Wavelets: A Student Guide

This text offers an excellent introduction to the mathematical theory of wavelets for senior undergraduate students. Despite the fact that this theory is intrinsically advanced, the author's elementary approach makes it accessible at the undergraduate level.

Beginning with thorough accounts of inner product spaces and Hilbert spaces, the book then shifts its focus to wavelets specifically, starting with the Haar wavelet, broadening to wavelets in general, and culminating in the construction of the Daubechies wavelets. All of this is done using only elementary methods, bypassing the use of the Fourier integral transform. Arguments using the Fourier transform are introduced in the final chapter, and this less elementary approach is used to outline a second and quite different construction of the Daubechies wavelets. The main text of the book is supplemented by more than 200 exercises ranging in difficulty and complexity.

AUSTRALIAN MATHEMATICAL SOCIETY LECTURE SERIES

Editor-in-chief: Professor C. Praeger, School of Mathematics and Statistics, University of Western Australia, Crawley, WA 6009, Australia

Editors:
Professor P. Broadbridge, School of Engineering and Mathematical Sciences, La Trobe University, Victoria 3086, Australia

Professor M. Murray, School of Mathematical Sciences, University of Adelaide, SA 5005, Australia

Professor J. Ramagge, Faculty of Science, University of Sydney, NSW 2006, Australia

Professor M. Wand, School of Mathematical Sciences, University of Technology, Sydney, NSW 2007, Australia

Australian Mathematical Society Lecture Series: 24

Wavelets: A Student Guide

PETER NICKOLAS

University of Wollongong, New South Wales

CAMBRIDGE
UNIVERSITY PRESS

University Printing House, Cambridge CB2 8BS, United Kingdom

One Liberty Plaza, 20th Floor, New York, NY 10006, USA

477 Williamstown Road, Port Melbourne, VIC 3207, Australia

4843/24, 2nd Floor, Ansari Road, Daryaganj, Delhi – 110002, India

79 Anson Road, #06–04/06, Singapore 079906

Cambridge University Press is part of the University of Cambridge.

It furthers the University's mission by disseminating knowledge in the pursuit of
education, learning, and research at the highest international levels of excellence.

www.cambridge.org
Information on this title: www.cambridge.org/9781107612518

First published 2017

Printed in the United Kingdom by Clays, St Ives plc

A catalogue record for this publication is available from the British Library

Library of Congress Cataloging-in-Publication data
Names: Nickolas, Peter.
Title: Wavelets : a student guide / Peter Nickolas, University of Wollongong,
New South Wales.
Description: Cambridge : Cambridge University Press, 2017. |
Series: Australian Mathematical Society lecture series ; 24 |
Includes bibliographical references and index.
Identifiers: LCCN 2016011212 | ISBN 9781107612518 (pbk.)
Subjects: LCSH: Wavelets (Mathematics)–Textbooks. | Inner product
spaces–Textbooks. | Hilbert space–Textbooks.
Classification: LCC QA403.3.N53 2016 | DDC 515/.2433–dc23 LC record available
at http://lccn.loc.gov/2016011212

ISBN 978-1-107-61251-8 Paperback

Contents

Preface

Overview

The overall aim of this book is to provide an introduction to the theory of wavelets for students with a mathematical background at senior undergraduate level. The text grew from a set of lecture notes that I developed while teaching a course on wavelets at that level over a number of years at the University of Wollongong.

Although the topic of wavelets is somewhat specialised and is certainly not a standard one in the typical undergraduate syllabus, it is nevertheless an attractive one for introduction to students at that level. This is for several reasons, including its topicality and the intrinsic interest of its fundamental ideas. Moreover, although a comprehensive study of the theory of wavelets makes the use of advanced mathematics unavoidable, it remains true that substantial parts of the theory can, with care, be made accessible at the undergraduate level.

The book assumes familiarity with finite-dimensional vector spaces and the elements of real analysis, but it does not assume exposure to analysis at an advanced level, to functional analysis, to the theory of Lebesgue integration and measure or to the theory of the Fourier integral transform. Knowledge of all these topics and more is assumed routinely in all full accounts of wavelet theory, which make heavy use of the Lebesgue and Fourier theories in particular.

The approach adopted here is therefore what is often referred to as 'elementary'. Broadly, full proofs of results are given precisely to the extent that they can be constructed in a form that is consistent with the relatively modest assumptions made about background knowledge. A number of central results in the theory of wavelets are by their nature deep and are not amenable in any straightforward way to an elementary approach, and a consequence is that while most results in the earlier parts of the book are supplied with complete proofs, a few of those in the later parts are given only partial proofs or are proved only in special cases or are stated without proof. While a degree of intellectual danger is inherent in giving an exposition that is incomplete in this way, I am careful to acknowledge gaps where they occur, and I think that any minor disadvantages are outweighed by the advantages of being able to introduce such an attractive topic at the undergraduate level.

Structure and Contents

If a unifying thread runs through the book, it is that of exploring how the fundamental ideas of an orthonormal basis in a finite-dimensional real inner product space, and the associated expression of a vector in terms of its projections onto the elements of such a basis, generalise naturally and elegantly to suitable infinite-dimensional spaces. Thus the work starts in the familiar and concrete territory of Euclidean space and moves towards the less familiar and more abstract domain of sequence and function spaces.

The structure and contents of the book are shown in some detail by the Contents, but brief comments on the first and last chapters specifically may be useful.

Chapter 1 is essentially a miniaturised version of the rest of the text. Its inclusion is intended to allow the reader to gain as early as possible some sense of what wavelets are, of how and why they are used and of the beauty and unity of the ideas involved, without having to wait for the more systematic development of wavelet theory that starts in Chapter 5.

As noted above, the Fourier transform is an indispensable technical tool for the rigorous and systematic study of wavelets, but its use has been bypassed in this text in favour of an elementary approach in order to make the material as accessible as possible. Chapter 8, however, is a partial exception to this principle, since we give there an overview of wavelet theory using the powerful extra insight provided by the use of Fourier analysis. For students with a deeper background in analysis than that assumed earlier in the text, this chapter brings the work of the book more into line with standard approaches to wavelet theory. In keeping with this, we work in this chapter with complex-valued rather than real-valued functions.

Pathways

The book contains considerably more material than could be covered in a typical one-semester course, but lends itself to use in a number of ways for such a course. It is likely that Chapters 3 and 4 on inner product spaces and Hilbert spaces would need to be included however the text is used, since this material is required for the work on wavelets that follows but is not usually covered in such depth in the undergraduate curriculum. Beyond this, coverage will depend on the knowledge that can be assumed. The following three pathways through the material are proposed, depending on the assumed level of mathematical preparation.

At the most elementary level, the book could be used as an introduction to Hilbert spaces with applications to wavelets, by covering just Chapters 1–5, perhaps with excursions, which could be quite brief, into Chapters 6 and 7. At a somewhat higher level of sophistication, the coverage could largely or completely bypass Chapter 1, survey the examples of Chapter 2 briefly and then proceed through to the end of Chapter 7. A third pathway through the material is possible for students with a more substantial background in analysis, including significant experience with the Fourier transform and, preferably, Lebesgue measure and integration. Here, coverage could begin comfortably with Chapter 3 and continue to the end of Chapter 8.

Exercises

The book contains about 230 exercises, and these should be regarded as an integral part of the text. Although they are referred to uniformly as 'exercises', they range substantially in difficulty and complexity: from short and simple problems to long and difficult ones, some of which extend the theory or provide proofs that are omitted from the body of the text. In the longer and more difficult cases, I have generally provided hints or outlined suggested approaches or broken possible arguments into steps that are individually more approachable.

Sources

Because the book is an introductory one, I decided against including references to the literature in the main text. At the same time, however, it is certainly desirable to provide such references for readers who want to consult source material or look more deeply into issues arising from the text, and the Appendix provides these.

Acknowledgements

The initial suggestion of converting my lecture notes on wavelets into a book came from Jacqui Ramagge. Adam Sierakowski read and commented on the notes in detail at an early stage and Rodney Nillsen likewise read and commented on substantial parts of the book when it was in something much closer to its final form. I am indebted to these colleagues, as well as to two anonymous referees.

1

An Overview

This chapter gives an overview of the entire book. Since our focus turns directly to wavelets only in Chapter 5, about halfway through, beginning with an overview is useful because it enables us early on to convey an idea of what wavelets are and of the mathematical setting within which we will study them.

The idea of the *orthonormality* of a collection of vectors, together with some closely related ideas, is central to the chapter. These ideas should be familiar at least in the finite-dimensional Euclidean spaces $\mathbb{R}^n$ (and certainly in $\mathbb{R}^2$ and $\mathbb{R}^3$ in particular), but after revising the Euclidean case we will go on to examine the same ideas in certain spaces of infinite dimension, especially spaces of functions. In many ways, the central chapters of the book constitute a systematic and general investigation of these ideas, and the theory of wavelets is, from one point of view, an application of this theory.

Because this chapter is only an overview, the discussion will be rather informal, and important details will be skated over quickly or suppressed altogether, but by the end of the chapter the reader should have a broad idea of the shape and content of the book.

1.1 Orthonormality in $\mathbb{R}^n$

Recall that the **inner product** $\langle a, b \rangle$ of two vectors

$$a = \begin{pmatrix} a_1 \\ a_2 \\ \vdots \\ a_n \end{pmatrix} \quad \text{and} \quad b = \begin{pmatrix} b_1 \\ b_2 \\ \vdots \\ b_n \end{pmatrix}$$

in $\mathbb{R}^n$ is defined by

$$\langle a, b \rangle = a_1 b_1 + a_2 b_2 + \cdots + a_n b_n = \sum_{i=1}^{n} a_i b_i.$$

Also, the **norm** $\|a\|$ of a is defined by

$$\|a\| = \sqrt{a_1^2 + a_2^2 + \cdots + a_n^2} = \left(\sum_{i=1}^{n} a_i^2 \right)^{\frac{1}{2}} = \langle a, a \rangle^{\frac{1}{2}}.$$

➤ A few remarks on notation and terminology are appropriate before we go further. First, in elementary linear algebra, special notation is often used for vectors, especially vectors in Euclidean space: a vector might, for example, be signalled by the use of bold face, as in $\boldsymbol{a}$, or by the use of a special symbol above or below the main symbol, as in $\vec{a}$ or $\underset{\sim}{a}$. This notational complication, however, is logically unnecessary, and is normally avoided in more advanced work.

Second, you may be used to the terminology *dot product* and the corresponding notation $a \cdot b$, but we will always use the phrase *inner product* and the 'angle-bracket' notation $\langle a, b \rangle$. Similarly, $\|a\|$ is sometimes referred to as the *length* or *magnitude* of a, but we will always refer to it as the *norm*.

The inner product can be used to define the component and the projection of one vector on another. Geometrically, the **component** of a on b is formed by projecting a perpendicularly onto b and measuring the length of the projection; informally, the component measures how far a 'sticks out' in the direction of b, or 'how long a looks' from the perspective of b. Since this quantity should depend only on the direction of b and not on its length, it is convenient to define it first when b is **normalised**, or a **unit vector**, that is, satisfies $\|b\| = 1$. In this case, the component of a on b is defined simply to be

$$\langle a, b \rangle.$$

If b is not normalised, then the component of a on b is obtained by applying the definition to its **normalisation** $(1/\|b\|) b$, which *is* a unit vector, giving the expression

$$\left\langle a, \frac{1}{\|b\|} b \right\rangle = \frac{1}{\|b\|} \langle a, b \rangle$$

for the component (note that we must assume that b is not the zero vector here, to ensure that $\|b\| \neq 0$).

Since the component is defined as an inner product, its value can be any real number – positive, negative or zero. This implies that our initial description of the component above was not quite accurate: the component represents not just the *length* of the first vector from the perspective of the second, but also, according to its sign, how the first vector is *oriented* with respect to the second. (For the special case when the component is 0, see below.)

It is now simple to define the **projection** of a on b; this is the vector whose direction is given by b and whose length and orientation are given by the component of a on b. Thus if $\|b\| = 1$, then the projection of a on b is given by

$$\langle a, b \rangle b,$$

and for general non-zero b by

$$\left\langle a, \frac{1}{\|b\|} b \right\rangle \frac{1}{\|b\|} b = \frac{1}{\|b\|^2} \langle a, b \rangle b.$$

Vectors a and b are said to be **orthogonal** if $\langle a, b \rangle = 0$, and a collection of vectors is **orthonormal** if each vector in the set has norm 1 and every two distinct elements from the set are orthogonal. If a set of vectors is indexed by the values of a subscript, as in $v_1, v_2, \ldots$, say, then the orthonormality of the set can be expressed conveniently using the Kronecker delta: the set is orthonormal if and only if

$$\langle v_i, v_j \rangle = \delta_{i,j} \quad \text{for all } i, j,$$

where the **Kronecker delta** $\delta_{i,j}$ is defined to be 1 when $i = j$ and 0 otherwise (over whatever is the relevant range of i and j).

All the above definitions are purely *algebraic*: they use nothing more than the algebraic operations permitted in a vector space and in the field of scalars $\mathbb{R}$. However, the definitions also have clear *geometric* interpretations, at least in $\mathbb{R}^2$ and $\mathbb{R}^3$, where we can visualise the geometry (see Exercise 1.2); indeed, we used this fact explicitly at a number of points in the discussion (speaking of 'how long' one vector looks from another's perspective and of the 'length' and 'orientation' of a vector, for example). We can attempt a somewhat more comprehensive list of such geometric interpretations as follows.

- A vector $a = \begin{pmatrix} a_1 \\ a_2 \\ \vdots \\ a_n \end{pmatrix}$ is just a point in $\mathbb{R}^n$ (whose coordinates are of course the numbers $a_1, a_2, \ldots, a_n$).

- The norm $\|a\|$ is the Euclidean distance of the point a from the origin, and more generally $\|a - b\|$ is the distance of the point a from the point b. (Given the formula defining the norm, this is effectively just a reformulation of Pythagoras' theorem; see Exercise 1.2.)

- For a unit vector b, the inner product $\langle a, b \rangle$ gives the magnitude and orientation of a when seen from b.

- The inner product and norm are related by the formula

$$\cos \theta = \frac{\langle a, b \rangle}{\|a\| \, \|b\|},$$

  where θ is the angle between the (non-zero) vectors a and b.
- Non-zero vectors a and b are orthogonal precisely when they are perpendicular, corresponding to the case $\cos \theta = 0$ above.
- A collection of vectors is orthonormal if its members have length 1 and are mutually perpendicular.

➤ It is reasonable to ask what we can make of these geometric interpretations in $\mathbb{R}^n$ when $n > 3$ and we can no longer visualise the situation. Can we sensibly talk about vectors being 'perpendicular', or about the 'lengths' of vectors, in spaces that we cannot visualise? The algebra works similarly in all dimensions, but does the geometry?

This question is perhaps as much philosophical as mathematical, but the experience and consensus of mathematicians is that the geometric terminology and intuition which are so central to our understanding in low dimensions are simply too valuable to discard in higher dimensions, and it is therefore used uniformly, whatever the dimension of the space. One of the remarkable and beautiful aspects of linear algebra is that the geometric ideas which are so obviously meaningful and useful in $\mathbb{R}^2$ and $\mathbb{R}^3$ play just as significant a role in higher-dimensional Euclidean spaces.

Further, as hinted earlier, an underlying theme of this book is to study how the same circle of ideas – of inner products, norms, orthogonality and orthonormality, and so on – plays a vital role in the study of certain spaces of infinite dimension, and in particular in the theory of wavelets.

Let us now look at two examples in $\mathbb{R}^3$.

The first, though extremely simple, is nevertheless fundamental. Consider the vectors

$$e_1 = \begin{pmatrix} 1 \\ 0 \\ 0 \end{pmatrix}, \quad e_2 = \begin{pmatrix} 0 \\ 1 \\ 0 \end{pmatrix} \quad \text{and} \quad e_3 = \begin{pmatrix} 0 \\ 0 \\ 1 \end{pmatrix}$$

in $\mathbb{R}^3$. These vectors form what is usually called the **standard basis** for $\mathbb{R}^3$; they form a basis, by the definition of that term, because they are linearly independent and every vector in $\mathbb{R}^3$ can be expressed uniquely as a linear combination of them. But what is especially of interest here, though it is mathematically trivial to check, is that e_1, e_2, e_3 form an *orthonormal basis*: $\langle e_i, e_j \rangle = \delta_{i,j}$ for $i, j = 1, 2, 3$. It follows from the orthonormality that in the expression for a given vector as a linear combination of e_1, e_2, e_3, the coefficients in the combination are the respective components of the vector on e_1, e_2, e_3. Specifically, if

$$a = \begin{pmatrix} a_1 \\ a_2 \\ a_3 \end{pmatrix},$$

then the component of a on e_i is $\langle a, e_i \rangle = a_i$ for $i = 1, 2, 3$, and we have

$$a = a_1 e_1 + a_2 e_2 + a_3 e_3 = \sum_{i=1}^{3} \langle a, e_i \rangle e_i.$$

For our second example, consider the three vectors

$$b_1 = \begin{pmatrix} \frac{1}{2} \\ -\frac{1}{2} \\ -\frac{1}{\sqrt{2}} \end{pmatrix}, \quad b_2 = \begin{pmatrix} \frac{1}{\sqrt{2}} \\ \frac{1}{\sqrt{2}} \\ 0 \end{pmatrix} \quad \text{and} \quad b_3 = \begin{pmatrix} \frac{1}{2} \\ -\frac{1}{2} \\ \frac{1}{\sqrt{2}} \end{pmatrix}.$$

It is simple to verify that these vectors are orthonormal. From orthonormality, it follows that the three vectors are linearly independent, and then, since $\mathbb{R}^3$ has dimension 3, that they form a basis for $\mathbb{R}^3$ (see Exercise 1.3). However, our main point here is to observe that we can express an arbitrary vector c as a linear combination of b_1, b_2 and b_3 nearly as easily as in the first example, by computing components.

For example, take

$$c = \begin{pmatrix} 1 \\ 2 \\ 3 \end{pmatrix}.$$

Projecting c onto each of b_1, b_2 and b_3, we obtain the components

$$\langle c, b_1 \rangle = -\frac{1}{2} - \frac{3}{\sqrt{2}}, \quad \langle c, b_2 \rangle = \frac{3}{\sqrt{2}} \quad \text{and} \quad \langle c, b_3 \rangle = -\frac{1}{2} + \frac{3}{\sqrt{2}},$$

respectively, and these quantities are the coefficients required to express c as a linear combination of b_1, b_2 and b_3. That is,

$$c = \langle c, b_1 \rangle b_1 + \langle c, b_2 \rangle b_2 + \langle c, b_3 \rangle b_3 = \sum_{i=1}^{3} \langle c, b_i \rangle b_i,$$

or, numerically,

$$\begin{pmatrix} 1 \\ 2 \\ 3 \end{pmatrix} = \left(-\frac{1}{2} - \frac{3}{\sqrt{2}}\right) \begin{pmatrix} \frac{1}{2} \\ -\frac{1}{2} \\ -\frac{1}{\sqrt{2}} \end{pmatrix} + \frac{3}{\sqrt{2}} \begin{pmatrix} \frac{1}{\sqrt{2}} \\ \frac{1}{\sqrt{2}} \\ 0 \end{pmatrix} + \left(-\frac{1}{2} + \frac{3}{\sqrt{2}}\right) \begin{pmatrix} \frac{1}{2} \\ -\frac{1}{2} \\ \frac{1}{\sqrt{2}} \end{pmatrix},$$

and of course we can check this easily by direct expansion.

➤ Although the algorithmic aspects of these ideas are not of primary interest here, it is worth making one observation about them in passing. If a basis is orthonormal, then the coefficient of a vector with respect to any given basis vector *depends only on that basis vector*. This is in marked contrast to the case when the basis is not orthonormal; the value of the coefficient then potentially involves *all* of the basis vectors, and finding the value may involve much more computation.

Thus orthonormality makes the *algebra* in the second example almost as simple as in the first example. Furthermore, the *geometry* is almost identical to that of the first example too, even though it is somewhat harder to visualise. The three vectors b_1, b_2 and b_3 are orthonormal, and hence define a system of mutually perpendicular axes in $\mathbb{R}^3$, just like the standard basis vectors e_1, e_2 and e_3, and the coefficients that we found for c by computing components are nothing other than the coordinates of the point c on these axes.

▶ The axes defined by b_1, b_2 and b_3, or any orthonormal basis in $\mathbb{R}^3$, can be obtained by rotation of the usual axes through some angle around some axis through the origin. However, while the usual coordinate system is a right-handed system, the coordinate system defined by an arbitrarily chosen orthogonal basis may be right- or left-handed. A linear algebra course typically explains how to compute the axis and the angle of rotation, as well as to determine the handedness of a system.

1.2 Some Infinite-Dimensional Spaces

1.2.1 Spaces of Sequences

A natural way of trying to translate our ideas so far into an infinite-dimensional space is to consider a vector space of 'infinity-tuples', instead of n-tuples for some $n \in \mathbb{N}$. Our vectors are thus infinitely long columns of the form

$$a = \begin{pmatrix} a_1 \\ a_2 \\ a_3 \\ \vdots \end{pmatrix},$$

and we will denote the resulting vector space by $\mathbb{R}^\infty$.

We make two simple points to begin with. First, column notation for vectors becomes increasingly inconvenient as the columns become longer, so we will switch henceforth to row notation instead. Second, we already have a standard name for infinitely long row vectors

$$a = (a_1, a_2, a_3, \dots) \in \mathbb{R}^\infty :$$

they are simply **sequences**. Thus, $\mathbb{R}^\infty$ is *the vector space of all sequences of real numbers*.

There is no difficulty in checking that $\mathbb{R}^\infty$ satisfies all the required conditions to be a vector space. We will not do this here, but we will glance briefly at a couple of illustrative cases and refer to Chapter 2 for details.

The axiom of closure under vector addition requires that when we add two vectors in the space, the result is again a vector in the space. This is clear for $\mathbb{R}^\infty$, provided that we define the sum of two vectors in the obvious way, following the definition in $\mathbb{R}^n$. Thus if

$$a = (a_1, a_2, a_3, \dots) \quad \text{and} \quad b = (b_1, b_2, b_3, \dots)$$

are in $\mathbb{R}^\infty$, then we define

$$a + b = (a_1 + b_1, a_2 + b_2, a_3 + b_3, \dots);$$

that is, addition of vectors in $\mathbb{R}^\infty$ is defined *entry by entry* or *entry-wise*. Closure is now obvious: each entry $a_n + b_n$ is a real number, by the axiom of closure for addition in $\mathbb{R}$, so $a + b$ is again an element of $\mathbb{R}^\infty$. For the axiom of commutativity of vector addition, which requires that $a + b = b + a$, we have

$$
\begin{aligned}
a + b &= (a_1, a_2, a_3, \dots) + (b_1, b_2, b_3, \dots) \\
&= (a_1 + b_1, a_2 + b_2, a_3 + b_3, \dots) \\
&= (b_1 + a_1, b_2 + a_2, b_3 + a_3, \dots) \\
&= (b_1, b_2, b_3, \dots) + (a_1, a_2, a_3, \dots) \\
&= b + a.
\end{aligned}
$$

Notice that the central step in this argument is an application of the law of commutativity of addition in $\mathbb{R}$, paralleling the way in which the argument for the closure axiom worked.

Now let us investigate how we might define an inner product and a norm in $\mathbb{R}^\infty$. Given vectors (that is, sequences) a and b as above, we would presumably wish, following our procedure in $\mathbb{R}^n$, to define their inner product by

$$\langle a, b \rangle = a_1 b_1 + a_2 b_2 + a_3 b_3 + \cdots = \sum_{n=1}^{\infty} a_n b_n,$$

and then to go on to say that a and b are orthogonal if $\langle a, b \rangle = 0$. But there is a problem: *the infinite series $\sum_{n=1}^{\infty} a_n b_n$ does not converge for all pairs of sequences a and b, so the proposed inner product is not defined for all pairs of vectors.*

It is reasonable to try to solve this problem pragmatically by simply removing all the troublesome vectors from the space, that is, by working in the largest subspace of $\mathbb{R}^\infty$ in which all of the required sums $\sum_{n=1}^{\infty} a_n b_n$ converge.

Now this description does not quite constitute a definition of the desired subspace, since it involves a condition on pairs of vectors and does not directly

give us a criterion for the membership of any single vector. However, there
is in fact such a criterion: the space can be defined directly as the set of all
sequences $a = (a_1, a_2, a_3, \dots)$ for which the sum

$$\sum_{n=1}^{\infty} a_n^2$$

converges. This new space is denoted by ℓ^2, so we have

$$\ell^2 = \left\{ (a_1, a_2, a_3, \dots) \in \mathbb{R}^\infty : \sum_{n=1}^{\infty} a_n^2 \text{ converges} \right\}.$$

▶ The name of the space is usually read as 'little-ℓ-2', the word 'little' being needed
because as we will see soon there are also 'capital-ℓ-2' or 'big-ℓ-2' spaces, which use
an 'L' rather than an 'ℓ'. (Note that in some sources the '2' is written as a subscript
rather than a superscript.)

It requires proof that this membership criterion for ℓ^2 solves our original
problem – that ℓ^2 is a vector space and that all the desired inner products now
lead to convergent sums – but this is left for Chapter 2. Notice, at least, that the
norm can certainly now be defined unproblematically by the formula

$$\|a\| = \sqrt{a_1^2 + a_2^2 + \cdots} = \left(\sum_{n=1}^{\infty} a_n^2 \right)^{\frac{1}{2}} = \langle a, a \rangle^{\frac{1}{2}}.$$

Thus we have found what, after checking, turn out to be satisfactory
definitions of an inner product and a norm in ℓ^2, and therefore also of the
notions of orthogonality and orthonormality. What can we do with all this?

Let $a = (a_1, a_2, a_3, \dots, a_n)$ be in $\mathbb{R}^n$ for any fixed n (for notational
convenience we now adopt row notation for vectors in $\mathbb{R}^n$). Then a can be
expressed as a linear combination of the n standard orthonormal basis vectors

$$e_1 = (1, 0, 0, \dots, 0),$$
$$e_2 = (0, 1, 0, \dots, 0),$$
$$e_3 = (0, 0, 1, \dots, 0),$$
$$\vdots$$
$$e_n = (0, 0, 0, \dots, 1),$$

in the form

$$a = a_1 e_1 + a_2 e_2 + a_3 e_3 + \cdots + a_n e_n = \sum_{i=1}^{n} a_i e_i,$$

as we noted in detail in the case of $\mathbb{R}^3$, in the first example of the previous
section.

It seems natural to try to do the same thing in ℓ^2, since one of our aims was to try to extend finite-dimensional ideas to the infinite-dimensional case. Thus we would hope to say that any $a = (a_1, a_2, a_3, \ldots) \in \ell^2$ can be written as a linear combination of *the infinite orthonormal sequence of vectors*

$$e_1 = (1, 0, 0, 0, \ldots), \quad e_2 = (0, 1, 0, 0, \ldots), \quad e_3 = (0, 0, 1, 0, \ldots), \ldots$$

in the form

$$a = a_1 e_1 + a_2 e_2 + a_3 e_3 + \cdots = \sum_{n=1}^{\infty} a_n e_n,$$

where the coefficient of e_n is the component

$$a_n = \langle a, e_n \rangle$$

of a on e_n, as in the finite-dimensional case.

But another problem arises: the axioms of a vector space only allow a *finite* number of vectors to added together, and hence only allow *finite linear combinations* of vectors to be formed. Therefore, if we want to justify the above very natural expression for a, we will have to find a way of giving meaning (at least in some circumstances) to 'infinite linear combinations' of vectors. This will be an important topic for detailed discussion in Chapter 4.

▶ The specific difficulty here is almost exactly the same as in the case of infinite series of real numbers. The ordinary laws of arithmetic only allow us to add together finitely many real numbers, and when we wish to add infinitely many, as in an infinite series, we have to develop appropriate definitions and results to justify the process. Specifically, we define partial sums and then take limits, and this will be exactly the route we follow in the present case as well when we return to the issue in detail in Chapter 4.

A related issue is raised by the phrase 'orthonormal basis', which we have used a number of times. For an integer n, it is correct to say (as we have) that the collection $e_1, e_2, \ldots, e_n$ is an orthonormal basis for $\mathbb{R}^n$, simply because the collection is both orthonormal and a basis. However, the collection $e_1, e_2, e_3, \ldots$ is *not* a basis for ℓ^2, and we consequently cannot correctly refer to this collection as an orthonormal basis for ℓ^2. (See Exercise 1.6 and Subsection 2.3.2 for the claim that $e_1, e_2, e_3, \ldots$ do not form a basis for ℓ^2.)

This is the case, at any rate, as long as we continue to use the term 'basis' in the ordinary sense of linear algebra, which only allows finite linear combinations. Once we have found a satisfactory definition of an infinite linear combination, however, we will be able to expand the scope of application of the term 'basis' in a way that will make it correct after all to call the collection $e_1, e_2, e_3, \ldots$ an orthonormal basis for ℓ^2.

Although the use of the phrase 'orthonormal basis' in this extended sense will not be formally justified until Chapter 4, we will nevertheless make informal use of it a few times in the remainder of this chapter.

1.2.2 Spaces of Functions

We now take a further step away from the familiar Euclidean spaces $\mathbb{R}^n$ and consider vector spaces of functions defined on an interval I of the real line. For most of the discussion, the interval I will either be $[-\pi, \pi]$ or the whole real line $\mathbb{R}$, but there is no need initially to restrict our choice. Our first attempt to specify a useful space of functions on I might be to consider $F(I)$, the collection of all functions $f : I \to \mathbb{R}$. This is indeed a vector space, if we define operations pointwise. Thus for $f, g \in F(I)$, we define $f + g$ by setting $(f + g)(x) = f(x) + g(x)$ for all $x \in I$, and it is clear that $f + g$ is again a function from I to $\mathbb{R}$, giving closure of $F(I)$ under addition. Scalar multiplication is handled similarly.

Further, we can easily write down what might appear to be a reasonable definition of an inner product on $F(I)$ by working in analogy to the inner product definitions that we introduced earlier in $\mathbb{R}^n$ and ℓ^2. In $\mathbb{R}^n$, for example, the definition

$$\langle a, b \rangle = \sum_{i=1}^{n} a_i b_i$$

multiplies corresponding entries of the n-tuples a and b and sums the resulting terms and so, given two functions f and g in $F(I)$, we might define

$$\langle f, g \rangle = \int_I fg,$$

since this definition multiplies corresponding values of the functions f and g and 'sums' the resulting terms, provided that we are prepared to think of integration as generalised summation.

➤ Note the two notational simplifications we have used here for integration. First, we can if we wish represent the region of integration in a definite integral as a subscript to the integral sign; thus

$$\int_{[-\pi,\pi]} \dots \quad \text{and} \quad \int_{-\pi}^{\pi} \dots$$

mean the same thing, as do

$$\int_{\mathbb{R}} \dots \quad \text{and} \quad \int_{-\infty}^{\infty} \dots .$$

Second, we can suppress the name of the variable of integration, the so-called dummy variable, when it is of no interest to us: thus

$$\int_I fg \quad \text{and} \quad \int_I f(x)g(x)\,dx$$

mean the same thing. The only situation in which we cannot do this is when we are dealing with a function for which we do not have a name; thus we cannot eliminate the use of the variable from an integral such as

$$\int_0^1 x^2\,dx.$$

After a moment's thought, we see that the formula

$$\langle f, g \rangle = \int_I fg$$

fails as the definition of an inner product on $F(I)$, because for arbitrary functions f and g on I, the integral $\int_I fg$ may not be defined. Observe that the failure occurs for essentially the same reason that our attempted definition earlier of an inner product on the space $\mathbb{R}^\infty$ using the formula $\langle a, b \rangle = \sum_{n=1}^\infty a_n b_n$ failed, where the problem was that the infinite series involved may not converge.

We can moreover rectify the problem in the case of $F(I)$ in a similar way to the case of $\mathbb{R}^\infty$, at least in principle: since the spaces are too large, we restrict to subspaces in which the problems vanish. In the case of $\mathbb{R}^\infty$, where we restricted to the subspace ℓ^2, finding the right subspace to which to restrict was fairly straightforward, but the solution is a little more elusive in the case of $F(I)$.

To discuss this in more detail, let us confine our attention for the time being to the specific case where $I = [-\pi, \pi]$. (We will enlarge our point of view again later, first to encompass the case $I = \mathbb{R}$ and then the general case.)

Let $C([-\pi, \pi])$ denote the set of continuous functions f that have as domain the interval $[-\pi, \pi]$. It is easy to check that $C([-\pi, \pi])$ is a vector subspace of $F([-\pi, \pi])$. Moreover, because of the added assumption of continuity, the integrals in the definitions

$$\langle f, g \rangle = \int_{-\pi}^\pi fg \quad \text{and} \quad \|f\| = \left(\int_{-\pi}^\pi f^2 \right)^{\frac{1}{2}} = \langle f, f \rangle^{\frac{1}{2}}$$

are now defined, and therefore give us a properly defined inner product and, consequently, norm on $C([-\pi, \pi])$.

This solves the immediate problem, but another less obvious and more difficult problem remains. We spoke above about trying to find the 'right' subspace to which to restrict, and while the choice of $C([-\pi, \pi])$ solves our original problem, a deeper analysis shows that it is actually too small a collection of functions to serve as an adequate basis for the development of the theory of later chapters. We cannot discuss this issue any further at this point, but see especially Subsection 4.4.4 below. We will also not try at this point to describe in any detail which extra functions need to be admitted to the space; it suffices for the moment to say that we need to add functions such as those that are piecewise continuous, as well as many others that are much more 'badly discontinuous'. For the moment, not too much is lost in thinking of the functions in question as, say, piecewise continuous.

The space that we end up with after this enlargement process is denoted by $L^2([-\pi, \pi])$.

▶ If f is continuous on the closed interval $[-\pi, \pi]$, then the integral $\int_{-\pi}^{\pi} f^2$ automatically has a finite value. Once we allow discontinuous functions, however, the corresponding integrals may not have finite values. Thus, a more precise statement is that $L^2([-\pi, \pi])$ consists of the functions f with the property that $\int_{-\pi}^{\pi} f^2$ exists and is finite. But, again, more needs to be said about this in later chapters.

Note that in accordance with our comments in Subsection 1.2.1, we read the notation $L^2([-\pi, \pi])$ as 'capital-ℓ-2 of $[-\pi, \pi]$' or 'big-ℓ-2 of $[-\pi, \pi]$'.

What corresponds in $L^2([-\pi, \pi])$ to the collection

$$e_1 = (1, 0, 0, 0, \dots), \quad e_2 = (0, 1, 0, 0, \dots), \quad e_3 = (0, 0, 1, 0, \dots), \dots$$

that we earlier referred to as an orthonormal basis for ℓ^2? One possible answer to this question (though there are many) should already be familiar: the constant function

$$\frac{1}{\sqrt{2\pi}}$$

together with the infinite family of trigonometric functions

$$\frac{1}{\sqrt{\pi}} \cos nx \quad \text{and} \quad \frac{1}{\sqrt{\pi}} \sin nx \quad \text{for } n \in \mathbb{N}.$$

Proving the orthonormality of this collection of functions is straightforward, and amounts to checking the following simple relations:

$$\int_{-\pi}^{\pi} \sin mx \cdot \cos nx \, dx = 0,$$

$$\int_{-\pi}^{\pi} \cos mx \cdot \cos nx \, dx = \pi \delta_{m,n},$$

$$\int_{-\pi}^{\pi} \sin mx \cdot \sin nx \, dx = \pi \delta_{m,n},$$

$$\int_{-\pi}^{\pi} 1 \cdot \cos nx \, dx = 0,$$

$$\int_{-\pi}^{\pi} 1 \cdot \sin nx \, dx = 0,$$

$$\int_{-\pi}^{\pi} 1 \cdot 1 \, dx = 2\pi,$$

for all $m, n \in \mathbb{N}$.

The really remarkable fact, however, is that these functions behave in the function space $L^2([-\pi, \pi])$ in exactly the same way as $e_1, e_2, e_3, \ldots$ behave in the sequence space ℓ^2: *every function in $L^2([-\pi, \pi])$ can be expressed as an infinite linear combination of them* – though as noted we have yet to make clear exactly what is meant by an infinite linear combination. Moreover, this remarkable fact should already be familiar, at least for, say, piecewise smooth functions f on $[-\pi, \pi]$: it is just the statement that *any function $f \in L^2([-\pi, \pi])$ can be expanded as a* **Fourier series**:

$$f(x) = a_0 \left(\frac{1}{\sqrt{2\pi}} \right) + \sum_{n=1}^{\infty} a_n \left(\frac{1}{\sqrt{\pi}} \cos nx \right) + \sum_{n=1}^{\infty} b_n \left(\frac{1}{\sqrt{\pi}} \sin nx \right), \qquad (1.1)$$

where $a_0, a_1, a_2, \ldots$ and $b_1, b_2, \ldots$ are the **Fourier coefficients** of f, defined by

$$a_0 = \frac{1}{\sqrt{2\pi}} \int_{-\pi}^{\pi} f(x) \, dx,$$

$$a_n = \frac{1}{\sqrt{\pi}} \int_{-\pi}^{\pi} f(x) \cos nx \, dx,$$

$$b_n = \frac{1}{\sqrt{\pi}} \int_{-\pi}^{\pi} f(x) \sin nx \, dx,$$

for $n \in \mathbb{N}$. Moreover, these Fourier coefficients are just components of the same kind that we used when we found the coordinates of vectors in $\mathbb{R}^n$ for $n \in \mathbb{N}$, or in ℓ^2, by projecting onto the elements of an orthonormal basis. For example, the formula

$$a_n = \frac{1}{\sqrt{\pi}} \int_{-\pi}^{\pi} f(x) \cos nx \, dx$$

is just a rearrangement of the formula

$$a_n = \left\langle f, \frac{1}{\sqrt{\pi}} \cos nx \right\rangle,$$

in which the right-hand side is the component of f on $(1/\sqrt{\pi}) \cos nx$.

➤ The notation used in the formula

$$a_n = \left\langle f, \frac{1}{\sqrt{\pi}} \cos nx \right\rangle$$

deserves comment.

In $L^2([-\pi,\pi])$ vectors are functions, and the inner product therefore requires two functions as arguments. Now in the expression for a_n, the first argument f certainly denotes a function, but the second argument, $(1/\sqrt{\pi}) \cos nx$, if we disregard the context in which it occurs, denotes the *value* of a function – a real number – rather than a function.

We adopt the convention here that since the expression $(1/\sqrt{\pi}) \cos nx$ occurs in a context where a function is required, the expression should be read as shorthand for *the function whose value at x is* $(1/\sqrt{\pi}) \cos nx$. On this convention, the arguments to the inner product are both functions, as they must be.

There are numerous comparable cases later that involve function notation, and we adopt the same convention consistently. In all cases, the context in which the notation is used will determine unambiguously how the notation is to be interpreted – as a function or as the value of a function.

In case of a_n above, we could avoid the need for the convention by introducing a new function name: we could write, say, $A_n(x) = (1/\sqrt{\pi}) \cos nx$ for $x \in [-\pi,\pi]$ and then define $a_n = \langle f, A_n \rangle$. However, there is little point in introducing a name for a function when the name will play little or no further role. Moreover, there are even cases later where notation for the function involved has already been introduced but it still seems preferable for clarity to apply the convention.

The discussion above implicitly described two distinct processes related to Fourier series. Given $f \in L^2([-\pi,\pi])$, we first compute the discrete collection of real numbers a_0 and a_n and b_n, for $n \in \mathbb{N}$; this is the process of **Fourier analysis**. Then, from this collection, Equation (1.1) tells us that f can be reconstructed; this is the process of **Fourier synthesis**. Diagrammatically:

$$f \xrightarrow{\text{analysis}} \{a_0, a_1, a_2, \ldots, b_1, b_2, \ldots\} \xrightarrow{\text{synthesis}} f.$$

The analysis phase could be thought of as the process of *encoding* f in the form of the collection of numbers a_0 and a_n and b_n, for $n \in \mathbb{N}$, and the synthesis phase as the process of *decoding* to recover f.

Many mathematical subtleties and details are concealed by this rapid account, and we will have to explore some of these later. We can begin to see already, however, how the concrete, geometrical ideas familiar to us in $\mathbb{R}^n$ extend to settings that are quite remote from the finite-dimensional one. A Fourier series in particular is essentially an infinite-dimensional analogue of the familiar decomposition of a vector in $\mathbb{R}^3$ into the sum of its projections onto the x-, y- and z-axes. We will see later that wavelets show the same phenomenon at work.

➤ The form in which you have met Fourier series previously is probably different in detail from the one outlined above. It is usual, purely for convenience, to work with the orthogonal (but not orthonormal) functions

$$1, \quad \cos nx \quad \text{and} \quad \sin nx,$$

for $n \in \mathbb{N}$, and to make up for the missing coefficients by adjusting the constant factors in the Fourier coefficients, so that

$$a_0 = \frac{1}{2\pi} \int_{-\pi}^{\pi} f(x)\,dx, \quad a_n = \frac{1}{\pi} \int_{-\pi}^{\pi} f(x) \cos nx\,dx \quad \text{and} \quad b_n = \frac{1}{\pi} \int_{-\pi}^{\pi} f(x) \sin nx\,dx.$$

1.3 Fourier Analysis and Synthesis

Fourier series are defined for functions f in the space $L^2([-\pi, \pi])$, and such functions therefore have as domain the interval $[-\pi, \pi]$. For a fixed such function f, however, the Fourier series of f has the form

$$a_0 \left(\frac{1}{\sqrt{2\pi}} \right) + \sum_{n=1}^{\infty} a_n \left(\frac{1}{\sqrt{\pi}} \cos nx \right) + \sum_{n=1}^{\infty} b_n \left(\frac{1}{\sqrt{\pi}} \sin nx \right),$$

and this function has *the whole real line* as its domain. However, the trigonometric functions involved are periodic, with periods that are integer fractions of 2π, and so the Fourier series represents not just the original function f but also the **periodic extension of** f to the whole real line. Informally, the graph of f on $[-\pi, \pi]$ is repeated in each interval

$$[(2n - 1)\pi, (2n + 1)\pi],$$

for $n \in \mathbb{Z}$; formally, $f(x + 2\pi) = f(x)$ for all $x \in \mathbb{R}$, where we re-use the symbol 'f' to denote the periodic extension of the original function f.

➤ The Fourier series above clearly yields the same numerical value when evaluated at $x = -\pi$ and $x = \pi$, but there is no reason to expect that $f(-\pi) = f(\pi)$, since f is an arbitrary member of $L^2([-\pi, \pi])$. This apparently contradictory finding is a sign that there are hidden subtleties involving the exact sense in which the series 'represents' the function. Having noted the existence of this problem, however, we propose to delay consideration and resolution of it until Chapter 4 (see especially Subsection 4.6.1).

Fourier series have numerous applications, in pure and applied mathematics and in many other disciplines. Let us consider one of these briefly. The application is to the analysis and synthesis of musical tones. A note of constant pitch played on a particular musical instrument has a characteristic waveform, which is periodic with a frequency corresponding to the pitch; the note A above middle C, for example, has frequency 440 Hz at the modern pitch standard, so-called concert pitch.

➤ An important issue in practice is that although the waveform is a major determinant of how the tone sounds to us, it is not the only one. Particularly important are the *transients* of the tone – the distinctive set of short-lived, very high-frequency oscillations that an instrument produces as each new note begins to sound. However, we will completely ignore this issue in the present short discussion.

Consider a tone of some kind which, by suitable scaling, we can take to have frequency of a multiple of 2π, with time t replacing x as the independent variable. We know that the function f on $[-\pi, \pi]$ representing one complete cycle of this sound can be decomposed, or analysed, into a sum of trigonometric functions – namely, its Fourier series.

➤ Fourier series are, of course, infinite series in general. But in this application, the series can be assumed to be finite for practical purposes, because there is a limit to the frequencies with which a musical instrument can oscillate, and to the frequencies to which the human ear can respond.

Therefore, if we had a set of pure sinusoidal oscillators whose waveforms corresponded to the graphs of $\cos nx$ and $\sin nx$ for each n, and if we set the oscillators in motion with amplitudes determined by the Fourier coefficients of f, then *we would hear exactly the original sound*, whose graph was given by f. We would thus have synthesised or recreated the original sound from its separate pure sinusoidal components.

➤ Note that we have ignored the constant term in the Fourier series here, because it does not seem meaningful to say that an oscillator can produce a constant waveform with any value other than zero. But this is not a serious issue: the constant term merely determines the absolute vertical position of the graph in the coordinate system, not its shape.

Here is a concrete illustration. Consider a sound, represented by a function f, which alternates between two pure sinusoidal oscillations, one given by the function $\sin kt$ over the time interval $[-\pi, 0)$ and the other given by the function $\sin \ell t$ over the time interval $[0, \pi]$, and which repeats this pattern indefinitely – that is, is the periodic extension of that pattern. Thus the formula for f is the periodic extension of

$$f(t) = \begin{cases} \sin kt, & -\pi \le t < 0, \\ \sin \ell t, & 0 \le t \le \pi, \end{cases}$$

and we find, after some calculation, that the Fourier series of f is given by

$$\frac{1}{2}\sin kt + \frac{1}{2}\sin \ell t - \frac{2}{\pi}\sum_{n=1}^{\infty}\left(\frac{k}{k^2 - (2n-1)^2} - \frac{\ell}{\ell^2 - (2n-1)^2}\right)\cos(2n-1)t.$$

Figure 1.1 shows graphs of the partial sums of this series for the specific parameter values $k = 10$ and $\ell = 20$, and for summation over the finite range $1 \le n \le N$, where $N = 5, 10, 15, 20, 25, 100$, respectively.

Notice that already when $N = 20$ – corresponding to 22 actual terms of the series – we have, at least according to visual inspection, quite accurate

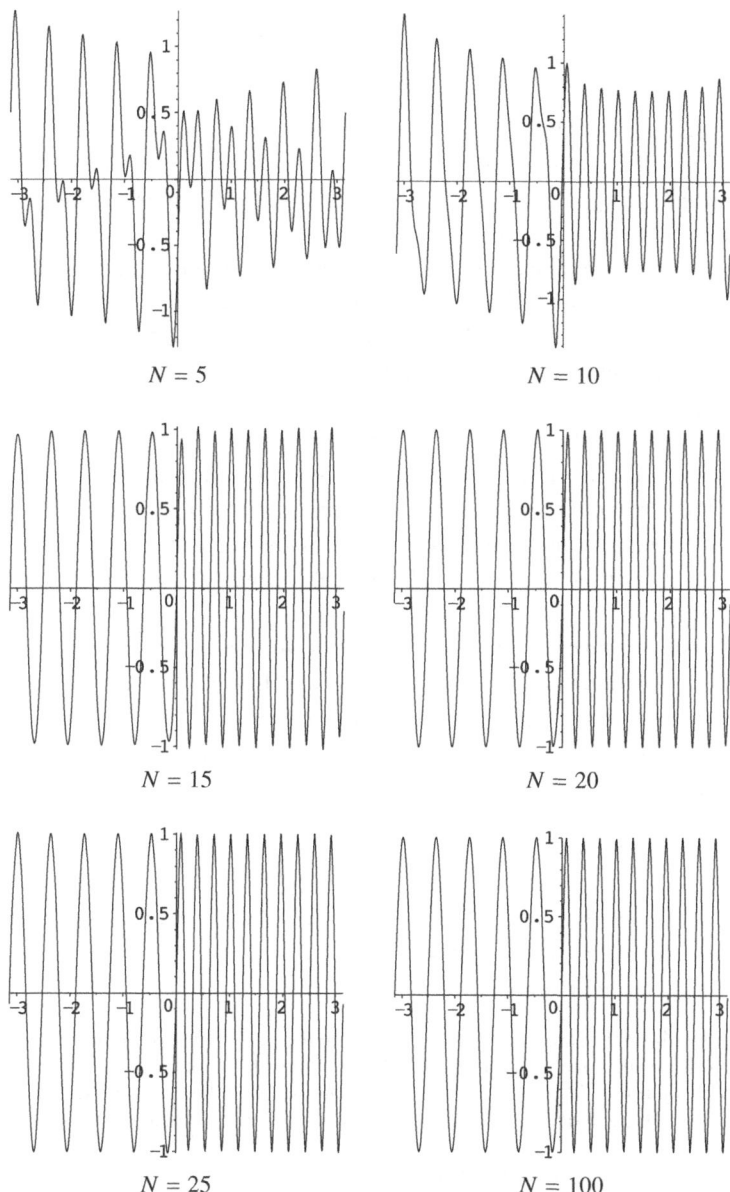

Figure 1.1 Graphs of some partial sums of the Fourier series for f

convergence. The physical implication is that if we switched on the appropriate 22 oscillators and adjusted the amplitudes to match the corresponding Fourier coefficients, then we would hear a pretty accurate rendition of the sound.

1.4 Wavelets

1.4.1 The Space $L^2(\mathbb{R})$

Fourier series can only ever represent *periodic* functions (even if by suitable
translation and scaling we can represent functions that are periodic on intervals
other than $[-\pi, \pi]$). Can we do something similar to represent *non-periodic*
functions on the whole real line?

The answer is that we can, and that there are several ways of doing it.
The classical technique is the use of the Fourier integral transform – a very
important topic which we will not discuss at all, except in the final chapter of
the book. Wavelets provide another, recently developed, technique. (There are
also hybrid techniques, sharing features of both.)

It is fair to say that the Fourier transform is one of the most important
concepts in mathematics, and one that has an enormous range of applications,
both within mathematics and outside. There are some applications, however,
where the Fourier transform is inadequate. Wavelet theory is partly the result
of attempts to find systematic mathematical techniques better suited to these
applications.

But first, what vector space of functions are we now dealing with? The
domain of our functions here is the whole real line $\mathbb{R}$, and we need an inner
product and a norm, defined in much the same way as they were on $L^2([-\pi, \pi])$.
Thus, the space we work in will be denoted by $L^2(\mathbb{R})$, and the inner product
and the norm will be defined by

$$\langle f, g \rangle = \int_{-\infty}^{\infty} fg \quad \text{and} \quad \|f\| = \left(\int_{-\infty}^{\infty} f^2 \right)^{\frac{1}{2}}.$$

As in the earlier case of $L^2([-\pi, \pi])$, the use of these formulae restricts the
functions eligible for membership of $L^2(\mathbb{R})$: a function f can only be in $L^2(\mathbb{R})$
if $\int_{-\infty}^{\infty} f^2$ exists and is finite.

▶ The condition that the integral be finite is much more restrictive in the case of $L^2(\mathbb{R})$
than it was in the case of $L^2([-\pi, \pi])$. For example, functions such as 1 and $\cos nx$
and $\sin nx$ are *not* in $L^2(\mathbb{R})$ – over the whole real line, the relevant integrals are infinite.
Very informally, a function in $L^2(\mathbb{R})$ must in some sense tend to 0 fairly rapidly as
$x \to \pm\infty$.

We have once again, unavoidably, been vague here about exactly which functions we
are dealing with; we will return to this point.

1.4.2 The Haar Wavelet

The **Haar wavelet** H is easy to define. Its simplicity makes it very useful for
illustrating some of the main ideas of the theory of wavelets; at the same time,

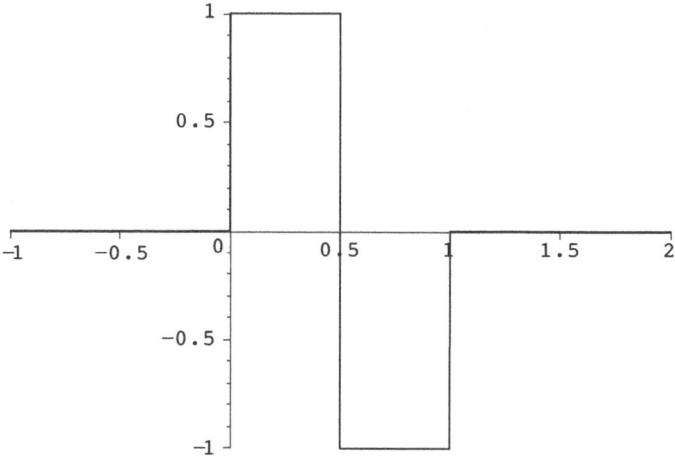

Figure 1.2 The graph of the Haar wavelet

however, its simplicity limits its use in practice: it is discontinuous, while most applications require wavelets that are smoother.

The definition of H is as follows, and its graph is shown in Figure 1.2.

$$H(x) = \begin{cases} 1, & 0 \le x < \frac{1}{2}, \\ -1, & \frac{1}{2} \le x < 1, \\ 0, & \text{otherwise.} \end{cases}$$

➤ The values we assign at the ends of the intervals in the definition are not particularly important; conventionally, the function is defined so as to be continuous from the right, but nothing of importance would change if we made it, say, continuous from the left instead. (It will become clear as discussion proceeds that this is essentially because changing the values of a function at isolated points does not affect the value of its integral.)

The vertical lines in Figure 1.2 are of course not part of the graph, but it is useful to include them as a guide to the eye.

If f is any function, we call the set

$$\{x \in \mathbb{R} : f(x) \neq 0\}$$

the **support** of f. Thus the support of the Haar wavelet H is the half-open interval $[0, 1)$. (Actually, the support is usually defined to be the smallest *closed* set containing the points where the function is non-zero; see Chapter 4. On this definition, the support of the Haar wavelet would be the closed interval $[0, 1]$, but the distinction will be not be of great importance here.)

To make use of the wavelet H, we work with its **scaled, dilated, translated** 'copies' $H_{j,k}$, defined by the formula

$$H_{j,k}(x) = 2^{j/2} H(2^j x - k),$$

for all $j, k \in \mathbb{Z}$. We will refer to the term $2^{j/2}$ as the **scaling factor**, the term 2^j as the **dilation factor** and the term k as the **translation factor**. Understanding the roles played by these factors both separately and together is vital. First, in a somewhat qualitative sense, their roles are as follows.

- The scaling factor $2^{j/2}$ is the simplest: its effect is merely to stretch or compress the graph in the vertical or y-direction, stretching in the case $j > 0$ and compressing in the case $j < 0$, and having no effect in the case $j = 0$.
- The dilation factor 2^j has the effect of stretching or compressing the graph in the horizontal or x-direction, compressing in the case $j > 0$ and stretching in the case $j < 0$, and having no effect in the case $j = 0$.
- The translation factor k has the effect of translating (that is, shifting) the graph in the horizontal or x-direction, to the right in the case $k > 0$ and to the left in the case $k < 0$, and having no effect in the case $k = 0$.

The different and seemingly contrary effects of the sign of j in the scaling and dilation factors should be noted carefully; drawing example graphs is strongly recommended (see Exercise 1.10).

Let us now examine analytically the roles of the three factors in combination. Given the original Haar wavelet H, and for fixed $j, k \in \mathbb{Z}$, consider in turn the functions

$$F(x) = H(2^j x)$$

and

$$G(x) = F(x - k/2^j).$$

Now the graph of F is the graph of H dilated by the factor 2^j, and the graph of G is the graph of F translated by the factor $k/2^j$. But

$$G(x) = F(x - k/2^j) = H(2^j(x - k/2^j)) = H(2^j x - k),$$

which is just an unscaled version of $H_{j,k}$. Thus, the graph of $H_{j,k}$ is the graph of H first dilated by the factor 2^j, then translated by the factor $k/2^j$, and finally scaled by the factor $2^{j/2}$. Thus, although we refer to k as the translation factor, the translation according to the analysis above is actually through a distance of $k/2^j$. (For more insight on the relationship between the dilation and the scaling, see Exercise 1.11.)

Since the dilations are always by a factor of 2^j for some $j \in \mathbb{Z}$, they are often referred to as **binary dilations**. Likewise, since the translations are through a distance of $k/2^j$ for some $j, k \in \mathbb{Z}$, and the numbers of this form are known as **dyadic numbers**, the translations are often referred to as **dyadic translations**. (See Chapter 7 for further discussion of the dyadic numbers.)

It follows from the discussion above that

$$\text{the support of } H_{j,k} \text{ is the interval } \left[\frac{k}{2^j}, \frac{k+1}{2^j}\right),$$

and that on that interval,

$$H_{j,k} \text{ takes the two non-zero values } \pm 2^{j/2},$$

the negative value in the left-hand half of the interval and the positive value in the right-hand half. In particular, for each fixed integer j, each real number lies in the support of *exactly one* of the functions

$$\{H_{j,k} : k \in \mathbb{Z}\}.$$

1.4.3 An Orthonormal Basis

What makes the Haar wavelet, and wavelets in general, of interest to us here, and part of what makes wavelets of importance in applications, is that the doubly infinite set

$$\{H_{j,k} : j, k \in \mathbb{Z}\},$$

consisting of all the scaled, dilated, translated copies of the Haar wavelet, behaves in $L^2(\mathbb{R})$ just as the trigonometric functions do in $L^2([-\pi, \pi])$, just as the vectors $e_1, e_2, e_3, \ldots$ do in ℓ^2, and just as any orthonormal set of n vectors in $\mathbb{R}^n$ does: the set is an orthonormal basis for the space $L^2(\mathbb{R})$, in the sense foreshadowed in Subsection 1.2.1.

Let us spell this claim out in detail. There are two parts to the claim. The first is that $\{H_{j,k} : j, k \in \mathbb{Z}\}$ is an orthonormal set. This means that

$$\langle H_{j,k}, H_{j',k'} \rangle = \int_{-\infty}^{\infty} H_{j,k} H_{j',k'} = \delta_{j,j'} \delta_{k,k'} = \begin{cases} 1, & j = j' \text{ and } k = k', \\ 0, & \text{otherwise,} \end{cases}$$

or, in words, that the inner product of any two functions from the collection is 0 if the functions are distinct and is 1 if the functions are the same.

The second part of the claim is that every function $f \in L^2(\mathbb{R})$ has a **Haar wavelet series expansion**, that is, an expansion of the form

$$f = \sum_{j=-\infty}^{\infty} \sum_{k=-\infty}^{\infty} w_{j,k} H_{j,k},$$

where the **Haar wavelet coefficient** $w_{j,k}$ is the component of f when projected onto $H_{j,k}$:

$$w_{j,k} = \langle f, H_{j,k} \rangle = \int_{-\infty}^{\infty} f \, H_{j,k}.$$

➤ The first property, orthonormality, is quite straightforward to prove – the case when $j = j'$, for example, is dealt with by the observation above about the supports of the $H_{j,k}$ for a fixed j. Here also is where the scaling factors are relevant: they simply ensure that when $j = j'$ and $k = k'$, the inner product has the specific value 1, rather than some other non-zero constant. (Detailed verification of orthonormality is left for Exercise 4.20, though that exercise could well be worked in the context of the present chapter.)

The second property is considerably harder to prove, and we postpone detailed discussion for later chapters (see especially Exercises 4.21, 4.22 and 4.23).

1.4.4 An Example

Let us now examine the Haar wavelet series for a specific function in some detail. We choose for the analysis a function which is simple enough to allow us to understand its wavelet series thoroughly but which is not so simple as to be trivial. Specifically, we consider

$$f(x) = \begin{cases} \sin 4\pi x, & -1 \le x < 1, \\ 0, & \text{otherwise,} \end{cases}$$

which is pictured, graphed on its support $[-1, 1)$, in Figure 1.3.

To begin the analysis, we note that the carefully chosen form of the function gives rise to two simplifications in the corresponding wavelet series expansion.

First, the choice of the coefficient 4π for x means that each of the intervals $[-1, 0]$ and $[0, 1]$ contains *exactly two* complete oscillations of the sine curve, which means in turn that *when $j \le 0$, the wavelet coefficients $w_{j,k}$ will be 0 for all k.* Similarly, we will also find that $w_{2,k} = 0$ for all k. (These claims should all be checked carefully; see Exercise 1.12.)

Second, recall that we found earlier that the support of $H_{j,k}$ is the interval $[k/2^j, (k+1)/2^j)$. Since the support of f is $[-1, 1)$, the integral defining the coefficient $w_{j,k}$ will be 0 unless these two supports intersect, and it is easy to check that this occurs for the values $j > 0$ (which by the previous paragraph are the only values of j that we need to consider) if and only if

$$-2^j \le k < 2^j.$$

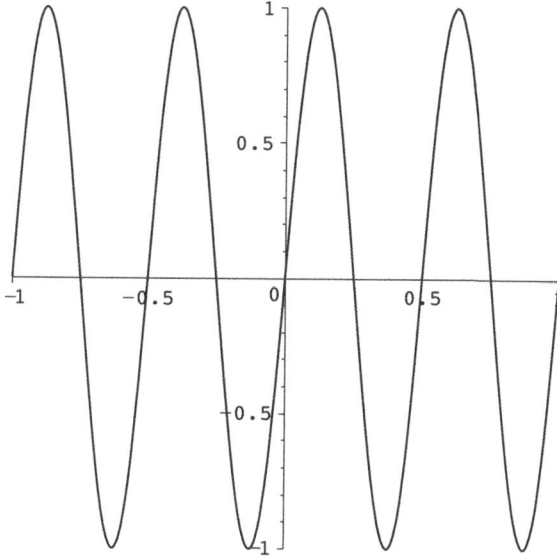

Figure 1.3 The graph of $\sin 4\pi x$

The general expression for the Haar wavelet series expansion of f, namely,

$$f = \sum_{j=-\infty}^{\infty} \sum_{k=-\infty}^{\infty} w_{j,k} H_{j,k},$$

therefore reduces to

$$f = \sum_{j=1}^{\infty} \sum_{k=-2^j}^{2^j-1} w_{j,k} H_{j,k}.$$

Figure 1.4 presents a series of graphs illustrating how the partial sums of this series approximate the original function better and better as more terms are included. Specifically, we graph the partial sums

$$\sum_{j=1}^{N} \sum_{k=-2^j}^{2^j-1} w_{j,k} H_{j,k},$$

for $N = 2, 3, \ldots, 7$. (Note that the graph for $N = 1$ is omitted. This is because the discussion above showed that $w_{2,k} = 0$ for all k, which implies that the graphs for $N = 1$ and $N = 2$ must be identical.)

24

1 An Overview

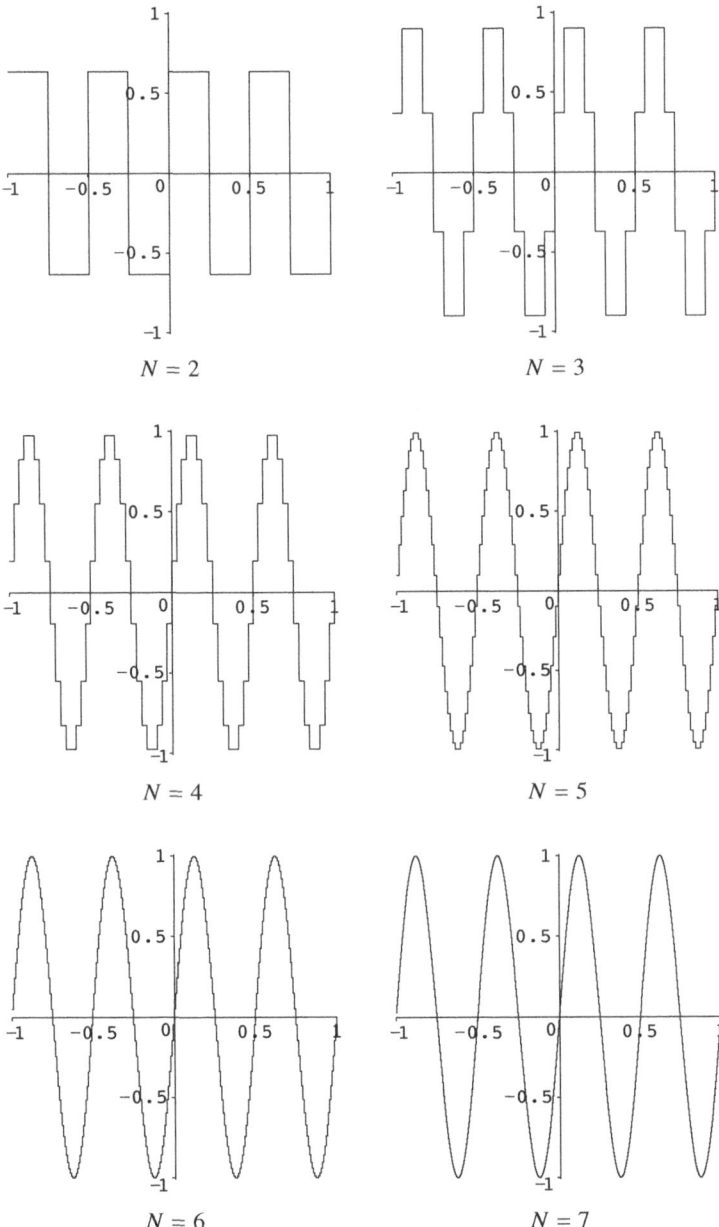

$N = 2$ 　　　 $N = 3$

$N = 4$ 　　　 $N = 5$

$N = 6$ 　　　 $N = 7$

Figure 1.4 Graphs of some partial sums of the Haar wavelet series for $\sin 4\pi x$

1.5 The Wavelet Phenomenon

Although the graphs in Figure 1.4 all relate to one carefully chosen function f, we can use them to illustrate some general points about the phenomenon of wavelets. These points are interesting and important mathematically and conceptually, and explain something of why wavelets are of great interest in applications.

The discussion will necessarily be, at this stage, rather informal, but we will see that the mathematical details of wavelet theory, when we later study them more thoroughly, justify it in a quite precise way.

1.5.1 Wavelets and Local Information

Let us focus for a moment on the information that one particular scaled, dilated, translated copy $H_{j,k}$ of the Haar wavelet H carries with it in the wavelet expansion of a given function f. This information, mathematically, is nothing other than the wavelet coefficient

$$w_{j,k} = \langle f, H_{j,k} \rangle = \int_{-\infty}^{\infty} f \, H_{j,k},$$

which is just the component of f on $H_{j,k}$.

What does this value tell us about f? Manipulating the integral, we find that

$$\int_{-\infty}^{\infty} f \, H_{j,k} = \int_{k/2^j}^{(k+1)/2^j} f \, H_{j,k}$$

(since the support of $H_{j,k}$ is $[k/2^j, (k+1)/2^j)$)

$$= 2^{j/2} \left[\int_{k/2^j}^{(k+1/2)/2^j} f \; - \; \int_{(k+1/2)/2^j}^{(k+1)/2^j} f \right]$$

(using the definition of $H_{j,k}$),

and it is reasonable to conclude from this that the quantity $w_{j,k}$ provides only a quite crude measure of the 'behaviour' or 'structure' of f. This is because:

- it carries no information whatsoever about f *outside* of $[k/2^j, (k+1)/2^j)$, the support of $H_{j,k}$; and
- the only information it gives us about f *inside* the support of $H_{j,k}$ is a kind of coarse 'average' of f, scaled by the factor $2^{j/2}$.

Within the support, the fine details about f are simply 'blurred' by the integration; in particular, many functions other than the given function f would yield exactly the same value for the integral $\int_{-\infty}^{\infty} f H_{j,k}$.

We might therefore say that in the wavelet expansion of f, a single coefficient $w_{j,k}$ *represents information about f down to a scale or resolution of* $1/2^j$ *on the support* $[k/2^j, (k+1)/2^j)$ *of* $H_{j,k}$, *but not in any other region, and not to any finer resolution.* Consequently, since for any fixed j the supports of the functions $\{H_{j,k} : k \in \mathbb{Z}\}$ cover the whole real line (without overlap), the corresponding family of terms in the wavelet series *represents information about f over the whole real line down to a scale or resolution of* $1/2^j$, *but not to any finer resolution.*

1.5.2 Wavelets and Global Information

Although we know that the local information carried by a single wavelet coefficient is only small, we also know that at the global level the situation is completely different. This is simply because we know that f can be expressed as the sum of its Haar wavelet series: the collection of all the coefficients taken together *carries enough information to reconstruct f.*

A closely related point is that the scaled, dilated, translated copies $H_{j,k}$ of H 'look the same', informally speaking, at every level of magnification. For the copies $H_{J,k}$ at any fixed scale $j = J$, there is an infinity of other copies $H_{j,k}$ both on a smaller scale (with $j > J$) and on a larger scale (with $j < J$). In particular, there is no 'privileged' scale, even though it is tempting to think of the scale corresponding to $j = 0$ as somehow special. Thus *a wavelet extracts information about a function uniformly at all scales or resolutions*, an observation that reinforces the thrust of our discussion above.

A straightforward consequence of this discussion is worth noting. Since, for each j, the functions $H_{j,k}$ for $k \in \mathbb{Z}$ extract all the detail about f at a resolution of $1/2^j$, it follows that the partial sum

$$\sum_{j=-\infty}^{J} \sum_{k=-\infty}^{\infty} w_{j,k} H_{j,k},$$

in which we include all the terms with $j \leq J$ but none with $j > J$, contains all the detail about f at all scales larger than or equal to $1/2^j$, and it is reasonable to refer to such a partial sum as giving a *coarse-grained image of f*, accurate to a resolution of $1/2^j$, but containing no information at finer levels of resolution. (In fact, we will see later that each such coarse-grained image is, in a precise sense, *the best possible approximation to f* at the chosen resolution.)

It should be noted how the graphs of Figure 1.4 demonstrate this phenomenon for the simple function f used in Subsection 1.4.4: each one shows an image of f accurate to the resolution $1/2^N$ determined by the parameter N, but no more accurate.

1.5.3 Wavelets and Applications

Our discussion above of the wavelet phenomenon helps to explain why wavelets are of importance in various applications. The extraction of detail on all scales simultaneously, which is at the root of the behaviour of wavelets, is often what is required in applications such as signal analysis and image compression. In signal analysis, one often wants to analyse signals in such a way that all frequency components are extracted and processed in the same way; similarly, in image compression one often wants to extract the details of an image at all levels of resolution simultaneously. (Image processing is of course a problem in a 2-dimensional domain rather than the 1-dimensional domain we have been considering, but the principle is exactly the same, and there is a theory of wavelets in two dimensions, or indeed in n dimensions for arbitrary $n \in \mathbb{N}$, similar to the one we have been informally exploring for $n = 1$.) The advantage of using wavelets for these sorts of applications is that wavelets exhibit the kind of behaviour we have just been discussing *as an intrinsic feature*. Other mathematical techniques have been devised to achieve somewhat similar effects, but none of them does so in the completely natural way that wavelets do.

Exercises

1.1   Show that the normalisation $(1/\|b\|)\, b$ of a non-zero vector $b \in \mathbb{R}^n$ is a unit vector.

1.2   This exercise involves confirming that various definitions in $\mathbb{R}^n$ from Section 1.1 are correct in $\mathbb{R}^2$ and $\mathbb{R}^3$, where we can use elementary geometry to check them. Let a and b be vectors in $\mathbb{R}^n$.

(a) If b is a unit vector, then the component of a on b is defined in Section 1.1 to be $\langle a, b \rangle$. Prove that this formula is correct in $\mathbb{R}^2$ and $\mathbb{R}^3$. (Include an analysis of how the sign of $\langle a, b \rangle$ is related to the orientation of a with respect to b.)

(b) It is claimed in Section 1.1 that the fact that the norm $\|a - b\|$ is the distance of the point a from the point b is just a reformulation of Pythagoras' theorem. Prove this claim in $\mathbb{R}^2$ and $\mathbb{R}^3$.

(c) Suppose that a and b are non-zero and let θ be the angle between a and b. The formula

$$\cos \theta = \frac{\langle a, b \rangle}{\|a\| \, \|b\|}$$

was noted in Section 1.1. Use trigonometry to show that this formula is correct in $\mathbb{R}^2$ and (though this is a bit more difficult) in $\mathbb{R}^3$.

1.3   Let $n \in \mathbb{N}$. Show, as claimed in Section 1.1, that any collection of orthonormal vectors in $\mathbb{R}^n$ must be linearly independent, and conclude that the collection is a basis for $\mathbb{R}^n$ if and only if it has n elements. (Also see Theorem 3.15 below for a more general result.)

1.4   Prove that the vectors

$$\left(\frac{1}{\sqrt{3}}, \frac{-1}{\sqrt{3}}, \frac{1}{\sqrt{3}}\right), \quad \left(\frac{1}{\sqrt{2}}, 0, \frac{-1}{\sqrt{2}}\right), \quad \left(\frac{1}{\sqrt{6}}, \frac{\sqrt{2}}{\sqrt{3}}, \frac{1}{\sqrt{6}}\right)$$

form an orthonormal basis for $\mathbb{R}^3$ (note that we are using row-vector notation here, rather than the column-vector notation of Section 1.1). Express the vector $c = (1, 2, 3)$ as a linear combination of these basis elements. If the coefficients in the combination are α_1, α_2 and α_3, calculate $(\alpha_1^2 + \alpha_2^2 + \alpha_3^2)^{1/2}$. Also calculate $\|(1, 2, 3)\|$, and explain your findings geometrically.

1.5   Consider the vectors

$$a_1 = (1, -1, 0), \quad a_2 = (2, 3, -1) \quad \text{and} \quad a_3 = (1, -1, -2).$$

(a) Show that a_1, a_2 and a_3 form a basis for $\mathbb{R}^3$, but not an orthonormal basis.

(b) Express the vector $(1, 2, 3)$ as a linear combination of a_1, a_2 and a_3.

1.6   Show that the vectors $e_1, e_2, e_3, \ldots$ in Subsection 1.2.1 form an orthonormal set in ℓ^2 but do not form a basis for ℓ^2.

1.7   (a) In the definition of the function spaces $F(I)$ at the beginning of Subsection 1.2.2, confirm that no use is made of the assumption that the set I is specifically an interval, and that we could therefore take it to be any chosen subset of $\mathbb{R}$.

(b) Confirm that $F(\{1, 2, \ldots, n\})$ can be naturally identified with $\mathbb{R}^n$ for each $n \in \mathbb{N}$ and that $F(\mathbb{N})$ can similarly be identified with $\mathbb{R}^\infty$.

1.8   Check the orthonormality relations given for the trigonometric functions in Subsection 1.2.2.

1.9   For the Fourier series example presented in Section 1.3, check the correctness of the series expansion given and use a computer algebra system to produce diagrams like those shown.

1.10 Sketch the graphs of the scaled, dilated, translated copies $H_{j,k}$ of the Haar wavelet H for a range of values of both j and k, using the same scale for all. Suitable choices might be

$$\begin{aligned} H_{-1,k} \quad &\text{for } k = -1, 0, \\ H_{0,k} \quad &\text{for } k = -2, -1, 0, 1, \\ H_{1,k} \quad &\text{for } k = -4, -3, -2, -1, 0, 1, 2, 3. \end{aligned}$$

1.11 In this exercise we study dilation and translation, and their interrelations, in a more systematic and general way than in the main text. Fix $b, c \in \mathbb{R}$ with $b \neq 0$. For $f \colon \mathbb{R} \to \mathbb{R}$, define the dilated function $D_b f$ by setting $D_b f(x) = f(bx)$ and the translated function $T_c f$ by setting $T_c f(x) = f(x - c)$, for all $x \in \mathbb{R}$.

(a) Show that $D_b \circ T_c = T_{c/b} \circ D_b$, that is, that $D_b(T_c f) = T_{c/b}(D_b f)$ for all f.

(b) If H is the Haar wavelet and $j, k \in \mathbb{Z}$, use part (a) to write down two expressions for $H_{j,k}$ in terms of H.

(c) If the support of f is the interval $[s, t]$, find the supports of $D_b f$ and $T_c f$.

(d) Assuming that $f, g \in L^2(\mathbb{R})$, show that $\langle f, D_b g \rangle = b^{-1} \langle D_{b^{-1}} f, g \rangle$ and $\langle f, T_c g \rangle = \langle T_{-c} f, g \rangle$.

(e) Find simplified expressions for $\langle D_b f, D_b g \rangle$ and $\langle T_c f, T_c g \rangle$.

1.12 This exercise refers to the example in Subsection 1.4.4.

(a) Check the various unproved assertions made in the discussion of the example.

(b) Calculate the coefficients $w_{j,k}$ for $j = 1, 2, 3$ and $k = -2^j, \ldots, 2^j - 1$, and hence graph by hand the partial sum of the wavelet series for $N = 2, 3$.

(c) Program a computer algebra system to produce the graphs shown in the main text.

1.13 Consider the function f given by

$$f(x) = \begin{cases} 1, & 0 \le x < 1, \\ 0, & \text{otherwise.} \end{cases}$$

(a) Sketch the graph of f.

(b) Show that the Haar wavelet coefficients $w_{j,k} = \langle f, H_{j,k} \rangle$ are given by

$$w_{j,k} = \begin{cases} 0, & \text{if } j \ge 0 \text{ or if } j < 0 \text{ and } k \neq 0, \\ 2^{j/2}, & \text{if } j < 0 \text{ and } k = 0, \end{cases}$$

and that the Haar wavelet series for f can therefore be written in the form

$$\sum_{j=-1}^{-\infty} 2^{j/2} H_{j,0}.$$

(c) Confirm that the value at $x \in \mathbb{R}$ of the function defined by this series can be written in the form

$$\sum_{j=-1}^{-\infty} 2^j H(2^j x) = \tfrac{1}{2} H(\tfrac{1}{2} x) + \tfrac{1}{4} H(\tfrac{1}{4} x) + \tfrac{1}{8} H(\tfrac{1}{8} x) + \cdots.$$

(d) Sketch the partial sums

$$\sum_{j=-1}^{-N} 2^j H(2^j x)$$

of this series for $N = 0, 1, 2, 3$, and compare them with the graph of f.

1.14 Consider the function g given by

$$g(x) = \begin{cases} \frac{1}{2} - x, & \text{if } 0 \le x < 1, \\ 0, & \text{otherwise.} \end{cases}$$

(a) Draw the graph of g.

(b) Show that the Haar wavelet coefficients $w_{j,k} = \langle g, H_{j,k} \rangle$ of g can only be non-zero when $j \ge 0$ and $0 \le k < 2^j$. (Consider the relevant supports.)

(c) Show further that for each fixed $j \ge 0$, the coefficients $w_{j,k}$ are equal for $0 \le k < 2^j$. (The fact that g is linear on its support is the key point here.)

(d) Compute the relevant coefficients, and hence sketch the partial Haar wavelet series sums

$$\sum_{j=0}^{N} \sum_{k=0}^{2^j-1} w_{j,k} H_{j,k}$$

for $N = 0, 1, 2, 3$.

2

Vector Spaces

This short chapter is about vector spaces, though it is by no means a self-contained introduction to the topic. It is assumed that the reader has already studied vector spaces and is familiar with the fundamentals of the theory – for example, the concepts of *linear independence*, *linear span*, *basis* and *dimension*.

We have three main aims in the chapter. The first is to give a brief but precise account of the definition of a vector space in 'axiomatic' style (see the discussion of this term later), partly as preparation for the definition of an inner product space in the same style in Chapter 3. The second is to develop a simple but very useful criterion for a subset of a vector space to be a subspace. The third is to survey more systematically than in Chapter 1 the vector spaces that will be most important for our later work, which are generally of infinite dimension. The three sections of the chapter correspond roughly to these three topics.

2.1 The Definition

We begin with the definition of a vector space.

A vector space consists of a set V, the elements of which are called **vectors**, and a field of **scalars**, which in this book (except for the final chapter) will always be the field of real numbers $\mathbb{R}$.

There are two operations involving these two sets. The first is the binary operation '+' on V of **addition** of vectors, and the second is the operation of **scalar multiplication**, or multiplication of vectors by scalars, which is not usually given its own symbol, but can be denoted by '$\cdot$' if necessary. Specifying that + is a binary operation on V is equivalent to requiring that + be a function

$$+ : V \times V \to V.$$

31

The operation of scalar multiplication $\cdot$ is required to be a function

$$\cdot : \mathbb{R} \times V \to V.$$

As usual, rather than writing $+(x, y)$ for two vectors $x, y \in V$, we write $x + y$, and rather than writing $\cdot (\alpha, x)$ for a scalar $\alpha \in \mathbb{R}$ and a vector $x \in V$, we write $\alpha \cdot x$ or just αx.

➤ Note that the operation of scalar multiplication indicated here by the dot '$\cdot$' bears no relation to the dot product mentioned in Chapter 1. As we noted there, we will always call the dot product the inner product and we will always represent it using angle-bracket notation. (In any case, we nearly always omit the symbol for scalar multiplication.)

The two operations are required to satisfy a certain collection of axioms, some pertaining to just one of the operations and others linking the two operations. Here is the list of axioms, in which it is assumed that x, y and z are arbitrary vectors in V and α and β are arbitrary scalars in $\mathbb{R}$.

First, we have a set of axioms involving only vector addition:

(V1)   *Commutativity of addition*: $x + y = y + x$.
(V2)   *Associativity of addition*: $(x + y) + z = x + (y + z)$.
(V3)   *Existence of an identity for addition*:   There exists a vector $0 \in V$ such that $0 + x = x + 0 = x$.
(V4)   *Existence of inverses for addition*:   For each vector $x \in V$, there exists a vector $-x \in V$ such that $x + (-x) = (-x) + x = 0$.

➤ If you have studied group theory, you will recognise that these four axioms can be summed up by saying that *V is an abelian group under the operation of addition*.

Second, we have two axioms involving only scalar multiplication:

(V5)   *Associativity of multiplication*: $\alpha(\beta x) = (\alpha\beta)x$.
(V6)   *Action of* 1: $1 \cdot x = x$.

Last, we have two axioms that specify the relations between vector addition and scalar multiplication:

(V7)   *Distributivity for scalar addition*: $(\alpha + \beta)x = \alpha x + \beta x$.
(V8)   *Distributivity for vector addition*: $\alpha(x + y) = \alpha x + \alpha y$.

We need to make a number of observations about the definition.

• We describe our presentation of the definition of a vector space as *axiomatic*, because it consists simply of a list of *axioms* (or *laws* or *rules*) that any vector space must satisfy. Although the axioms are derived from our experience with familiar examples, they are not tied to any of the particulars of those examples. Thus they make no assumptions about what the vectors are as objects: whether the vectors are n-tuples in $\mathbb{R}^n$ or sequences in ℓ^2 or functions in an L^2 space or something else again plays no role in the definition. (Our definition of an inner product space in the next chapter will be formulated in a similar way.)

- The vector space axioms can be formulated in a number of slightly different ways, which are all ultimately equivalent in the sense that any structure which is a vector space under one set of axioms will also be a vector space under any of the other sets.

  For example, two *laws of closure* are often listed among the axioms for vector spaces. The *law of closure under vector addition* would read

  $$if\ x, y \in V,\ then\ x + y \in V,$$

  while the *law of closure under scalar multiplication* would read

  $$if\ \alpha \in \mathbb{R}\ and\ x \in V,\ then\ \alpha \cdot x \in V.$$

  However, we have stated the vector space axioms in such a way that *explicit* closure axioms are not needed: they are equivalent to our description of $+$ and $\cdot$ as functions. For example, the specification of $+$ as a function

  $$+ : V \times V \to V$$

  tells us that when two vectors in V are added, then the result is again in V, and this is precisely the law of closure under vector addition.

  Although closure laws do not play a formal role in our definition of a vector space, it is useful to be able to refer to them, and we will do so at several points below (see for example Theorem 2.1; also note that we have already referred to the laws in passing in Chapter 1).

- If it is necessary for clarification, we can speak of our vector space as a vector space *over the field* $\mathbb{R}$, meaning that the scalars involved are elements of $\mathbb{R}$. There are many fields other than $\mathbb{R}$, and a systematic study of vector spaces would examine vector spaces over fields in general. The field $\mathbb{C}$ of complex numbers is probably the most familiar field after $\mathbb{R}$, followed by the field $\mathbb{Q}$ of rational numbers, but there are many others, including a whole family of finite fields. There is an entire branch of mathematics devoted to the study of fields in general, in which an axiomatic definition of a field would be given. However, the field $\mathbb{R}$ is the only field we consider in this book (except in the final chapter), and it is therefore unnecessary for us to list the general field axioms.

2.2 Subspaces

A **subspace** of a vector space V is a subset of V which is a vector space in its own right, using the operations $+$ and $\cdot$ inherited from V. Our aim in this section is to investigate which of the vector space axioms need to be checked in order to show that a given subset of V is a subspace.

Suppose that $X \subseteq V$ is the subset in question. Now since we can add any two vectors in V, and since X is a subset of V, we can certainly add any two vectors in X; but although we know that their sum is in V, we have no guarantee that their sum is *in X*, which is what we require for closure of X under vector addition. A similar comment applies to scalar multiplication. Thus the closure laws are two criteria that we will need to check if we wish to show that X is a subspace.

What about axioms (V1) to (V8)? It is easy to see that axioms (V1) and (V2) and axioms (V5) to (V8) *hold automatically in X*. This is because they make no demands on *where* particular vectors lie – specifically, on whether or not they are in X – but require only that certain equations hold. Since these equations already hold throughout V, they hold automatically in X too.

Therefore, the only axioms that X might fail to satisfy are axioms (V3) and (V4), which *do* make demands on where certain vectors lie: the zero vector should be *in X*, not just in V, and elements of X should have additive inverses *in X*, not just in V.

However, it turns out, most conveniently, that if X is non-empty and the two closure requirements are established – that X is closed under vector addition and scalar multiplication – then axioms (V3) and (V4) hold automatically.

Theorem 2.1 *A non-empty subset of a vector space is a vector subspace if and only if it is closed under vector addition and scalar multiplication.*

To prove this, we need the following two simple auxiliary facts.

Lemma 2.2 *If V is a vector space and $v \in V$, then*

(i) $0 \cdot v = 0$ *and*
(ii) $(-1) \cdot v = -v$.

Proof   First, since $0 + 0 = 0$, axioms (V3) and (V7) give

$$0 \cdot v = (0 + 0) \cdot v = (0 \cdot v) + (0 \cdot v),$$

and then subtracting $0 \cdot v$ from both sides (or, more precisely, using (V4) and *adding* $-(0 \cdot v)$ to both sides) gives $0 \cdot v = 0$, proving (i).

Second, since $1 + (-1) = 0$, we have

$$(1 + (-1)) \cdot v = 0 \cdot v = 0.$$

Therefore, axiom (V7) gives $1 \cdot v + (-1) \cdot v = 0$, and so by (V6) we have $v + (-1) \cdot v = 0$. Subtracting v from both sides then gives $(-1) \cdot v = -v$, proving (ii).   □

The proof of the theorem is now easy.

Proof of Theorem 2.1   Let X be a non-empty subset of a vector space V, and suppose that the two closure laws hold in X. Let x be any vector in X (there is one, since X is non-empty). Then using closure under scalar multiplication and the lemma, the vector $0 \cdot x = 0$ must be in X, proving (V3). Further, by the lemma again, $(-1) \cdot x = -x \in X$, proving (V4).   □

➤ It is worth noting in passing that although it is necessary for X to be closed under vector addition for it to be a subspace, the above argument did not actually make use of closure under vector addition: (V3) and (V4) follow from closure under scalar multiplication alone.

2.3 Examples

Here we survey our standard collection of vector spaces. We have looked at each of these already in Chapter 1, and we will generally not stop here to check that the vector space axioms hold; in most cases, this is straightforward.

2.3.1 Spaces of n-Tuples

For each $n \in \mathbb{N}$, the **Euclidean space** $\mathbb{R}^n$ is the vector space of all n-tuples of real numbers, with both vector addition and scalar multiplication defined entry by entry. This space is a vector space of dimension n, and has as its **standard basis** the n vectors

$$e_1 = (1, 0, 0, \ldots, 0),$$
$$e_2 = (0, 1, 0, \ldots, 0),$$
$$e_3 = (0, 0, 1, \ldots, 0),$$
$$\vdots$$
$$e_n = (0, 0, 0, \ldots, 1).$$

2.3.2 Spaces of Sequences

Generalising from $\mathbb{R}^n$ for $n \in \mathbb{N}$, we define $\mathbb{R}^\infty$ to be the vector space of all infinite sequences of real numbers, with both vector addition and scalar multiplication defined entry by entry. The straightforward formal verification that $\mathbb{R}^\infty$ is a vector space is left for the exercises (representative parts of the argument were of course outlined in Chapter 1).

As we noted in Chapter 1, $\mathbb{R}^\infty$ is an obvious generalisation of $\mathbb{R}^n$ for finite n, but for many purposes it is not the most useful one. Usefulness depends, of course, on the intended application, and there are a number of subspaces of $\mathbb{R}^\infty$ that have applications in various areas, but the most useful for us will be the following space.

Definition 2.3 The space ℓ^2 is defined by

$$\ell^2 = \left\{ (a_1, a_2, a_3, \ldots) \in \mathbb{R}^\infty : \sum_{n=1}^{\infty} a_n^2 \text{ converges} \right\}.$$

Thus, ℓ^2 consists of those sequences $(a_1, a_2, a_3, \ldots)$ of real numbers for which the infinite series $\sum_{n=1}^{\infty} a_n^2$ converges. We often describe ℓ^2 in words as the vector space of **square-summable sequences**.

We will prove that ℓ^2 is a vector space by showing that it is a vector subspace of $\mathbb{R}^\infty$. Since we know that $\mathbb{R}^\infty$ is a vector space and ℓ^2 is obviously a non-empty subset, the appropriate tool to use is Theorem 2.1, and only the two closure laws need to be checked.

Theorem 2.4 *The space ℓ^2 is a vector space.*

Proof For closure under addition, let

$$a = (a_1, a_2, a_3, \ldots), \ b = (b_1, b_2, b_3, \ldots) \in \ell^2,$$

so that the series

$$\sum_{n=1}^{\infty} a_n^2 \quad \text{and} \quad \sum_{n=1}^{\infty} b_n^2$$

converge. We have to prove that

$$a + b = (a_1 + b_1, a_2 + b_2, a_3 + b_3, \ldots) \in \ell^2,$$

that is, that the series

$$\sum_{n=1}^{\infty} (a_n + b_n)^2$$

converges.

We claim that $(a + b)^2 \leq 2(a^2 + b^2)$ for any real numbers a and b. To see this, we start with the inequality $(a - b)^2 \geq 0$; this immediately gives $2ab \leq a^2 + b^2$, and the desired inequality then follows from the identity $(a+b)^2 = a^2 + 2ab + b^2$.

Now, by the inequality, we have

$$(a_n + b_n)^2 \leq 2(a_n^2 + b_n^2)$$

for each n. Therefore, since $\sum_{n=1}^{\infty} 2(a_n^2 + b_n^2)$ converges, so does $\sum_{n=1}^{\infty}(a_n + b_n)^2$, by the Comparison Test for infinite series. Hence $a + b \in \ell^2$, and we conclude that ℓ^2 is closed under addition. The proof that ℓ^2 is closed under scalar multiplication – that is, that if $a \in \ell^2$ then $\alpha a \in \ell^2$ for all scalars α – is straightforward and is left as an exercise. $\qquad\square$

Note carefully that the infinite set of vectors

$$e_1 = (1, 0, 0, \dots), \quad e_2 = (0, 1, 0, \dots), \quad e_3 = (0, 0, 1, \dots), \dots$$

is *not* a basis for ℓ^2 (or for $\mathbb{R}^{\infty}$), contrary to what our experience with $\mathbb{R}^n$ for finite n might lead us to expect. This is because if a set of vectors is a basis for a space, then each vector in the space must be expressible as a linear combination of a *finite* number of vectors from that set; but any finite linear combination of vectors from the set $e_1, e_2, e_3, \dots$ can have only *a finite number of non-zero entries*, and the set of all such combinations therefore cannot exhaust the whole of ℓ^2. (Recall the brief discussion of this and closely related issues near the end of Subsection 1.2.1, and see Exercises 1.6 and 2.4.)

The space ℓ^2 as defined above consists of all square-summable sequences indexed by the set $\mathbb{N}$ of natural numbers. Sometimes it is useful to allow indexing by $\mathbb{Z}$, the set of all integers. This gives us the space $\ell^2(\mathbb{Z})$ of *doubly infinite square-summable sequences*, that is, sequences of the form

$$a = (\dots, a_{-2}, a_{-1}, a_0, a_1, a_2, \dots)$$

with the property that $\sum_{n=-\infty}^{\infty} a_n^2$ converges. An infinite sum must always be defined as the limit of a sequence of partial sums, and for concreteness we take our partial sums in this case to be over a symmetric finite range, so that $\sum_{n=-\infty}^{\infty} a_n^2$ is defined to be

$$\lim_{n \to \infty} \sum_{k=-n}^{n} a_k^2.$$

Note that for clarity or emphasis, we could denote the space ℓ^2 by $\ell^2(\mathbb{N})$.

2.3.3 Spaces of Functions

Let I be an interval on the real line, of any form, and finite or infinite in length; thus, cases such as $I = [0, 1]$, $I = [-\pi, \pi]$, $I = [0, \infty)$ and $I = \mathbb{R}$ are all permitted. We wish to define vector spaces of functions with the interval I as their domain. (It is quite possible, in fact, to allow I to be a set other than an interval – something like $[0, 1] \cup [2, 3]$ is perfectly admissible, for example, and see also Exercise 1.7 – but we will not need to consider such cases.)

Now the set $F(I)$ of all functions from I into $\mathbb{R}$ becomes a vector space if we define addition and scalar multiplication of functions pointwise, that is, by setting

$$(f + g)(x) = f(x) + g(x) \quad \text{and} \quad (\alpha f)(x) = \alpha f(x)$$

for all $f, g \in F(I)$, all $\alpha \in \mathbb{R}$ and all $x \in I$; checking that the vector space axioms hold is easy. Like $\mathbb{R}^\infty$, this space is not of much use to us here in itself, since we generally want our functions to have additional properties such as continuity or differentiability or integrability, as below. It is useful nevertheless for us to know that $F(I)$ is a vector space, for the same reason as for $\mathbb{R}^\infty$: it can greatly simplify the task of checking that various other more useful sets of functions are vector spaces.

We denote by $C(I)$ the vector space of continuous real-valued functions on the interval I. Using Theorem 2.1 to check that $C(I)$ is indeed a vector space is straightforward, and is left as an exercise.

Various subspaces of $C(I)$ in turn are often encountered. For example, for any $k \in \mathbb{N}$, we have the space $C^k(I)$ of functions that have continuous derivatives up to and including order k. It is also usual to define $C^0(I)$ to be $C(I)$, and to define $C^\infty(I)$ to be the space of all infinitely differentiable functions on I – that is, the functions that have (necessarily continuous) derivatives of all orders – and we then have the sequence of inclusions

$$C(I) = C^0(I) \supseteq C^1(I) \supseteq C^2(I) \supseteq \cdots \supseteq C^\infty(I).$$

The function spaces that we deal with most frequently in this book, however, will be spaces of integrable functions. Specifically, we will later study the space $L^2(I)$ for an interval I, which consists of the functions f on I for which the integral

$$\int_I f^2$$

exists and is finite; $L^2(I)$ is often referred to as the space of **square-integrable functions** on I. The two commonest cases in our discussion are those for $I = [-\pi, \pi]$ and $I = \mathbb{R}$.

We must note, at this point, that the integral in the expression defining $L^2(I)$ is not the integral that is usually encountered in undergraduate mathematics courses. The usual integral is called the **Riemann integral** when we are working in contexts such as the present one where it is necessary to give it a distinguishing name. Unfortunately, there are a number of respects in which the Riemann integral is inadequate for advanced work, and it is then substituted by the more general **Lebesgue integral**. (It is from this integral that the L^2 spaces get their 'L'.)

To say that the Lebesgue integral is a generalisation of the Riemann integral means that every Riemann-integrable function is Lebesgue-integrable and that the two integrals have the same numerical value when they are both defined; but the Lebesgue integral is *strictly* more general than the Riemann integral, in the sense that there are functions that are Lebesgue-integrable but not Riemann-integrable.

As an illustration, consider the function f defined on $[0, 1]$ by

$$f(x) = \begin{cases} 1, & x \text{ rational,} \\ 0, & x \text{ irrational.} \end{cases}$$

This function is not Riemann-integrable on $[0, 1]$, but is Lebesgue-integrable, with $\int_0^1 f = 0$. Further, since $f^2 = f$, we have $\int_0^1 f^2 = 0$, and so $f \in L^2([0, 1])$. Now this function f is a rather peculiar one – it would certainly be unusual to encounter a function like it in an application – but there turn out to be good reasons for admitting it into the collection of integrable functions, along with other functions that are even more exotic.

Since we have not attempted to explain how the Lebesgue integral is defined, we cannot give a complete or rigorous proof (via Theorem 2.1) that $L^2(I)$ is indeed a vector space. However, once a few elementary and quite unsurprising facts are known about Lebesgue integration, a proof can be constructed by mimicking the steps of the earlier proof for ℓ^2.

To prove that $L^2(I)$ is closed under addition, for example, we could argue as follows. Given f and g in $L^2(I)$, we need to show that $f + g \in L^2(I)$. Now the algebra used in the ℓ^2 argument shows that for each $x \in I$,

$$(f(x) + g(x))^2 \leq 2(f(x)^2 + g(x)^2).$$

But the fact that $f, g \in L^2(I)$ implies that the integral of the right-hand side over I is finite, and so the inequality tells us that the integral of the left-hand side is also finite. Thus $f + g \in L^2(I)$, as required.

We will say a little more in Chapters 3 and 4 to help clarify the distinction between Riemann and Lebesgue integration and hence to clarify which functions belong to $L^2(I)$. In Subsection 4.4.5 in particular we discuss some of the inadequacies of the Riemann integral that were referred to above.

However, this book is not the place for a detailed or rigorous discussion of these questions, which for proper treatment need a book in their own right. For our purposes, little harm is done by thinking of the elements of $L^2(I)$ as, say, piecewise continuous, even though, as the function f defined above shows, some elements of $L^2(I)$ are in fact considerably 'wilder' than that.

Exercises

2.1 Write out proofs of the vector space axioms for $\mathbb{R}^n$, $\mathbb{R}^\infty$ and $F(I)$ for an interval I. Confirm in particular how in each case the arguments are ultimately reflections of corresponding laws in $\mathbb{R}$.

2.2 Write out the arguments of Lemma 2.2 in full detail, justifying every step by reference to a vector space axiom or a fact about the real numbers.

2.3 Let V be a vector space, and let $\alpha \in \mathbb{R}$ and $x \in V$. Prove that $\alpha \cdot 0 = 0$ and that $\alpha \cdot x = 0$ implies that either $\alpha = 0$ or $x = 0$. (The proofs should be in the spirit of those of Lemma 2.2, using only the vector space axioms and the properties of the real numbers. Note that as in Lemma 2.2 some occurrences of the symbol '0' above refer to $0 \in \mathbb{R}$ and some to $0 \in V$.)

2.4 (a) For each $n \in \mathbb{N} \cup \{0\}$, find a subspace of ℓ^2 of dimension n and write down a basis.

   (b) Show that the collection of sequences of real numbers that have only a finite number of non-zero entries forms a proper subspace of ℓ^2. Confirm that the set $\{e_1, e_2, e_3, \ldots\}$ forms a basis for the subspace, which therefore has infinite dimension. (Compare with the discussion in Subsection 2.3.2.)

2.5 For which real numbers α does the sequence whose nth term is n^α for all n belong to ℓ^2?

2.6 Denote by c the set of all convergent sequences of real numbers and by c_0 the set of all sequences of real numbers that converge to 0. Show using Theorem 2.1 that c and c_0 are subspaces of $\mathbb{R}^\infty$.

2.7 Define ℓ^∞ to be the set of bounded sequences of real numbers. (A sequence $(a_1, a_2, \ldots)$ is **bounded** if there exists $M \geq 0$ such that $|a_n| \leq M$ for all n.) Show that ℓ^∞ is a subspace of $\mathbb{R}^\infty$. What is the relation between ℓ^∞, the space ℓ^2 and the spaces c and c_0 of the previous exercise?

2.8 Define $\ell^p = \{(a_1, a_2, a_3, \ldots) : \sum_{n=1}^\infty |a_n|^p \text{ converges}\}$ for any p such that $1 \leq p < \infty$. (Note that $p = 2$ gives us the space ℓ^2 already defined.) Show that ℓ^p is a subspace of $\mathbb{R}^\infty$. What is the relation between ℓ^p and ℓ^q when $p < q$, and between ℓ^p and the space ℓ^∞ of the previous exercise? (To prove closure under addition in ℓ^p, you will need an inequality of the form $(a + b)^p \leq c(a^p + b^p)$ for $a, b \geq 0$ and a constant c that depends only upon p. You should be able to show without too much trouble that $(a + b)^p \leq 2^p(a^p + b^p)$, and that suffices. The tightest inequality, however, is $(a+b)^p \le 2^{p-1}(a^p+b^p)$. We proved this inequality within the proof of Theorem 2.4 for the case $p = 2$, but the general case is harder; the easiest proof is probably by application of the standard and very important inequality known as **Hölder's inequality**. We will not state this result, but it is easy to find in analysis and functional analysis texts.)

2.9 Show that $C^\infty(I) = \bigcap_{k=0}^\infty C^k(I)$.

2.10 Show that the set of polynomial functions $1, x, x^2, x^3, \ldots$ is linearly independent in $C(I)$ for any non-trivial interval I. (Note that in this exercise and the next we make use of the notational convention explained near the end of Subsection 1.2.2: where appropriate, x^k (for each fixed value of k) denotes the function whose value at x is x^k, and $e^{\alpha x}$ similarly denotes the function whose value at x is $e^{\alpha x}$. See also the comments in Subsection 3.5.4 below.)

2.11 Show that the set of all functions $e^{\alpha x}$ for $\alpha \in \mathbb{R}$ is linearly independent in $C(I)$ for any non-trivial interval I. (If you are familiar with the idea of the *cardinality* of a set, go on to deduce that the dimension of $C(I)$ as a vector space is equal to the cardinality of $\mathbb{R}$.) Show further that the collection does not span $C(I)$, and is therefore not a basis for $C(I)$.

2.12 Use your knowledge of differential equations to show that the set of solutions to the equation $y'' + y = 0$ is a subspace of $C^\infty(\mathbb{R})$ of dimension 2. Write down a basis for the subspace.

2.13 (a) Show that the function f on $\mathbb{R}$ defined by

$$f(x) = \begin{cases} 1, & x \text{ rational,} \\ 0, & x \text{ irrational} \end{cases}$$

is not continuous at any point of $\mathbb{R}$. (The Lebesgue-integrable but non-Riemann-integrable function defined in Subsection 2.3.3 and denoted there also by f is of course just the restriction of the new f to the interval $[0, 1]$.)

(b) Adapt the definition of f to produce a function that is continuous at and only at n chosen points $x_1, \ldots, x_n$ in $\mathbb{R}$.

3

Inner Product Spaces

In the first two sections of this chapter, we give an axiomatic definition of inner product spaces and survey our standard examples of such spaces. In the remaining sections, we examine the basic properties of inner product spaces and explore their geometry in some detail.

At all points in the development, the familiar and fundamental example of $\mathbb{R}^n$ with its usual inner product should be kept in mind, to motivate and illustrate the ideas.

3.1 The Definition

Definition 3.1   Let V be a vector space. We say that V is an **inner product space** if there is a function

$$\langle\,\cdot\,,\cdot\,\rangle \colon V \times V \to \mathbb{R},$$

called the **inner product**, with properties as follows, for all $x, y, z \in V$ and all $\alpha \in \mathbb{R}$:

(IP1) $\langle x, y \rangle = \langle y, x \rangle$;

(IP2) $\langle x + y, z \rangle = \langle x, z \rangle + \langle y, z \rangle$;

(IP3) $\langle \alpha x, y \rangle = \alpha \langle x, y \rangle$;

(IP4) $\langle x, x \rangle \geq 0$, and $\langle x, x \rangle = 0$ if and only if $x = 0$.

The properties described by these axioms have standard names: (IP1) is the property of **symmetry**; (IP2) and (IP3) together are the property of **linearity** (in the first argument position); and (IP4) is the property of **positive definiteness**.

➤ The notation '$\langle\,\cdot\,,\cdot\,\rangle$' is a standard and convenient way of indicating that the inner product is a function of two arguments, and that the arguments are written in the places occupied by the dots, as in '$\langle x, y\rangle$'.

The simple properties below follow immediately from the definition.

Theorem 3.2 *For all $x, y, z \in V$ and all $\alpha \in \mathbb{R}$,*

(i) $\langle x, y + z\rangle = \langle x, y\rangle + \langle x, z\rangle$,

(ii) $\langle x, \alpha y\rangle = \alpha\langle x, y\rangle$ *and*

(iii) $\langle 0, x\rangle = \langle x, 0\rangle = 0$.

Proof (i) $\langle x, y + z\rangle = \langle y + z, x\rangle = \langle y, x\rangle + \langle z, x\rangle = \langle x, y\rangle + \langle x, z\rangle$.

(ii) $\langle x, \alpha y\rangle = \langle \alpha y, x\rangle = \alpha\langle y, x\rangle = \alpha\langle x, y\rangle$.

(iii) $\langle 0, x\rangle = \langle 0x, x\rangle = 0\langle x, x\rangle = 0$. Hence $\langle x, 0\rangle = \langle 0, x\rangle = 0$. □

3.2 Examples

Here we will cycle through a sequence of vector spaces that will be familiar from Chapters 1 and 2, noting how each one can be made into an inner product space and formalising some of the informal discussion of Chapter 1.

3.2.1 The Euclidean Spaces $\mathbb{R}^n$

For $x = (x_1, x_2, \ldots, x_n), y = (y_1, y_2, \ldots, y_n) \in \mathbb{R}^n$, we define

$$\langle x, y\rangle = \sum_{i=1}^{n} x_i y_i.$$

This definition makes $\mathbb{R}^n$ into an inner product space; all the axioms are easy to check (see Exercise 3.1).

3.2.2 The Sequence Space ℓ^2

Recall that the space ℓ^2 consists of the square-summable sequences, that is, those sequences $(x_1, x_2, x_3, \ldots) \in \mathbb{R}^\infty$ for which the infinite series $\sum_{n=1}^{\infty} x_n^2$ converges. We define an inner product on this space by specifying that for $x = (x_1, x_2, x_3, \ldots)$ and $y = (y_1, y_2, y_3, \ldots)$ in ℓ^2,

$$\langle x, y\rangle = \sum_{n=1}^{\infty} x_n y_n.$$

Before we can check that the inner product space axioms hold here, we must check that the infinite sum defining $\langle x, y\rangle$ converges. Consider vectors $x, y \in \ell^2$, as above. The definition of ℓ^2 tells us that the series $\sum_{n=1}^{\infty} x_n^2$ and $\sum_{n=1}^{\infty} y_n^2$ both converge. Also, since we have already proved that ℓ^2 is a vector space, we know that $x + y \in \ell^2$, meaning that the series $\sum_{n=1}^{\infty} (x_n + y_n)^2$ converges. But for each $n \in \mathbb{N}$, we have the algebraic identity

$$x_n y_n = \tfrac{1}{2}((x_n + y_n)^2 - x_n^2 - y_n^2),$$

and it follows that

$$\langle x, y \rangle = \sum_{n=1}^{\infty} x_n y_n$$

converges, as required. (See also Exercise 3.3.)

With this established, the inner product space properties themselves are now nearly as easy to check as in the case of $\mathbb{R}^n$, the only minor complication being that we must deal with infinite rather than finite sums. As an example, here is a proof of inner product space axiom (IP2): for $x = (x_1, x_2, x_3, \ldots)$, $y = (y_1, y_2, y_3, \ldots)$ and $z = (z_1, z_2, z_3, \ldots)$ in ℓ^2,

$$\begin{aligned}
\langle x + y, z \rangle &= \sum_{n=1}^{\infty} (x_n + y_n) z_n \\
&= \sum_{n=1}^{\infty} (x_n z_n + y_n z_n) \\
&= \sum_{n=1}^{\infty} x_n z_n + \sum_{n=1}^{\infty} y_n z_n \\
&= \langle x, z \rangle + \langle y, z \rangle.
\end{aligned}$$

3.2.3 Spaces of Functions

Let I be a closed, bounded interval $[a, b]$, for some $a, b \in \mathbb{R}$ such that $a < b$. In the space of continuous functions $C(I)$, we define an inner product by setting

$$\langle f, g \rangle = \int_I fg$$

for all $f, g \in C(I)$. Note that this definition gives well-defined (finite) values for $\langle f, g \rangle$, because if f and g are continuous, then so is fg, and a continuous function on a closed, bounded interval is always integrable and its integral has a finite value.

The following proofs of the inner product properties are straightforward, depending only on elementary facts about integration and continuous functions. Let $f, g, h \in C(I)$ and $\alpha \in \mathbb{R}$.

(i) $\langle f, g \rangle = \int_I fg = \int_I gf = \langle g, f \rangle$.

(ii) $\langle f + g, h \rangle = \int_I (f + g)h = \int_I (fh + gh) = \int_I fh + \int_I gh = \langle f, h \rangle + \langle g, h \rangle$.

(iii) $\langle \alpha f, g \rangle = \int_I (\alpha f)g = \int_I \alpha(fg) = \alpha \int_I fg = \alpha \langle f, g \rangle$.

(iv) Since $\langle f, f \rangle = \int_I f^2$ and $f^2 \geq 0$ on I, we have $\langle f, f \rangle \geq 0$. If $f = 0$ (meaning that f is the zero vector in $C(I)$, that is, the function that has value 0 everywhere on I), then clearly $\langle f, f \rangle = 0$.

Finally, suppose that $\langle f, f \rangle = 0$, for some continuous function f. Then the function $g = f^2$ is a *non-negative* continuous function whose integral over I is 0, and this implies that g, and hence f, has value 0 at all points of I. (The last sentence is not so much a proof as a statement of what needs to be proved; Exercise 3.4 outlines a way of constructing a rigorous proof of the statement.)

➤ Although the collection $C(I)$ of continuous functions on I forms a vector space for all intervals I, it does not form an inner product space for arbitrary I, at least with the inner product defined above. This is why we restricted consideration to closed, bounded intervals I.

The explanation for this is simple. As noted, on a closed, bounded interval, the formula $\langle f, g \rangle$ for the inner product always yields well-defined finite values. But on intervals that are open at an endpoint or are infinite in length, there always exist continuous functions f with the property that the integral of f^2 is infinite, so that the formula that we want to use for the inner product will not always yield finite values (see Exercise 3.5).

The spaces of functions that we wish to spend most time with later are the L^2 spaces, and we will say as much now as we can about these spaces as inner product spaces in the absence of a detailed discussion of Lebesgue integration.

Here, we allow our interval I to range without restriction, in contrast to the case of $C(I)$: it may be closed, open or half-open and it may be of finite or infinite length. We begin by defining our proposed inner product on $L^2(I)$ using exactly the same formula that we used for the spaces $C(I)$:

$$\langle f, g \rangle = \int_I fg,$$

for $f, g \in L^2(I)$.

The first question that arises is whether this proposed inner product is well defined. That is, given $f, g \in L^2(I)$, we wish to know whether the integral $\int_I fg$ that defines $\langle f, g \rangle$ exists and is finite.

Although we have not defined the Lebesgue integral, we can still carry out the required argument on the basis of some reasonable assumptions. (A similar point was made in Subsection 2.3.3 in the discussion of the vector space axioms in $L^2(I)$.) It is easy enough to accept that *any* reasonable integration theory, whatever its exact details, will have properties such as the following:

- if f and g are integrable over I, then so is $f + g$, and $\int_I (f + g) = \int_I f + \int_I g$;
- if f is integrable over I, then so is αf, and $\int_I (\alpha f) = \alpha \int_I f$, for every $\alpha \in \mathbb{R}$.

If we accept that Lebesgue integration has these properties, as it does, then we can prove easily that the inner product is well defined.

We argue in direct analogy to the earlier case of ℓ^2. The following purely algebraic identity holds for all x:

$$(fg)(x) = f(x)g(x) = \tfrac{1}{2}((f(x) + g(x))^2 - f(x)^2 - g(x)^2).$$

If now $f, g \in L^2(I)$, then $f + g \in L^2(I)$ also, by closure under addition. Hence, by the criterion for membership of $L^2(I)$, the integrals $\int_I f^2$, $\int_I g^2$ and $\int_I (f+g)^2$ all exist and are finite, and it follows using the algebraic identity and the above two properties of the Lebesgue integral that the integral $\int_I fg$ is also defined and finite, as required.

Moreover, assuming the same two properties, we can also verify straightforwardly, along the lines used above for $C(I)$, that the four inner product space axioms hold in $L^2(I)$, *except* for the last part of the fourth inner product space axiom, which requires that if $\langle f, f \rangle = 0$ then $f = 0$. The proof of this statement for $C(I)$, which we relegated to Exercise 3.4, certainly fails if we try to carry it out for $L^2(I)$, because it depends specifically on continuity, which we can no longer assume. However, more seriously, the statement is in fact false for $L^2(I)$. That is, our proposed inner product does not obey the inner product space axioms in full, and so we have not succeeded in making the vector space $L^2(I)$ into an inner product space.

We will find a way around this difficulty shortly, but let us first discuss two examples, to see how the last part of the fourth axiom fails.

Consider the function f defined on $[0, 1]$ by

$$f(x) = \begin{cases} 1, & x = \tfrac{1}{2}, \\ 0, & \text{otherwise.} \end{cases}$$

This function is Riemann-integrable, with $\int_0^1 f = 0$. Since f is Riemann-integrable, it is also Lebesgue-integrable (the Lebesgue theory is a generalisation of the Riemann theory), and so we have $f \in L^2([0, 1])$. Also, since $f^2 = f$, we have $\langle f, f \rangle = \int_0^1 f^2 = \int_0^1 f = 0$. But of course f does *not* satisfy $f = 0$, because to say that $f = 0$ means that f is the zero vector in $L^2([0, 1])$, the function whose value is 0 at *all* points in $[0, 1]$.

A more extreme illustration is provided by the function f defined on $[0, 1]$ by

$$f(x) = \begin{cases} 1, & x \text{ rational}, \\ 0, & x \text{ irrational}. \end{cases}$$

This function, as we noted in Chapter 2, is not Riemann-integrable, but is Lebesgue-integrable, with $\int_0^1 f = 0$; further, since $f^2 = f$, we have $\int_0^1 f^2 = 0$, and hence $f \in L^2([0, 1])$ and $\langle f, f \rangle = 0$. But, as for the first example, f is not the zero vector, because there are infinitely many points at which its value is non-zero.

Notice that both these examples are of functions f which, while not identically 0, are 'almost 0', in the sense that *the integral of f^2 in each case is* 0. This observation is the key to making $L^2(I)$ into an inner product space after all: we simply declare that any function f such that $\int_I f^2 = 0$ is *identical to the zero function*. The more general statement follows that any *two* functions f and g such that $\int_I (f-g)^2 = 0$ are identical. A simpler analytic formulation of the condition under which f and g are identified can be obtained by invoking the Lebesgue integration theory, which shows that $\int_I (f - g)^2 = 0$ if and only if $\int_I |f - g| = 0$. (The following section gives another way of looking at this condition.)

In formal terms, what we have done here is to redefine the space $L^2(I)$ *by making its elements the equivalence classes of the functions from our earlier definition* under the equivalence relation given by:

$$f \text{ and } g \text{ are equivalent if } \int_I |f - g| = 0.$$

At first sight, it may appear rather complicated and cumbersome to have to deal with spaces whose elements are equivalence classes of functions rather than functions, but the definition works in a straightforward way in practice: most of the time, we continue to think of $L^2(I)$ as a space of functions, as originally defined, but we exercise great caution if we have to make a statement about the value of a function in $L^2(I)$ at any particular point.

It is useful to reconsider some comments made in Chapter 1 in the light of our revised definition of $L^2(I)$. When defining the Haar wavelet H in Subsection 1.4.2, we commented that the values chosen for H at 0, $\frac{1}{2}$ and 1 were not particularly important, and that nothing significant would change if we made different choices. We can now be precise about why this is the case. If we defined a variant of the Haar wavelet, say H', with differently chosen values at a few isolated points, it would nevertheless be the case that

$$\int_{-\infty}^{\infty} |H - H'| = 0,$$

and in $L^2(\mathbb{R})$, as just redefined, H and H' represent *the same element of* $L^2(\mathbb{R})$; from a standpoint within this space, H and H' are indistinguishable.

3.3 Sets of Measure Zero

The length of an interval $[a, b]$ (with $a \leq b$) is given by $b - a$; further, whether the interval contains its own endpoints or not does not affect its length, so the intervals $[a, b]$, $[a, b)$, $(a, b]$ and (a, b) all have length $b - a$. Generalising the idea of length to slightly more complicated sets is easy enough; for example, we would presumably say that the length of $[-12, 2] \cup [17, 28]$ is $14 + 11 = 25$. Generalising the idea of length to its furthest possible extent, however, involves a substantial theoretical development, and would take us into the topic of *measure theory*. In measure theory, the (generalised) length of a set on the real line is called its *Lebesgue measure* or just its *measure*.

A proper study of Lebesgue integration requires a proper study of Lebesgue measure, but we do not attempt either in this book. What we do wish to discuss here, though only briefly, is one special but very important case: the case when a set has *(Lebesgue) measure* 0.

Consider again the case of a bounded interval. Since the length or measure of the interval $[a, b]$ is $b - a$, the interval will have measure 0 only when $a = b$, in which case the interval is just the singleton set $\{a\}$. Also, presumably, we will want to say that the empty set has measure 0. The formal definition below allows us to say that many other larger and far more complicated sets of real numbers also have measure 0.

The informal idea of the definition is straightforward: a set A has measure 0 if its length (or measure) is less than every positive real number.

Definition 3.3   If $A \subseteq \mathbb{R}$, then A has **measure** 0 if for every $\epsilon > 0$ there exists a sequence $\{I_n\}_{n=1}^{\infty}$ of bounded intervals such that

$$A \subseteq \bigcup_{n=1}^{\infty} I_n$$

and the sum of the lengths of the I_n is less than ϵ.

We make a few minor observations about the definition. First, from our discussion above, it does not matter whether the intervals contain their endpoints or not, so the definition is phrased in a way that avoids the issue. Second, although we allow ourselves an infinite number of intervals, we only allow a *countable* infinity – that is, we must be able to arrange the intervals in a sequence indexed by the natural numbers. Third, the definition in effect allows the use of a finite number of intervals, or even just a single interval, since 'degenerate' intervals of the form $[a, a]$ containing only one point and even empty intervals such as (a, a) can be included in the sequence. Fourth, the order in which we list the intervals I_n does not affect the sum of their lengths: their lengths are non-negative real numbers, and so the convergent series formed by summing their lengths is absolutely convergent and therefore has a sum which is independent of the order of summation. (Compare the comments just preceding Subsection 4.6.1 below, and see Exercise 4.36.)

Example 3.4   (i) A singleton set $\{x\}$ has measure 0, since it can be covered by the single degenerate interval $[x, x]$ of length 0, and 0 is less than any chosen $\epsilon > 0$.

 (ii) A countable set of points $\{x_n : n = 1, 2, 3, \ldots\}$ also has measure 0, since it can be covered by the sequence $\{[x_n, x_n] : n = 1, 2, 3, \ldots\}$ of degenerate intervals.

(iii) A consequence is that the set of rational numbers has measure 0, since the rationals can be arranged into a sequence, that is, form a countable set.

We note a couple of simple results about sets of measure 0. (The proofs are left for Exercise 3.7.)

Theorem 3.5   (i) *Any subset of a set of measure 0 has measure 0.*

(ii) *If* $\{A_n : n = 1, 2, 3, \ldots\}$ *is a sequence of sets of measure 0, then the union* $\cup_{n=1}^{\infty} A_n$ *also has measure 0.*

There are many uncountable sets that are still of measure 0. We do not need to examine any such cases here, but we note that the well-known **Cantor set** is one example (see Exercise 3.8).

It is also important to note that 'not too many' sets have measure 0. For example, since $[a, b]$ has length $b - a$, and this quantity is non-zero if $a < b$, our definition of measure 0, if it is satisfactory, should not assign this set measure 0, and this is indeed the case (see Exercise 3.9). (If we were to introduce Lebesgue measure in the general case, then $[a, b]$ would, as expected, have Lebesgue measure $b - a$ and the sets we have defined as having measure 0 would have Lebesgue measure 0.)

➤ Sets of measure 0 are, intuitively, 'thin' or 'sparse' or 'negligible', though we need to be careful about where our intuition may lead. Certainly, the rationals between 0 and 1 seem thinly spread throughout the interval [0, 1], because there are infinitely many irrational numbers between any two rational numbers, no matter how close together (see Exercise 4.5), and the rationals do indeed have measure 0, as we have shown. On the other hand the *irrationals* between 0 and 1 might also seem to be thinly spread for a similar reason – there are infinitely many rational numbers between any two irrational numbers, no matter how close together (Exercise 4.5 again). But the irrationals between

0 and 1 are *not* of measure 0. One way of explaining this is to argue that since the interval [0, 1] has length 1, and given that the rationals have measure (or generalised length) 0, the measure of the irrationals must be 1 to make up the difference. The general theory of Lebesgue measure shows that this argument is correct: the irrationals between 0 and 1 have Lebesgue measure 1. (Observe also that while the set of rationals is countable, the set of irrationals in [0, 1] is uncountable, so part (ii) of Example 3.4 does not lead us to a contradiction.)

3.3.1 Properties That Hold Almost Everywhere

A property that may or may not hold for any given real number x is said to hold **almost everywhere** or **for almost all** x if it holds for all x *except* those in some set of measure 0. (It is very common to see the phrase 'almost everywhere' abbreviated to 'a. e.', but we will use the longer form.) The phrase is generally used in situations where it is not important or not possible to identify the specific measure 0 set involved.

For example, recall the following function on $\mathbb{R}$, which we discussed in Subsection 3.2.3 and in earlier chapters:

$$f(x) = \begin{cases} 1, & x \text{ rational}, \\ 0, & x \text{ irrational}. \end{cases}$$

Since the rationals form a set of measure 0 in $\mathbb{R}$, we can say that $f(x) = 0$ *for almost all* x, or simply that f *is* 0 *almost everywhere*.

For a second example, recall our definition of the space $L^2(I)$ for an interval I. The elements of these spaces, strictly, are equivalence classes of functions under the equivalence relation that identifies any two functions f and g such that $\int_I |f - g| = 0$. The theory of Lebesgue integration shows that the equation $\int_I |f - g| = 0$ holds if and only if f and g are equal almost everywhere, that is, $f(x) = g(x)$ except for x in some set of measure 0.

In the last two chapters of the book we will encounter many statements involving conditions that hold almost everywhere.

3.3.2 Measurability of Functions

This is an appropriate point at which to include a brief comment about a technical issue in the definition of the L^2 spaces. A glance at any text on the subject will show that $L^2(I)$ is defined as the set of *measurable* or *Lebesgue measurable* functions f for which $\int_I f^2$ is finite. Measurability of functions is one of the fundamental concepts in the theory of Lebesgue integration and measure, but since we do not assume this theory is known, and we certainly cannot cover it in this text, we have simply ignored the condition of measurability altogether above, and we will continue to ignore it below as well. It suffices to say here that the condition is satisfied in all the cases where it is needed. In particular, all the standard functions of everyday mathematics are measurable, all functions that are continuous or piecewise continuous are measurable, and many far more exotic functions, such as the one mentioned above that is 1 on the rationals and 0 on the irrationals, are measurable.

3.4 The Norm and the Notion of Distance

The inner product space axioms tell us that $\langle x, x \rangle$ is always non-negative, and this implies that we can properly make the following definition.

Definition 3.6 If V is an inner product space, then the **norm** in V is the function $\| \cdot \| : V \to \mathbb{R}$ defined by

$$\|x\| = \langle x, x \rangle^{\frac{1}{2}},$$

for all $x \in V$.

We can immediately prove the following very important result, which relates the inner product and the norm. The result will be used directly below to prove another simple but fundamental inequality involving the norm, and will reappear at other significant points later.

Theorem 3.7 (The Cauchy–Schwarz inequality) *If V is an inner product space, then*

$$\left| \langle x, y \rangle \right| \leq \|x\| \, \|y\|,$$

for all $x, y \in V$.

Proof If $y = 0$, then the inequality holds, since both sides are 0. Therefore, let us assume that $y \neq 0$. Then $\|y\|$ is a non-zero real number (this is immediate from the definition, but see also part (iii) of Theorem 3.8), and so it will be enough if we prove the inequality

$$\left| \left\langle x, \frac{y}{\|y\|} \right\rangle \right| \leq \|x\|,$$

since multiplying through this inequality by $\|y\|$ yields the inequality we are trying to prove. Now the vector $y/\|y\|$ is a unit vector (that is, has norm 1), so it will be enough if we prove the formally simpler inequality

$$\left| \langle x, u \rangle \right| \leq \|x\|,$$

for every unit vector u, and this inequality is immediate from the following simple calculation:

$$0 \le \left\| x - \langle x, u \rangle u \right\|^2$$
$$= \langle x - \langle x, u \rangle u, x - \langle x, u \rangle u \rangle$$
$$= \langle x, x - \langle x, u \rangle u \rangle - \langle \langle x, u \rangle u, x - \langle x, u \rangle u \rangle$$
$$= \langle x, x \rangle - 2\langle x, u \rangle \langle x, u \rangle + \langle x, u \rangle \langle x, u \rangle \langle u, u \rangle$$
$$= \langle x, x \rangle - 2\langle x, u \rangle \langle x, u \rangle + \langle x, u \rangle \langle x, u \rangle$$
$$= \langle x, x \rangle - \langle x, u \rangle \langle x, u \rangle$$
$$= \|x\|^2 - \langle x, u \rangle^2. \qquad \qquad \square$$

We can define components and projections in a general inner product space exactly as we did in Chapter 1 in the case of $\mathbb{R}^n$ and other spaces. Thus if u is a unit vector, then the **component of** x **on** u is defined to be $\langle x, u \rangle$ and the **projection of** x **on** u is defined to be $\langle x, u \rangle u$, and the case when the second vector is not a unit vector is obtained as in Section 1.1 by normalisation. (See Section 3.6 below for discussion of a notion of projection that generalises the one just introduced.)

The statement, extracted from the proof of the Cauchy–Schwarz inequality, that

$$\left| \langle x, u \rangle \right| \le \|x\|$$

for every unit vector u therefore has the geometrically very plausible interpretation that the absolute value of the component of x in any direction, that is, the norm of the projection of x in any direction, is less than or equal to the norm of x itself.

There is a set of three standard properties of the norm.

Theorem 3.8   *For all vectors x and y, and all scalars α,*

(i) $\|x + y\| \le \|x\| + \|y\|$ *(the **triangle inequality**),*
(ii) $\|\alpha x\| = |\alpha| \|x\|$ *and*
(iii) $\|x\| \ge 0$, *and* $\|x\| = 0$ *if and only if* $x = 0$.

Proof   Properties (ii) and (iii) follow immediately from the inner product space axioms. For (i), we need the Cauchy–Schwarz inequality. We have:

$$\|x + y\|^2 = \langle x + y, x + y \rangle$$
$$= \langle x, x + y \rangle + \langle y, x + y \rangle$$
$$= \langle x, x \rangle + \langle x, y \rangle + \langle y, x \rangle + \langle y, y \rangle$$
$$= \|x\|^2 + 2\langle x, y \rangle + \|y\|^2$$

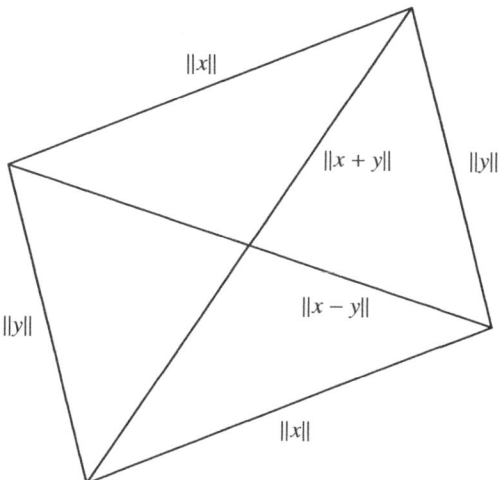

Figure 3.1 Illustration of the parallelogram law

$$\leq \|x\|^2 + 2|\langle x, y\rangle| + \|y\|^2$$
$$\leq \|x\|^2 + 2\|x\|\,\|y\| + \|y\|^2$$
$$= (\|x\| + \|y\|)^2,$$

from which the inequality is immediate.   □

The following result is also worth noting in passing, though it is of much less significance than the other theorems of this section and the proof is left as an exercise. The geometric content of the result is summed up clearly enough in Figure 3.1.

Theorem 3.9   *The norm in an inner product space satisfies the **parallelogram** law:*

$$\|x + y\|^2 + \|x - y\|^2 = 2\|x\|^2 + 2\|y\|^2,$$

for all x and y.

In Euclidean space, the norm $\|x\|$ of a vector x is the **distance** (or **Euclidean distance**) of x from the origin. In detail, if $x = (x_1, x_2, \ldots, x_n) \in \mathbb{R}^n$, then

$$\|x\| = \sqrt{\langle x, x\rangle} = \sqrt{x_1^2 + x_2^2 + \cdots + x_n^2},$$

and of course the most familiar examples are when n is 1, 2 or 3, when the space is $\mathbb{R}$, $\mathbb{R}^2$ or $\mathbb{R}^3$ and the distance of x from the origin is

$$\sqrt{x_1^2} = |x_1|, \qquad \sqrt{x_1^2 + x_2^2} \quad \text{or} \quad \sqrt{x_1^2 + x_2^2 + x_3^2},$$

respectively. More generally, the distance between two points x and y in $\mathbb{R}^n$ is $\|x - y\|$, and again this idea is very familiar in $\mathbb{R}$, $\mathbb{R}^2$ and $\mathbb{R}^3$. All this, of course, is as already discussed in Chapter 1.

In accordance with our strategy of trying to carry the geometry of $\mathbb{R}^n$ into inner product spaces in general, we make an analogous definition in any inner product space.

Definition 3.10   If V is an inner product space and $x, y \in V$, then **the distance of x from y** is $\|x - y\|$.

Note that this distance $\|x - y\|$ is always *a non-negative real number*, in conformity with our expectations of the nature of distance.

We can now speak, for example, of the distance between two vectors or sequences x and y in ℓ^2: if $x = (x_1, x_2, x_3, \ldots)$ and $y = (y_1, y_2, y_3, \ldots)$, then the distance between x and y is

$$\|x - y\| = \left(\sum_{n=1}^{\infty} |x_n - y_n|^2 \right)^{\frac{1}{2}}$$
$$= \left(|x_1 - y_1|^2 + |x_2 - y_2|^2 + |x_3 - y_3|^2 + \cdots \right)^{\frac{1}{2}}.$$

Similarly, we can speak of the distance between two functions f and g in $L^2(I)$: the distance between f and g is

$$\|f - g\| = \left(\int_I |f - g|^2 \right)^{\frac{1}{2}}.$$

➤ Recall that for $f, g, \in L^2(I)$ we regard f and g as equal if $\int_I |f - g|^2 = 0$. Since $\|f - g\|$ is just the square root of this integral, we can rephrase the criterion for the identification of f and g in the following more geometric form: f and g are identical if the distance between them is 0.

We note several elementary but important properties of distance; the proofs follow directly from Theorem 3.8 (checking them is left for Exercise 3.14).

Theorem 3.11   *For all vectors x, y and z,*

(i) $\|x - y\| \leq \|x - z\| + \|z - y\|$ *(the **triangle inequality**),*
(ii) $\|x - y\| = \|y - x\|$, *and, more generally,* $\|\alpha x - \alpha y\| = |\alpha| \|x - y\|$ *for all scalars α, and*
(iii) $\|x - y\| \geq 0$, *and* $\|x - y\| = 0$ *if and only if $x = y$.*

➤ If you are familiar with metric spaces, you will recognise these three properties (with the omission of the generalised form of (ii)) as the metric space axioms; the theorem says that an inner product space is a metric space, with $\|x - y\|$ as the metric or distance function.

Note that in Theorems 3.8 and 3.11 we have given the name 'triangle inequality' to two different inequalities. But it is easy to see that the two inequalities are really equivalent, and both are simply abstract expressions of the elementary geometric fact that the lengths of two sides of a triangle always have a sum greater than or equal to the length of the remaining side (see Exercise 3.15).

Notice that the ideas and results of this section suggest strongly that our definitions have captured axiomatically some of the essential features of the familiar geometry of $\mathbb{R}^3$ (or of $\mathbb{R}^n$ in general). The Cauchy–Schwarz inequality, the triangle inequalities and the parallelogram law, for example, are standard results in $\mathbb{R}^3$ (where the Cauchy–Schwarz inequality is usually known as **Cauchy's inequality**), but we now have proofs of all three which proceed *purely axiomatically* in an arbitrary inner product space, making no use of any assumptions about the internal structure of vectors in the space.

It thus seems clear that we are on the way to a successful extension of familiar finite-dimensional ideas to much more general, frequently infinite-dimensional, contexts, a point that will be reinforced by the remainder of the chapter. We noted in Chapter 1 that the study of this extension was one of our main themes.

3.5 Orthogonality and Orthonormality

Now that we have surveyed a range of basic facts about inner products and norms, we will begin to explore their deeper properties, a process which we could describe as an exploration of *the geometry of inner product spaces*.

The Cauchy–Schwarz inequality tells us that

$$\left| \langle x, y \rangle \right| \leq \|x\| \, \|y\|,$$

for all $x, y \in V$, where V is any inner product space. Hence for all x and y, we have

$$-\|x\| \, \|y\| \leq \langle x, y \rangle \leq \|x\| \, \|y\|,$$

and therefore (provided that neither x nor y is 0)

$$-1 \leq \frac{\langle x, y \rangle}{\|x\| \, \|y\|} \leq 1. \tag{3.1}$$

It is a standard geometric fact that in $\mathbb{R}^2$ and $\mathbb{R}^3$ (see Section 1.1 and Exercise 1.2) the angle θ between two vectors x and y satisfies the relation

$$\cos \theta = \frac{\langle x, y \rangle}{\|x\| \, \|y\|},$$

and by (3.1) we could, if it proved to be useful, define the angle between two vectors in any inner product space by the same formula. In practice, it seems not to be particularly useful to discuss angles in general in spaces other than Euclidean spaces, but the special case when $\cos\theta = 0$, or when θ is a right angle, is of central importance in all inner product spaces, as we have seen already in Chapter 1.

Definition 3.12 Vectors x and y in an inner product space are **orthogonal** if $\langle x, y \rangle = 0$; we then write $x \perp y$ and we may say 'x perp y'.

The properties below follow directly from the definition of orthogonality and the inner product, and the proofs are left for Exercise 3.17.

Theorem 3.13 *For all vectors x, y and $y_1, y_2, \ldots, y_n$,*

(i) *$0 \perp x$ and $x \perp 0$,*
(ii) *$x \perp y$ if and only if $y \perp x$,*
(iii) *$x \perp x$ if and only if $x = 0$, and*
(iv) *if $x \perp y_1, x \perp y_2, \ldots, x \perp y_n$, then $x \perp \alpha_1 y_1 + \alpha_2 y_2 + \cdots + \alpha_n y_n$, for all scalars $\alpha_1, \alpha_2, \ldots, \alpha_n$.*

Note that an alternative way of expressing the result of (iv) is to say that if x is orthogonal to each of $y_1, y_2, \ldots, y_n$, then it is also orthogonal to the subspace spanned by $y_1, y_2, \ldots, y_n$.

Definition 3.14 Let A be a subset of an inner product space. Then A is **orthogonal** if $\langle x, y \rangle = 0$ for all distinct $x, y \in A$, and A is **orthonormal** if in addition $\langle x, x \rangle = 1$ for all $x \in A$.

Thus, A is orthonormal if

$$\langle x, y \rangle = \begin{cases} 1, & x = y, \\ 0, & x \neq y, \end{cases}$$

for all $x, y \in A$. As we already noted in Chapter 1, if the elements of the set A are indexed then we can express the orthonormality of A succinctly using the Kronecker delta $\delta_{i,j}$. Specifically, if $A = \{x_i : i \in I\}$, where I is some index set, then A is orthonormal if

$$\langle x_i, x_j \rangle = \delta_{i,j}$$

for all $i, j \in I$.

The following result was noted for the particular case of the Euclidean spaces in Section 1.1 and Exercise 1.3.

Theorem 3.15 *Every orthonormal set is linearly independent.*

Proof Let $x_1, x_2, \ldots, x_n$ be a finite collection of vectors from an orthonormal set (which itself may be finite or infinite), and suppose that

$$\sum_{i=1}^{n} \alpha_i x_i = 0$$

for some scalars $\alpha_1, \alpha_2, \ldots, \alpha_n$. Then for any j such that $1 \leq j \leq n$, we have

$$\left\langle \sum_{i=1}^{n} \alpha_i x_i, x_j \right\rangle = \langle 0, x_j \rangle = 0,$$

and also

$$\left\langle \sum_{i=1}^{n} \alpha_i x_i, x_j \right\rangle = \sum_{i=1}^{n} \alpha_i \langle x_i, x_j \rangle = \alpha_j,$$

giving $\alpha_j = 0$. Since this holds for $j = 1, 2, \ldots, n$, we conclude that the vectors $x_1, x_2, \ldots, x_n$ are linearly independent. $\square$

3.5.1 The Gram–Schmidt Process

The Gram–Schmidt process is an algorithm for converting a finite or infinite sequence of linearly independent vectors in an inner product space into an orthonormal sequence which has the same length and spans the same subspace.

Suppose that $x_1, x_2, \ldots$ is a given linearly independent sequence, which, as mentioned, may be finite or infinite. If it is finite with, say, N elements, then the Gram–Schmidt process will stop after exactly N steps, and if it is infinite, the process will continue through an infinite number of steps – there is in either case exactly one step for each term of the sequence.

The following theorem specifies precisely the steps involved in applying the Gram–Schmidt process and states precisely the mathematical properties of the process.

Theorem 3.16 (The Gram–Schmidt process) *Let $x_1, x_2, \ldots$ be a finite or infinite sequence of linearly independent vectors in an inner product space. For $n = 1, 2, \ldots,$ define*

$$d_n = x_n - \sum_{i=1}^{n-1} \langle x_n, e_i \rangle e_i \qquad \text{and} \qquad e_n = \frac{d_n}{\|d_n\|}.$$

Then the sequence $e_1, e_2, \ldots$ has the following properties:

(i) *for each meaningful value of n, the vectors $e_1, e_2, \ldots, e_n$ form an orthonormal set that spans the same subspace as $x_1, x_2, \ldots, x_n$, and*

(ii) *the entire sequence $e_1, e_2, \ldots$ forms an orthonormal set that spans the same subspace as the entire sequence $x_1, x_2, \ldots$.*

➤ Note that we have tried to deal with the cases of a finite and an infinite sequence in a single statement here. If $x_1, x_2, \ldots$ is a finite sequence with last element x_N, then n will run from 1 to N in the steps that construct the vectors d_n and e_n, while if $x_1, x_2, \ldots$ is an infinite sequence, n will run through the natural numbers $\mathbb{N}$, and vectors d_n and e_n will be constructed for all $n \in \mathbb{N}$. Thus in the finite case, n in part (i) will be a 'meaningful value' if $1 \le n \le N$, while any $n \in \mathbb{N}$ will be 'meaningful' in the infinite case.

The Gram–Schmidt algorithm is recursive (or inductive), in the sense that the output of each step depends upon the output of the previous steps. Proofs about recursive constructions are typically carried out by induction, and this is the form our proof of part (i) will take, although the induction will be a finite induction, terminating at the Nth step, in the first case above, and a full induction over $\mathbb{N}$ in the second case.

Proof   For the base case of our inductive proof of part (i), consider the case $n = 1$. Note that the range of the summation in the definition of d_n is empty, so that the formula simply gives $d_1 = x_1$. Since the sequence $x_1, x_2, \ldots$ is linearly independent, we have $x_1 \ne 0$ and hence $\|x_1\| \ne 0$, so we can divide by $\|x_1\|$ and obtain a vector e_1, which clearly satisfies $\|e_1\| = 1$. Also, the definition of e_1 shows immediately that e_1 is in the subspace spanned by x_1, and the equation $x_1 = \|x_1\| e_1$ likewise shows that x_1 is in the subspace spanned by e_1. The singleton set $\{e_1\}$ is clearly orthonormal.

For the inductive step, fix any value $n > 1$ lying within the range of meaningful values and assume that the claims of part (i) hold for $n-1$. Define d_n by the given formula. We claim that $d_n \ne 0$. Indeed, if $d_n = 0$, then the formula for d_n gives

$$x_n = \sum_{i=1}^{n-1} \langle x_n, e_i \rangle e_i,$$

expressing x_n as a linear combination of $e_1, \ldots, e_{n-1}$. But our assumption tells us that $e_1, \ldots, e_{n-1}$ span the same subspace as $x_1, \ldots, x_{n-1}$, so it follows that x_n is expressible as a linear combination of $x_1, \ldots, x_{n-1}$, and this contradicts the linear independence of the sequence $x_1, \ldots, x_{n-1}, x_n$. Thus, $d_n \ne 0$.

We can therefore legitimately define $e_n = d_n / \|d_n\|$. Now the definitions of d_n and e_n imply that x_n lies in the space spanned by $e_1, \ldots, e_n$, and since the inductive assumption implies that $x_1, \ldots, x_{n-1}$ lie in the space spanned by $e_1, \ldots, e_{n-1}$, it follows that $x_1, \ldots, x_n$ lie in the space spanned by $e_1, \ldots, e_n$.

On the other hand, the inductive assumption also tells us that $e_1, \ldots, e_{n-1}$ lie in the space spanned by $x_1, \ldots, x_{n-1}$, and the definitions of d_n and e_n imply that e_n can be written as a linear combination of $e_1, \ldots, e_{n-1}$ and x_n, and hence of $x_1, \ldots, x_{n-1}, x_n$, so we conclude that $e_1, \ldots, e_n$ lie in the space spanned by $x_1, \ldots, x_n$.

For orthonormality, we have assumed inductively that $\{e_1, \ldots, e_{n-1}\}$ is orthonormal, a simple calculation shows that $e_n \perp e_i$ for $i = 1, \ldots, n - 1$, and the fact that $\|e_n\| = 1$ is immediate from the definition of e_n. The fact that $\{e_1, \ldots, e_n\}$ is orthonormal follows.

Part (i) therefore holds by induction. The easy deduction of part (ii) is left as an exercise (see Exercise 3.21). □

It is sometimes useful to employ a variant of the Gram–Schmidt process which yields orthogonal but not necessarily orthonormal vectors. Now the vectors required are just the vectors d_n in the theorem, but we defined the d_n inductively in terms of the e_n, and it would be helpful to have an expression for the d_n that avoids mention of the e_n if the d_n are all we want. We claim that the expression defining the d_n directly is

$$d_n = x_n - \sum_{i=1}^{n-1} \frac{\langle x_n, d_i \rangle}{\|d_i\|^2} d_i,$$

for $n \geq 1$.

It is straightforward to confirm this. From the theorem, we have

$$d_n = x_n - \sum_{i=1}^{n-1} \langle x_n, e_i \rangle e_i$$

for each n and

$$e_i = \frac{d_i}{\|d_i\|}$$

for each i. Therefore,

$$d_n = x_n - \sum_{i=1}^{n-1} \langle x_n, e_i \rangle e_i = x_n - \sum_{i=1}^{n-1} \left\langle x_n, \frac{d_i}{\|d_i\|} \right\rangle \frac{d_i}{\|d_i\|} = x_n - \sum_{i=1}^{n-1} \frac{\langle x_n, d_i \rangle}{\|d_i\|^2} d_i,$$

as claimed. (Of course, we are still free to orthonormalise if we wish, by setting $e_n = d_n / \|d_n\|$ for each n.)

Not surprisingly, we can visualise the Gram–Schmidt process geometrically, making use of the ideas of components and projections. Now the general expression for d_n in the Gram–Schmidt process is

$$d_n = x_n - \sum_{i=1}^{n-1} \langle x_n, e_i \rangle e_i,$$

and we observe that the ith term $\langle x_n, e_i \rangle e_i$ in the sum is just the projection of x_n onto the ith unit vector e_i. Thus d_n is obtained by subtracting from x_n all of its projections into the $(n - 1)$-dimensional subspace spanned by $e_1, e_2, \ldots, e_{n-1}$; and e_n is then obtained by normalising d_n. We should therefore expect, *on purely geometric grounds*, what Theorem 3.16 already tells us algebraically: that e_n is perpendicular to the subspace spanned by $e_1, e_2, \ldots, e_{n-1}$.

Thus, for example, in $\mathbb{R}^3$:

- e_2 is perpendicular to the 1-dimensional subspace, or straight line through the origin, defined by e_1 (or, equivalently, by x_1);
- e_3 is perpendicular to the 2-dimensional subspace, or plane through the origin, defined by e_1 and e_2 (or, equivalently, by x_1 and x_2);

and similarly in higher dimensions.

3.5.2 An Example in $\mathbb{R}^n$

We begin with a simple example of an application of the Gram–Schmidt process in Euclidean space. Consider the vectors

$$x_1 = (-1, 2, 0, 2), \quad x_2 = (2, -4, 1, -4) \quad \text{and} \quad x_3 = (-1, 3, 1, 1)$$

in $\mathbb{R}^4$. These vectors are linearly independent. We can easily enough confirm this directly, of course, but Exercise 3.22 shows that if they were dependent, the dependence would show up in the Gram–Schmidt process. Hence we will omit direct confirmation of independence and go straight to the Gram–Schmidt steps, which proceed as follows (though the details of the calculations are left for Exercise 3.23):

- $d_1 = x_1 = (-1, 2, 0, 2)$, and hence $e_1 = \frac{1}{3}(-1, 2, 0, 2)$;
- $d_2 = x_2 - \langle x_2, e_1 \rangle e_1 = (0, 0, 1, 0)$, and hence $e_2 = d_2 = (0, 0, 1, 0)$; and
- $d_3 = x_3 - \langle x_3, e_1 \rangle e_1 - \langle x_3, e_2 \rangle e_2 = (0, 1, 0, -1)$, and hence $e_3 = \frac{1}{\sqrt{2}}(0, 1, 0, -1)$.

We now know from the theory that the three vectors e_1, e_2 and e_3 are orthonormal (though this is also obvious by direct inspection), and that they span the same 3-dimensional subspace of $\mathbb{R}^4$ as x_1, x_2 and x_3.

3.5.3 An Example in ℓ^2

In the space ℓ^2, consider the infinite sequence of vectors

$$x_1 = (\alpha_1, 0, 0, 0, \ldots), \quad x_2 = (\alpha_1, \alpha_2, 0, 0, \ldots), \quad x_3 = (\alpha_1, \alpha_2, \alpha_3, 0, \ldots), \ldots .$$

If $\alpha_n \neq 0$ for all n, then these vectors are independent. If also $\alpha_n > 0$ for all n, then the Gram–Schmidt process applied to them yields the familiar sequence

$$e_1 = (1, 0, 0, 0, \ldots), \quad e_2 = (0, 1, 0, 0, \ldots), \quad e_3 = (0, 0, 1, 0, \ldots), \ldots$$

(see Exercise 3.24).

3.5.4 An Example in $C([-1, 1])$: The Legendre Polynomials

Let I be any non-trivial interval, and consider the infinite sequence of functions

$$1, x, x^2, x^3, \ldots,$$

regarded as vectors in the vector space $C(I)$. This sequence forms a linearly independent set in $C(I)$. (Exercise 2.10 already asked for a proof of this fact, but we now supply a proof. Let us also briefly recall the notational convention established near the end of Subsection 1.2.2, under which the notation x^k, for any given k, may be taken, as the context demands here, to denote a function – the function whose value at x is the real number x^k.)

Suppose that some finite linear combination of the functions $1, x, x^2, x^3, \ldots$ is the zero function (which is the zero vector in $C(I)$). Now the functions involved in the linear combination can be written as

$$x^{n_1}, x^{n_2}, \ldots, x^{n_k},$$

for some $k \geq 1$ and for some integers $n_1 < n_2 < \cdots < n_k$, so our assumption is that the linear combination

$$a_1 x^{n_1} + a_2 x^{n_2} + \cdots + a_k x^{n_k}$$

is identically 0 for some scalars $a_1, a_2, \ldots, a_k$. But if any of these scalars is non-zero, then $a_1 x^{n_1} + a_2 x^{n_2} + \cdots + a_k x^{n_k}$ is a non-zero polynomial, which can have only a finite number of roots in I, whereas our assumption that it is the zero function says that every number x in I is a root. Hence every coefficient must be 0, which proves independence.

Let us now consider the specific case of $C([-1, 1])$; since $[-1, 1]$ is closed and bounded, this vector space is an inner product space with the standard inner product (see Subsection 3.2.3).

If we use the Gram–Schmidt process to orthogonalise (rather than orthonormalise) the functions $1, x, x^2, x^3, \ldots$, then we obtain a sequence of polynomials beginning

$$p_0(x) = 1, \quad p_1(x) = x, \quad p_2(x) = x^2 - \tfrac{1}{3}, \quad p_3(x) = x^3 - \tfrac{3}{5}x, \ldots.$$

Although calculating additional terms of this sequence is not hard, it is not at all obvious how to write down a general expression for the nth term of the sequence in any simple way. It is possible, however, to show that

$$p_n(x) = \frac{n!}{(2n)!} \frac{d^n}{dx^n} (x^2 - 1)^n$$

(see Exercise 3.28).

These polynomials, with a different normalisation, are very well known in analysis. The more or less standard procedure, used largely for historical reasons, is to normalise the polynomials so that their value at $x = 1$ is 1. They are then known as the **Legendre polynomials**, and the sequence begins

$$P_0(x) = 1, \quad P_1(x) = x, \quad P_2(x) = \tfrac{3}{2}x^2 - \tfrac{1}{2}, \quad P_3(x) = \tfrac{5}{2}x^3 - \tfrac{3}{2}x, \dots .$$

The nth term is now given by

$$P_n(x) = \frac{1}{2^n n!} \frac{d^n}{dx^n} (x^2 - 1)^n,$$

an expression known as **Rodrigues' formula**. The norm of P_n in $C([-1, 1])$ is

$$\|P_n\| = \sqrt{\int_{-1}^{1} P_n^2} = \sqrt{\frac{2}{2n + 1}},$$

and if we divide P_n by $\|P_n\|$, we obtain the orthonormalised Legendre polynomials:

$$Q_0(x) = \frac{1}{\sqrt{2}},$$

$$Q_1(x) = \frac{\sqrt{3}}{\sqrt{2}} x,$$

$$Q_2(x) = \frac{3\sqrt{5}}{2\sqrt{2}} x^2 - \frac{\sqrt{5}}{2\sqrt{2}},$$

$$Q_3(x) = \frac{5\sqrt{7}}{2\sqrt{2}} x^3 - \frac{3\sqrt{7}}{2\sqrt{2}} x,$$

$$\vdots$$

The graphs of the polynomials Q_n for $n = 0, 1, \dots, 5$ are shown in Figure 3.2.

A whole sub-discipline of classical analysis is devoted to the general study of orthogonal polynomials, which have many fascinating properties and important applications. Among the many families of such polynomials in addition to the Legendre family are the **Hermite**, **Laguerre** and **Chebyshev** **polynomials**; these are all obtained by applying the Gram–Schmidt process to the sequence $1, x, x^2, x^3, \dots$ in spaces $C(I)$, for various intervals I, or in closely related spaces.

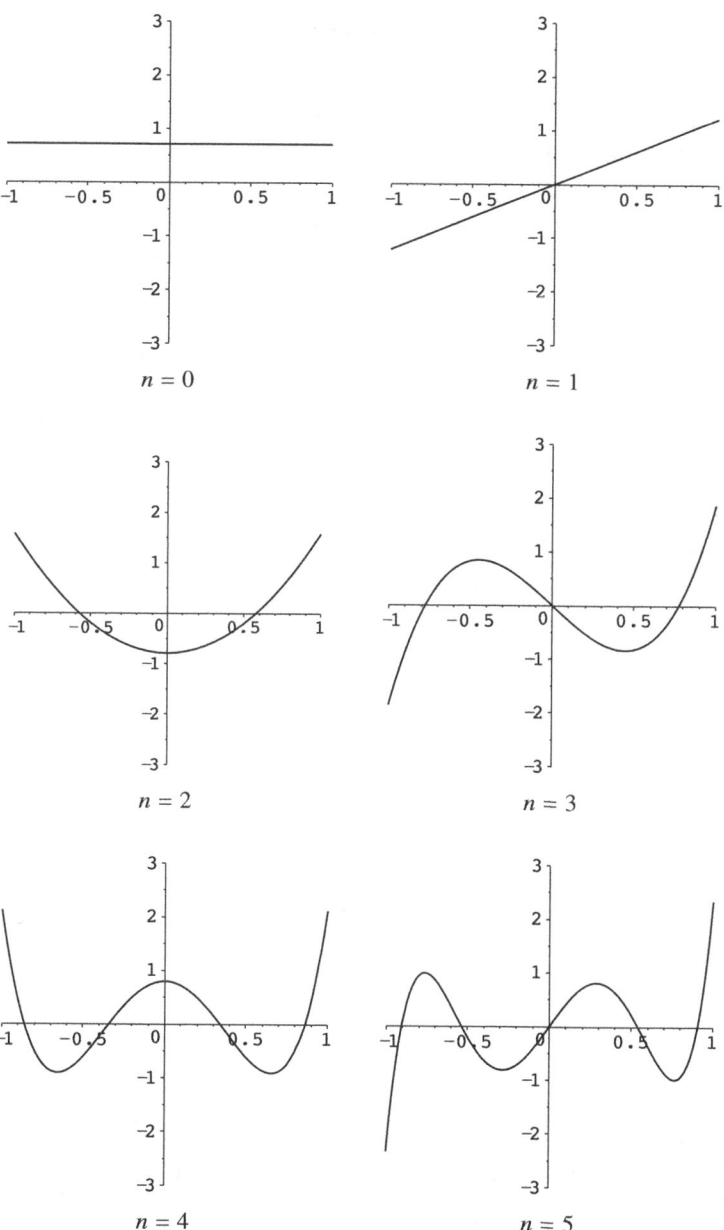

Figure 3.2 Graphs of the orthonormalised Legendre polynomials Q_n

3.6 Orthogonal Projection

We saw a number of times in Chapter 1, at least informally, how a vector in an inner product space may be 'decomposed' into a linear combination of the members of a finite orthonormal basis by projecting the vector onto each member of the basis.

The first example of this was in Section 1.1, where we were given a basis b_1, b_2, b_3 of three orthonormal vectors for $\mathbb{R}^3$ and saw how a specific vector c was expressible in the form

$$c = \langle c, b_1 \rangle b_1 + \langle c, b_2 \rangle b_2 + \langle c, b_3 \rangle b_3 = \sum_{i=1}^{3} \langle c, b_i \rangle b_i :$$

the coefficient $\langle c, b_i \rangle$ is of course the component of c on b_i for each i. We now both formalise and generalise this process in the context of an arbitrary inner product space.

Theorem 3.17 *Let $\{e_1, e_2, \ldots, e_n\}$ be a finite orthonormal set in an inner product space V, and let x be any vector in V. Then*

$$\sum_{i=1}^{n} \langle x, e_i \rangle^2 \leq \|x\|^2 \tag{3.1}$$

and, for $j = 1, \ldots, n$,

$$x - \sum_{i=1}^{n} \langle x, e_i \rangle e_i \perp e_j. \tag{3.2}$$

Further, if x is in the n-dimensional subspace spanned by $\{e_1, e_2, \ldots, e_n\}$, then

$$x = \sum_{i=1}^{n} \langle x, e_i \rangle e_i, \tag{3.3}$$

and

$$\sum_{i=1}^{n} \langle x, e_i \rangle^2 = \|x\|^2. \tag{3.4}$$

➤ It is worth emphasising that V here may be an inner product space of any, possibly infinite, dimension, even though the orthonormal set $\{e_1, e_2, \ldots, e_n\}$ is finite. Also note carefully that (3.1) and (3.2) hold for any vector $x \in V$, while (3.3) and (3.4), which give stronger information than (3.1) and (3.2), require the stronger assumption that x is in the subspace spanned by the e_i.

Here, as always in inner product space work, it is important and instructive to note the geometric interpretation of all of these statements. The first statement says that the sum of the squares of the lengths of the projections of x in various perpendicular directions is less than or equal to the square of the length of x

itself; the inequality is called **Bessel's inequality**. The second statement can be interpreted as saying that if we subtract from a vector its projections in several perpendicular directions, then the result has no component left in any of those directions – that is, it is orthogonal to all of them. Similarly for the remaining statements: if x is in the subspace, then both it and its length are *exactly* expressible in terms of its projections and their lengths. The expression (3.3) for x is a **finite Fourier series for** x with respect to the orthonormal basis $\{e_1, e_2, \ldots, e_n\}$ for the n-dimensional subspace involved, and the terms $\langle x, e_i \rangle$ are the **Fourier coefficients of** x with respect to that basis. The equation (3.4) is called **Parseval's relation**. (The comments at the end of Subsection 3.6.2 below shed further light on all the issues raised here.)

In the next chapter, we will be able to generalise all the parts of the above theorem to *infinite* Fourier series with respect to certain *infinite* orthonormal sets, encompassing the usual trigonometric Fourier series and wavelet series as special cases.

Proof of Theorem 3.17   The main part of the proof is a straightforward computation. For any $x \in V$,

$$0 \leq \left\| x - \sum_{i=1}^{n} \langle x, e_i \rangle\, e_i \right\|^2$$

$$= \left\langle x - \sum_{i=1}^{n} \langle x, e_i \rangle\, e_i, x - \sum_{j=1}^{n} \langle x, e_j \rangle\, e_j \right\rangle$$

$$= \langle x, x \rangle - \left\langle x, \sum_{j=1}^{n} \langle x, e_j \rangle\, e_j \right\rangle - \left\langle \sum_{i=1}^{n} \langle x, e_i \rangle\, e_i, x \right\rangle + \left\langle \sum_{i=1}^{n} \langle x, e_i \rangle\, e_i, \sum_{j=1}^{n} \langle x, e_j \rangle\, e_j \right\rangle$$

$$= \langle x, x \rangle - \sum_{j=1}^{n} \langle x, e_j \rangle \langle x, e_j \rangle - \sum_{i=1}^{n} \langle x, e_i \rangle \langle x, e_i \rangle + \sum_{i=1}^{n} \sum_{j=1}^{n} \langle x, e_i \rangle \langle x, e_j \rangle \langle e_i, e_j \rangle$$

$$= \|x\|^2 - \sum_{i=1}^{n} \langle x, e_i \rangle^2,$$

which gives (3.1). Also, for each j,

$$\left\langle x - \sum_{i=1}^{n} \langle x, e_i \rangle\, e_i, e_j \right\rangle = \langle x, e_j \rangle - \sum_{i=1}^{n} \langle x, e_i \rangle \langle e_i, e_j \rangle = \langle x, e_j \rangle - \langle x, e_j \rangle = 0,$$

which gives (3.2). Now if x is in the subspace spanned by $\{e_1, e_2, \ldots, e_n\}$, then we have

$$x = \sum_{i=1}^{n} \alpha_i e_i,$$

for some collection of scalars $\alpha_1, \alpha_2, \ldots, \alpha_n$. Therefore, for any j,

$$\langle x, e_j \rangle = \left\langle \sum_{i=1}^{n} \alpha_i e_i, e_j \right\rangle = \sum_{i=1}^{n} \alpha_i \langle e_i, e_j \rangle = \alpha_j,$$

and replacing α_j by $\langle x, e_j \rangle$ in the expression for x gives (3.3). Equation (3.4) follows easily from (3.3) by expanding $\|x\|^2$ as an inner product (verification of this is left for Exercise 3.38). □

In Theorem 3.17, in the case when x belonged to the subspace spanned by the orthonormal set $\{e_1, e_2, \ldots, e_n\}$, we referred to the expansion $\sum_{i=1}^{n} \langle x, e_i \rangle e_i$ as a finite Fourier series for x. Sums of this form turn out to be of great importance whether or not x is a member of the subspace, and we introduce the following definition to describe them.

Definition 3.18   Let $\{e_1, e_2, \ldots, e_n\}$ be a finite orthonormal set in an inner product space V, and let x be any vector in V. Then the vector

$$\sum_{i=1}^{n} \langle x, e_i \rangle e_i$$

is called the **orthogonal projection**, or just the **projection**, of x into the subspace spanned by the set $\{e_1, e_2, \ldots, e_n\}$.

We will typically use the symbol 'P' (perhaps with appropriate subscripts) for the function that maps a vector to its projection into a given subspace, so that we could write $Px = \sum_{i=1}^{n} \langle x, e_i \rangle e_i$ in Definition 3.18. At present, the subspace involved is restricted to being finite-dimensional, but a significant task in the next chapter is to discover how to remove that constraint.

Observe that the definition is a generalisation of the notion of the projection of a vector onto a single unit vector, as discussed in Section 3.4: if e is a unit vector, then the projection of x into the 1-dimensional subspace spanned by e is given by $\langle x, e \rangle e$, and this is the same as the projection, in the sense of Section 3.4, of x onto the vector e. (See Subsection 3.6.2 for further discussion of the geometric properties of projection.)

3.6.1 An Example

We have seen that the first three orthonormalised Legendre polynomials Q_0, Q_1 and Q_2 are given by the formulae

$$Q_0(x) = \frac{1}{\sqrt{2}}, \quad Q_1(x) = \frac{\sqrt{3}}{\sqrt{2}} x \quad \text{and} \quad Q_2(x) = \frac{3\sqrt{5}}{2\sqrt{2}} x^2 - \frac{\sqrt{5}}{2\sqrt{2}}.$$

We also know from the Gram–Schmidt process (see Theorem 3.16) that the set of functions $\{Q_0, Q_1, \ldots, Q_n\}$ spans the same subspace of $C([-1, 1])$ as the set $\{1, x, x^2, \ldots, x^n\}$ for each n, and that subspace is therefore the space of all polynomials on $[-1, 1]$ of degree at most n. Hence, by Theorem 3.17, the polynomial x^2 must be expressible as a finite Fourier series in the orthonormal polynomials Q_0, Q_1 and Q_2. Moreover, the Fourier coefficients will simply be the components of x^2 on those three vectors. Computing these (and applying once again the convention of Subsection 1.2.2), we find that

$$\langle x^2, Q_0 \rangle = \int_{-1}^{1} x^2 \tfrac{1}{\sqrt{2}} \, dx = \tfrac{\sqrt{2}}{3},$$

$$\langle x^2, Q_1 \rangle = \int_{-1}^{1} x^2 \tfrac{\sqrt{3}}{\sqrt{2}} x \, dx = 0$$

(we obtain 0 in the second case because we are integrating an odd function over a symmetric interval; this is obvious, but see Exercise 3.30), and

$$\langle x^2, Q_2 \rangle = \int_{-1}^{1} x^2 \left(\tfrac{3\sqrt{5}}{2\sqrt{2}} x^2 - \tfrac{\sqrt{5}}{2\sqrt{2}} \right) dx = \tfrac{2\sqrt{2}}{3\sqrt{5}}.$$

Hence, by Theorem 3.17,

$$x^2 = \tfrac{\sqrt{2}}{3} Q_0(x) + \tfrac{2\sqrt{2}}{3\sqrt{5}} Q_2(x),$$

which can be verified directly by expanding the right-hand side.

In this example, the function x^2 was an element of the subspace spanned by Q_0, Q_1 and Q_2. The expansion we obtained for x^2 was therefore an *equation* – it gave an *exact* expression for x^2 as a combination of $Q_0(x)$, $Q_1(x)$ and $Q_2(x)$.

However, Theorem 3.17 also shows how we can project a vector (that is, function) that is *outside* of the subspace into the subspace. As an example, let us compute the projection of the function x^4 (which is certainly not in the subspace). Now x^4 is even, and so $\langle x^4, Q_1 \rangle$ will be 0 exactly as for $\langle x^2, Q_1 \rangle$ above, and for the other two components, we find that

$$\langle x^4, Q_0 \rangle = \frac{\sqrt{2}}{5} \quad \text{and} \quad \langle x^4, Q_2 \rangle = \frac{4\sqrt{2}}{7\sqrt{5}},$$

giving

$$\frac{\sqrt{2}}{5} Q_0(x) + \frac{4\sqrt{2}}{7\sqrt{5}} Q_2(x) = -\frac{3}{35} + \frac{6}{7} x^2$$

as the projected function. (It is a useful exercise to verify directly the properties of this expansion that are predicted by Theorem 3.17; see Exercise 3.39. Some further comments on this example are made in Subsection 3.6.3.)

3.6.2 Projection and Best Approximation

Our intuitions about orthogonal projection in Euclidean space are well developed. In $\mathbb{R}^3$, for example, we might talk about projecting a point x into a plane through the origin (the latter being a geometric description of a 2-dimensional subspace of $\mathbb{R}^3$). We would probably envisage carrying out the projection process by 'dropping a perpendicular' from x to the plane; the projection Px of x would then be the point where the perpendicular met the plane. We would further expect two other properties: that Px was the closest point to x in the plane, and that it was the unique such point, that is, was strictly closer to x than any other point.

We will prove below that exactly these properties hold quite generally for an orthogonal projection into a finite-dimensional subspace of an inner product space. The classical context for these ideas is *the problem of best approximation*. This is the following problem: Given a finite set of vectors $x_1, x_2, \ldots, x_n$ in an inner product space V, and another vector $x \in V$, find scalars $\alpha_1, \alpha_2, \ldots, \alpha_n$ such that the distance

$$\left\| x - (\alpha_1 x_1 + \alpha_2 x_2 + \cdots + \alpha_n x_n) \right\|$$

is minimised (if possible). Now the set of all linear combinations of the form

$$\alpha_1 x_1 + \alpha_2 x_2 + \cdots + \alpha_n x_n$$

for $\alpha_1, \alpha_2, \ldots, \alpha_n \in \mathbb{R}$ is just the subspace of V spanned by $x_1, x_2, \ldots, x_n$ (this may not be of dimension n, but will be if $x_1, x_2, \ldots, x_n$ are linearly independent). So the analytic problem of best approximation is equivalent to the geometric problem: Find the vector Px in the subspace spanned by $x_1, x_2, \ldots, x_n$ *which is as close as possible to* x, or *whose distance* $\|Px - x\|$ *from x is a minimum*, if such a point exists. The problem is comprehensively resolved by the following theorem.

Theorem 3.19 *Let V be an inner product space. For every finite set of linearly independent vectors $x_1, x_2, \ldots, x_n$ in V, and every vector $x \in V$, the problem of best approximation has a unique solution. More specifically, the projection Px of x into the subspace spanned by $x_1, x_2, \ldots, x_n$ solves the problem, and is the unique vector in the subspace with that property. Finally, $Px - x$ is orthogonal to the subspace and Px is the unique vector y in the subspace such that $y - x$ is orthogonal to the subspace.*

Proof Using the Gram–Schmidt process, we can construct an orthonormal set of vectors $e_1, e_2, \ldots, e_n$ which span the same subspace as $x_1, x_2, \ldots, x_n$. We now work exclusively with the orthonormal set, and we carry out a computation similar to the main part of the proof of Theorem 3.17.

We claim that for any set of scalars $\alpha_1, \alpha_2, \ldots, \alpha_n$,

$$\left\| x - \sum_{i=1}^{n} \langle x, e_i \rangle e_i \right\| \leq \left\| x - \sum_{i=1}^{n} \alpha_i e_i \right\|,$$

with equality between the left- and right-hand sides if and only if $\alpha_i = \langle x, e_i \rangle$ for $i = 1, 2, \ldots, n$.

To prove this, we observe, using the orthonormality of $e_1, e_2, \ldots, e_n$, that

$$\left\| x - \sum_{i=1}^{n} \alpha_i e_i \right\|^2 = \left\langle x - \sum_{i=1}^{n} \alpha_i e_i, x - \sum_{j=1}^{n} \alpha_j e_j \right\rangle$$

$$= \langle x, x \rangle - \left\langle x, \sum_{j=1}^{n} \alpha_j e_j \right\rangle - \left\langle \sum_{i=1}^{n} \alpha_i e_i, x \right\rangle + \left\langle \sum_{i=1}^{n} \alpha_i e_i, \sum_{j=1}^{n} \alpha_j e_j \right\rangle$$

$$= \|x\|^2 - 2 \sum_{i=1}^{n} \alpha_i \langle x, e_i \rangle + \sum_{i=1}^{n} \alpha_i^2$$

$$= \|x\|^2 + \sum_{i=1}^{n} (\alpha_i - \langle x, e_i \rangle)^2 - \sum_{i=1}^{n} \langle x, e_i \rangle^2.$$

Now the only part of the final expression above that depends on the scalars $\alpha_1, \alpha_2, \ldots, \alpha_n$ is the middle term,

$$\sum_{i=1}^{n} (\alpha_i - \langle x, e_i \rangle)^2;$$

moreover, this term is a sum of squares, and so takes on its minimum value when and only when all the squared terms are 0, and this is precisely when $\alpha_i = \langle x, e_i \rangle$ for $i = 1, 2, \ldots, n$, which proves our claim.

Hence the vector $Px = \sum_{i=1}^{n} \langle x, e_i \rangle, e_i$ is the unique vector solving the problem: it is in the subspace by definition, and the claim just proved shows that it both minimises the distance $\|Px - x\|$ and does so uniquely.

To confirm that $Px - x$ is orthogonal to the subspace, it is enough to confirm that it is orthogonal to each member e_i of the orthonormal basis for the subspace, and we have

$$\langle Px - x, e_i \rangle = \left\langle \sum_{j} \langle x, e_j \rangle e_j - x, e_i \right\rangle$$

$$= \left\langle \sum_{j} \langle x, e_j \rangle e_j, e_i \right\rangle - \langle x, e_i \rangle$$

$$= \sum_j \langle x, e_j \rangle \langle e_j, e_i \rangle - \langle x, e_i \rangle$$
$$= \langle x, e_i \rangle - \langle x, e_i \rangle$$
$$= 0.$$

The uniqueness assertion is left for Exercise 3.41.   □

Note that it is not strictly necessary to specify that the initial set of vectors $x_1, x_2, \ldots, x_n$ in the theorem is linearly independent; if they are dependent, then the Gram–Schmidt process will uncover this fact for us (see Subsection 3.5.2 and Exercise 3.22), and can be adapted to output an orthonormal (and therefore independent) set $e_1, e_2, \ldots, e_m$ for some $m < n$ which spans the same subspace as $x_1, x_2, \ldots, x_n$. The rest of the argument can then proceed as before.

When we introduced the Gram–Schmidt process, we made informal mention of projection to motivate the steps of the process. Now that we have formally introduced the idea of projection into a subspace, it is useful to revisit this briefly. Continuing the notation used in the discussion of the Gram–Schmidt process, recall that at the nth step we produce the next element e_n of the orthonormal set by defining

$$d_n = x_n - \sum_{i=1}^{n-1} \langle x_n, e_i \rangle e_i \qquad \text{and} \qquad e_n = \frac{d_n}{\|d_n\|}.$$

The expression $\sum_{i=1}^{n-1} \langle x_n, e_i \rangle e_i$ of course represents precisely the projection, say $P_n x_n$, of x_n into the subspace spanned by $x_1, \ldots, x_{n-1}$ or, equivalently, $e_1, \ldots, e_{n-1}$, and d_n is just $x_n - P_n x_n$, which by the last theorem is orthogonal to that subspace, and in particular to $e_1, \ldots, e_{n-1}$. Normalising d_n must therefore give us the vector e_n that we require.

3.6.3 Examples

We discuss three examples, building upon work done earlier in the chapter.

First, consider again the example in Section 3.5.2. We were given the vectors

$$x_1 = (-1, 2, 0, 2), \quad x_2 = (2, -4, 1, -4) \quad \text{and} \quad x_3 = (-1, 3, 1, 1)$$

in $\mathbb{R}^4$, and the Gram–Schmidt process yielded the corresponding orthonormal set

$$e_1 = \tfrac{1}{3}(-1, 2, 0, 2), \quad e_2 = (0, 0, 1, 0) \quad \text{and} \quad e_3 = \tfrac{1}{\sqrt{2}}(0, 1, 0, -1),$$

which spans the same 3-dimensional subspace of $\mathbb{R}^4$ as x_1, x_2 and x_3. The closest point in this 3-dimensional subspace to the vector $y = (1, 2, 3, 4)$ is therefore the projection

$$\sum_{i=1}^{3} \langle y, e_i \rangle e_i = \tfrac{11}{3} \times \tfrac{1}{3}(-1, 2, 0, 2) + 3 \times (0, 0, 1, 0) - \sqrt{2} \times \tfrac{1}{\sqrt{2}}(0, 1, 0, -1)$$

$$= (-\tfrac{11}{9}, \tfrac{13}{9}, 3, \tfrac{31}{9})$$

$$= \tfrac{1}{9}(-11, 13, 27, 31).$$

Second, refer to Example 3.6.1, where we computed the projection of x^4 into the 3-dimensional subspace of $C([-1, 1])$ of polynomials of degree at most 2 to be the function given by $-\tfrac{3}{35} + \tfrac{6}{7}x^2$. From Theorem 3.19 above, we can now say that this function is the unique closest element in the subspace to the function x^4 (with respect to the distance defined by the norm arising from the inner product)

For our third example, we start from the fact that the first $n + 1$ orthonormalised Legendre polynomials span the subspace of all polynomials of degree less than or equal to n in $C([-1, 1])$. Let us find the cubic polynomial that is closest to e^x, considered as a function on $[-1, 1]$. We know that the required polynomial is the projection of e^x into the subspace spanned by the first four orthonormal Legendre polynomials Q_0, Q_1, Q_2 and Q_3. After much calculation (or assistance from a computer algebra package), we find

$$\langle e^x, Q_0 \rangle = \int_{-1}^{1} e^x Q_0(x)\,dx = \tfrac{1}{\sqrt{2}}\left(e - e^{-1}\right),$$

$$\langle e^x, Q_1 \rangle = \int_{-1}^{1} e^x Q_1(x)\,dx = \sqrt{6}e^{-1},$$

$$\langle e^x, Q_2 \rangle = \int_{-1}^{1} e^x Q_2(x)\,dx = \tfrac{\sqrt{5}}{\sqrt{2}}\left(e - 7e^{-1}\right)$$

and

$$\langle e^x, Q_3 \rangle = \int_{-1}^{1} e^x Q_3(x)\,dx = \tfrac{\sqrt{7}}{\sqrt{2}}\left(-5e + 37e^{-1}\right).$$

By the theorem on best approximation, the cubic polynomial

$$\sum_{i=0}^{3} \langle e^x, Q_i \rangle Q_i(x)$$

is therefore the best approximation of degree 3 to e^x on $[-1, 1]$. Expanding this sum, we find (again after some calculation) that it equals

$$\left(-\frac{3e}{4} + \frac{33}{4e}\right) + \left(\frac{105e}{4} - \frac{765}{4e}\right)x + \left(\frac{15e}{4} - \frac{105}{4e}\right)x^2 + \left(-\frac{175e}{4} + \frac{1295}{4e}\right)x^3,$$

or approximately

$$0.9962940183 + 0.9979548730x + 0.5367215259x^2 + 0.1761390841x^3.$$

3 Inner Product Spaces

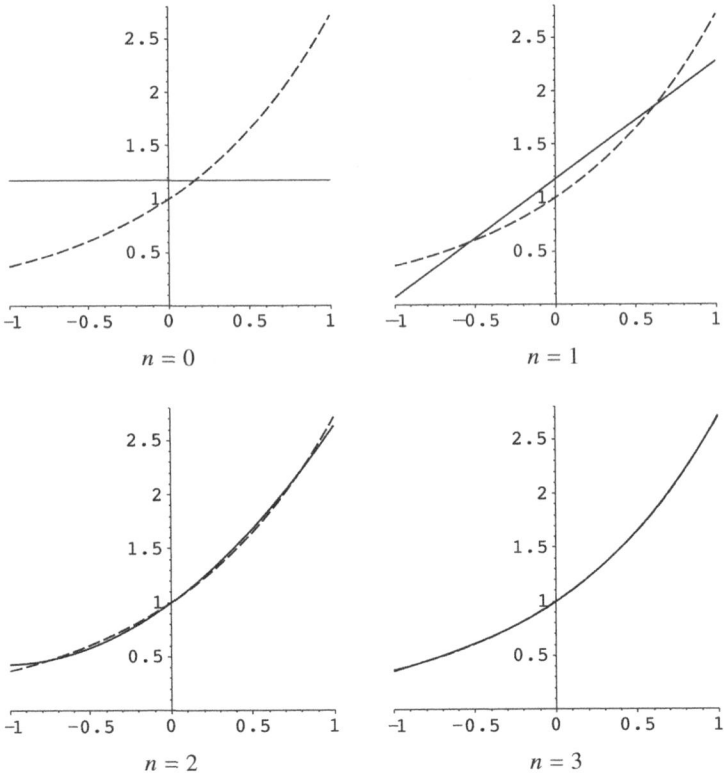

Figure 3.3 Graphs of Legendre polynomial approximations to e^x

Figure 3.3 shows graphs comparing the sums $\sum_{i=0}^{n}\langle e^x, Q_i\rangle Q_i(x)$ (shown as solid lines) with e^x (shown as a dashed line) on $[-1, 1]$ for $0 \le n \le 3$; the polynomial computed above corresponds to $n = 3$. The graphs indicate that the approximations approach e^x rather rapidly as n increases: at the scale used, the graph of the degree 3 approximation is hardly distinguishable from that of e^x.

It is interesting to make a comparison between the Legendre polynomial approximations to e^x and another set of polynomial approximations to e^x that we know well, the Maclaurin polynomials for e^x, which are the partial sums of the Maclaurin series

$$1 + x + \tfrac{1}{2}x^2 + \tfrac{1}{6}x^3 + \cdots .$$

(Note that the coefficients here are numerically quite close to those of the Legendre polynomial approximation.) We know by Theorem 3.19 that for each specific degree n, the Legendre polynomial approximation gives a *strictly* better approximation on $[-1, 1]$ than the corresponding Maclaurin polynomial approximation, using the norm in $L^2([-1, 1])$ as our measure of distance. For brevity, write L_n for the Legendre polynomial approximation of degree n and write M_n for the Maclaurin polynomial approximation of degree n. Then we have, to 10 places:

n	$\|L_n(x) - e^x\|$	$\|M_n(x) - e^x\|$
0	0.9298734950	0.9623178442
1	0.2294624530	0.3481444172
2	0.0379548881	0.0947975702
3	0.0047211090	0.0205090218
4	0.0004704933	0.0036677905

– which numerically confirms the results of our analysis above, and indeed appears to suggest that the L_n are significantly better approximations than the M_n as n increases.

➤ Note carefully, however, that our analysis tells us nothing whatsoever about the comparative merits of the L_n and the M_n as approximations to e^x *outside* of the interval $[-1, 1]$. Strictly speaking, in fact, the functions L_n do not even exist outside of $[-1, 1]$, even though the formulae that define them make sense for all real x, because they and the Legendre polynomials Q_n are elements of $L^2([-1, 1])$.

Exercises

3.1 Check that the inner product space axioms hold in $\mathbb{R}^n$.

3.2 Check the validity of the inner product space axioms in ℓ^2 that were not checked in Section 3.2.

3.3 Give an alternative proof of the fact from Subsection 3.2.2 that the infinite sum defining the inner product in ℓ^2 converges. (Base the argument on the inequality $2ab \leq a^2 + b^2$, which was used in Theorem 2.4 to prove that ℓ^2 is a vector space.)

3.4 Use the ϵ, δ-definition of continuity to confirm the claim in Section 3.2 that a non-negative continuous function g that has integral 0 over a non-trivial interval I must be identically 0. (If g is not identically 0, then g is positive at some point $x_0 \in I$; use its continuity, via the ϵ, δ-definition, to show that it is greater than some positive constant over some non-trivial interval within I containing x_0; then deduce that its integral is greater than 0, contrary to assumption.)

3.5 Let I be any interval that is not of the form $[a, b]$ for $a, b \in \mathbb{R}$. Show that there is a continuous function f on I such that $\int_I f^2$ is infinite.

3.6   Suppose that u and v are vectors in an inner product space and that $\langle x, u \rangle = \langle x, v \rangle$ for all x. Show that $u = v$.

3.7   Prove Theorem 3.5.

3.8   Use a real analysis text to investigate the construction of the Cantor set, which was mentioned just following Theorem 3.5, and read one of the proofs that the set is uncountable (there are at least two quite different arguments). Show that the Cantor set has measure 0; this follows quite easily from its definition.

3.9   Let $a < b$ and let $\mathcal{I}$ be a collection of intervals.

(a) Suppose that $\mathcal{I}$ is a finite collection of open intervals and that the interval $[a, b]$ is covered by $\mathcal{I}$ (that is, $[a, b]$ is contained in the union of the elements of $\mathcal{I}$). Show that the sum of the lengths of the elements of $\mathcal{I}$ is greater than $b - a$.

(b) Suppose that $\mathcal{I}$ is a countable collection of open intervals and that the interval $[a, b]$ is covered by $\mathcal{I}$. The Heine–Borel Theorem from real analysis implies that $[a, b]$ is covered by some finite sub-collection of $\mathcal{I}$. Deduce from this that the sum of the lengths of the elements of $\mathcal{I}$ is greater than $b - a$.

(c) Let I be any one of the intervals $[a, b]$, $[a, b)$, $(a, b]$ and (a, b) and let $\mathcal{I}$ be a countable collection of intervals of any type that covers I. Use part (b) to show that the sum of the lengths of the elements of $\mathcal{I}$ is greater than or equal to $b - a$.

(d) Deduce that, as noted in the main text, the interval I in part (c) is not of measure 0.

3.10 Show that if $\langle x, y \rangle = \|x\| \|y\|$ for some x and y in an inner product space, then $x = \alpha y$ or $y = \alpha x$ for some $\alpha \in \mathbb{R}$. (Hint: Examine the proof of the Cauchy–Schwarz inequality.)

3.11 (a) Prove that if x is any element of an inner product space, then

$$\|x\| = \max\{\langle x, u \rangle : \|u\| = 1\}.$$

(Hint: Prove the inequalities '$\leq$' and '$\geq$' separately.)

(b) Give an informal, geometrically plausible interpretation of this statement in $\mathbb{R}^n$.

3.12 Prove the parallelogram law (Theorem 3.9).

3.13 The norm in an inner product space is defined in terms of the inner product. Show that the inner product can also be expressed in terms of the norm (that is, without explicit mention of the inner product).

3.14 Prove Theorem 3.11.

3.15 Verify that, as claimed near the end of Section 3.4, the two 'triangle inequalities' of that section are equivalent, in the sense that either can be derived from the other by substituting suitably chosen vectors.

3.16 (a) Let $x, y \in \mathbb{R}^2$. Sketch and describe the set of points

$$\{\alpha x + (1 - \alpha)y : 0 \le \alpha \le 1\}.$$

(b) A subset C of a vector space is called **convex** if whenever $x, y \in C$ and $0 \le \alpha \le 1$, we have $\alpha x + (1 - \alpha)y \in C$. Explain informally what a set which is convex according to this definition looks like, and why it is reasonable to use the word 'convex' in this way.

(c) Specify and sketch three non-trivial sets in the plane that are convex and three that are not. Give a brief explanation of your claim in each case.

(d) Let V be an inner product space. Prove that the **closed unit ball**

$$B = \{x \in V : \|x\| \le 1\}$$

in V is convex. (You will probably need to use the Cauchy–Schwarz inequality.)

(e) Indicate how to modify the proof to show that the **open unit ball**

$$B^\circ = \{x \in V : \|x\| < 1\}$$

is convex.

(f) Show that $rB(x, 1) = B(rx, r)$ for any $r \ge 0$. What happens for $r < 0$? Also show that $B(x, r) + y = B(x + y, r)$ for any vector y. Check that things work similarly in the case of open balls. (Before proving the required statements, write down suitable definitions of $rB(x, 1)$ and $B(x, r) + y$, or of what is meant in general by the multiplication of a set of vectors by a scalar and by the addition of a vector to a set of vectors.)

3.17 Prove the properties of orthogonality given in Theorem 3.13.

3.18 Show that if $x_1, x_2, \ldots, x_n$ form an orthonormal basis for an inner product space, then $x = \sum_{i=1}^{n} \langle x, x_i \rangle x_i$ for all x in the space. (The statement actually follows immediately from Theorem 3.17, but the idea of the exercise is to give a direct proof.)

3.19 Show that if $x_1, x_2, \ldots, x_n$ are orthogonal vectors in an inner product space, then

$$\|x_1 + x_2 + \cdots + x_n\|^2 = \|x_1\|^2 + \|x_2\|^2 + \cdots + \|x_n\|^2.$$

Comment on the geometrical significance of the result.

3.20 Suppose that the vectors $e_1, e_2, \ldots, e_n$ form an orthonormal set in an inner product space and let $\alpha_1, \alpha_2, \ldots, \alpha_n$ and $\beta_1, \beta_2, \ldots, \beta_n$ be arbitrary scalars. Prove that

$$\left\langle \sum_{i=1}^{n} \alpha_i e_i, e_j \right\rangle = \alpha_j \quad \text{for } j = 1, 2, \ldots, n,$$

$$\left\langle \sum_{i=1}^{n} \alpha_i e_i, \sum_{i=1}^{n} \beta_i e_i \right\rangle = \sum_{i=1}^{n} \alpha_i \beta_i$$

and

$$\left\| \sum_{i=1}^{n} \alpha_i e_i \right\|^2 = \sum_{i=1}^{n} \alpha_i^2.$$

3.21 (a) Confirm the claim, near the end of the inductive argument in the proof of Theorem 3.16, that $e_n \perp e_i$ for $i = 1, \ldots, n - 1$.

(b) Complete the proof of Theorem 3.16 by deducing part (ii) from part (i). (A minor but relevant observation is that a vector is in the span of some given set of vectors if it is a *finite* linear combination of vectors from the set; finiteness is simply part of the definition of a linear combination. It might also be noted that if the original sequence $x_1, x_2, x_3, \ldots$ in the theorem is finite, then the assertion made by part (ii) is included among the assertions made by part (i), and so requires no proof at all; but a suitably formulated proof, as above, will deal uniformly with both the finite and the infinite case.)

3.22 Let $x_1, x_2, \ldots$ be a linearly dependent sequence of vectors in an inner product space.

(a) Determine what happens when the Gram–Schmidt process is applied to the sequence. (Suppose that for some $k \geq 0$, the vectors $x_1, \ldots, x_k$ are linearly independent but that x_{k+1} is a linear combination of $x_1, \ldots, x_k$. What exactly happens on the $(k + 1)$th step?)

(b) Show that by a slight adaptation of the process, we can arrange for it to produce an orthonormal (and hence independent) sequence $e_1, e_2, \ldots$ that spans the same subspace as the original dependent sequence.

3.23 Write out the details of the example of Subsection 3.5.2.

3.24 Confirm the claims made about the example of Subsection 3.5.3, and investigate what can be said if we do not make the assumption that $\alpha_n > 0$ for all n.

3.25 Confirm the expressions given in Subsection 3.5.4 for the unnormalised Legendre polynomials $p_0(x)$, $p_1(x)$, $p_2(x)$ and $p_3(x)$, the classical Legendre polynomials $P_0(x)$, $P_1(x)$, $P_2(x)$ and $P_3(x)$ and the orthonormalised Legendre polynomials $Q_0(x)$, $Q_1(x)$, $Q_2(x)$ and $Q_3(x)$. Also, compute at least one more term of each sequence.

3.26 Show that if $x_1, \ldots, x_n$ are independent vectors in an inner product space, and $y \neq 0$ is such that $y \perp x_i$ for each i, then y is not a linear combination of $x_1, \ldots, x_n$ (or, equivalently, is not in the subspace spanned by $x_1, \ldots, x_n$).

3.27 Deduce from the previous exercise that there is only one polynomial of degree n with leading term x^n that is orthogonal in $C([-1, 1])$ (or $C(I)$ for any non-trivial interval I) to all polynomials of degree less than n.

3.28 Let $p_0(x), p_1(x), p_2(x), \ldots$ be the sequence of unnormalised Legendre polynomials. Thus, for each n, $p_n(x)$ is a polynomial of degree n with leading term x^n. Show that Rodrigues' formula (Subsection 3.5.4) holds. (Hint: Show that the function on the right-hand side in the formula is orthogonal to all polynomials of degree less than n, and use the previous exercise.)

3.29 Consider $Q_0(x), Q_1(x), Q_2(x), \ldots$, the orthonormalised Legendre polynomials. Without reference to the formulae for the polynomials or the details of the Gram–Schmidt calculations, show how we can infer from Theorem 3.16 that Q_n must be a polynomial of degree exactly n for each n.

3.30 Show by induction that the Legendre polynomials are even functions for even subscripts and odd functions for odd subscripts.

3.31 For vectors $y_1, y_2, \ldots, y_n$ in an inner product space, the matrix

$$G \equiv G(y_1, y_2, \ldots, y_n) = \begin{pmatrix} \langle y_1, y_1 \rangle & \langle y_1, y_2 \rangle & \cdots & \langle y_1, y_n \rangle \\ \langle y_2, y_1 \rangle & \langle y_2, y_2 \rangle & \cdots & \langle y_2, y_n \rangle \\ \vdots & \vdots & \ddots & \vdots \\ \langle y_n, y_1 \rangle & \langle y_n, y_2 \rangle & \cdots & \langle y_n, y_n \rangle \end{pmatrix}$$

is called the **Gram matrix** of $y_1, y_2, \ldots, y_n$, and the corresponding determinant, $|G| \equiv |G(y_1, y_2, \ldots, y_n)|$, is called the **Gram determinant** of $y_1, y_2, \ldots, y_n$. Show that $y_1, y_2, \ldots, y_n$ are linearly independent if and only if $|G| \neq 0$. (Hint: Consider the existence and significance of non-zero solutions to the system of linear equations $G\alpha = 0$.)

3.32 Show that the Gram determinant $|G|$ of the previous exercise is always non-negative. (Hint: When applied to $y_1, y_2, \ldots, y_n$, the Gram–Schmidt process yields orthonormal vectors $v_1, v_2, \ldots, v_n$ such that

$$v_1 = a_{1,1} y_1,$$
$$v_2 = a_{2,1} y_1 + a_{2,2} y_2,$$
$$v_3 = a_{3,1} y_1 + a_{3,2} y_2 + a_{3,3} y_3,$$
$$\vdots$$
$$v_n = a_{n,1} y_1 + a_{n,2} y_2 + a_{n,3} y_3 + \cdots + a_{n,n} y_n,$$

for suitable scalars $\{a_{i,j}\}$. Express this information in matrix form, and apply your knowledge of determinants.)

3.33 What familiar and important facts about inner product spaces are represented by the cases $n = 1$ and $n = 2$ in the two previous exercises?

3.34 Recall that an $n \times n$ real matrix A is called **orthogonal** if $AA^T = I$ (that is, the inverse of A is its transpose).

(a) Show that an $n \times n$ matrix A is orthogonal if and only if its rows form an orthonormal set (and hence an orthonormal basis) in $\mathbb{R}^n$.

(b) Deduce that if the n vectors

$$(x_{1,1}, x_{1,2}, \ldots, x_{1,n}), (x_{2,1}, x_{2,2}, \ldots, x_{2,n}), \ldots, (x_{n,1}, x_{n,2}, \ldots, x_{n,n})$$

are orthonormal in $\mathbb{R}^n$, then so are the n vectors

$$(x_{1,1}, x_{2,1}, \ldots, x_{n,1}), (x_{1,2}, x_{2,2}, \ldots, x_{n,2}), \ldots, (x_{1,n}, x_{2,n}, \ldots, x_{n,n}).$$

(c) Confirm the result of part (b) directly for the vectors in Exercise 1.4 of Chapter 1.

(d) Show that an $n \times n$ orthogonal matrix A *preserves the inner product and the norm* in $\mathbb{R}^n$, in the sense that $\langle Ax, Ay \rangle = \langle x, y \rangle$ and $\|Ax\| = \|x\|$ for all $x, y \in \mathbb{R}^n$.

3.35 The **cross product in $\mathbb{R}^n$** is a product of $n-1$ vectors, defined as follows. Let

$$x_1 = (x_{1,1}, x_{1,2}, \ldots, x_{1,n}),$$
$$x_2 = (x_{2,1}, x_{2,2}, \ldots, x_{2,n}),$$
$$\vdots$$
$$x_{n-1} = (x_{n-1,1}, x_{n-1,2}, \ldots, x_{n-1,n})$$

be $n-1$ arbitrary vectors in $\mathbb{R}^n$ and let $e_1, e_2, \ldots, e_n$ be the standard unit vectors in $\mathbb{R}^n$. Then the cross product $x_1 \times x_2 \times \cdots \times x_{n-1}$ of $x_1, x_2, \ldots, x_{n-1}$ is defined to be

$$x_1 \times x_2 \times \cdots \times x_{n-1} = (-1)^{n-1} \begin{vmatrix} e_1 & e_2 & \cdots & e_n \\ x_{1,1} & x_{1,2} & \cdots & x_{1,n} \\ x_{2,1} & x_{2,2} & \cdots & x_{2,n} \\ \vdots & \vdots & \ddots & \vdots \\ x_{n-1,1} & x_{n-1,2} & \cdots & x_{n-1,n} \end{vmatrix}.$$

(Note that the 'determinant' here is not, strictly speaking, a determinant at all, since the elements of its first row are vectors rather than scalars. But as in the case of the cross product in $\mathbb{R}^3$, it is very convenient to use determinant notation: expansion of the 'determinant' along the first row, using the usual expansion rules, yields the vector value that is required for the definition.)

(a) Check that the case $n = 3$ does indeed yield the usual cross product in $\mathbb{R}^3$.

(b) Use properties of determinants to show that

$$x_1 \times x_2 \times \cdots \times x_{n-1} \perp x_i, \quad \text{for } i = 1, 2, \ldots, n - 1.$$

(c) What is $e_1 \times e_2 \times \cdots \times e_{n-1}$? Generalise!

(d) Investigate the effect on the cross product of (i) interchanging two of its arguments, (ii) having two of its arguments equal and (iii) having one of its arguments a linear combination of the others.

3.36 Find a vector in $\mathbb{R}^4$ that is orthogonal to each of $(1, 2, 3, 4)$, $(-1, -3, 1, 0)$ and $(3, -8, 0, 1)$. Do this in two ways:

(a) using the cross product in $\mathbb{R}^4$; and

(b) using the Gram–Schmidt process.

3.37 Adapting ideas from the two previous exercises (or otherwise), construct a quadratic polynomial in $C([-1, 1])$ that is orthogonal to both $1 + 2x$ and $2 + x - 3x^2$. It may be helpful to use the following expressions for the first three powers of x as linear combinations of the polynomials $Q_n(x)$:

$$x^0 = \sqrt{2}Q_0(x), \quad x^1 = \frac{\sqrt{2}}{\sqrt{3}}Q_1(x), \quad x^2 = \frac{2\sqrt{2}}{3\sqrt{5}}Q_2(x) + \frac{\sqrt{2}}{3}Q_0(x).$$

3.38 Prove the final statement in Theorem 3.17 (Parseval's relation).

3.39 For the examples in Subsection 3.6.1, verify directly that the properties predicted by Theorem 3.17 hold.

3.40 Let V be an inner product space, let $e_1, e_2, \ldots, e_n$ be orthonormal vectors in V, let P denote orthogonal projection into the subspace spanned by $e_1, e_2, \ldots, e_n$, and let $x \in V$. Show that

$$\|x - Px\|^2 = \left\|x - \sum_{i=1}^n \langle x, e_i \rangle e_i\right\|^2 = \|x\|^2 - \sum_{i=1}^n \langle x, e_i \rangle^2.$$

Why should this result be expected geometrically? (Hint: Think about Theorem 3.19 and Pythagoras' theorem.)

3.41 Prove the uniqueness assertion at the end of Theorem 3.19.

3.42 In $\mathbb{R}^3$, find the vector x in the plane (that is, the 2-dimensional subspace) spanned by $x_1 = (1, 0, -3)$ and $x_2 = (2, -2, 1)$ which is closest to the vector $y = (4, 8, -5)$. Calculate the distance from x to y.

3.43 Find the cubic polynomial that best approximates the function $\sin x$ in the space $C([-1, 1])$.

4

Hilbert Spaces

Although we have already seen that significant parts of Euclidean geometry can be generalised to arbitrary inner product spaces, there is another extremely important set of properties of Euclidean spaces that we have not yet attempted to generalise. Specifically, we have not yet made any examination at all of *limiting processes* in inner product spaces; these are processes, not explicitly geometric in character, such as forming the limits of infinite sequences and summing infinite series. Such processes are fundamental in all deeper work on the real line $\mathbb{R}$, which usually goes under the general heading of *analysis*.

Our task in this chapter is to develop these ideas in general inner product spaces, and an alternative title for the chapter might therefore be 'Analysis in inner product spaces'. One of the most fundamental of all these ideas is that of the convergence of a sequence, and we turn to this first.

4.1 Sequences in Inner Product Spaces

4.1.1 The Definition of Convergence

We use the definition of convergence of a sequence on the real line $\mathbb{R}$, which we understand well, as a model for the corresponding definition in a general inner product space.

Recall that if $\{x_n\}$ is a sequence of real numbers, then we write

$$\lim_{n \to \infty} x_n = x \qquad \text{or} \qquad x_n \to x \text{ as } n \to \infty,$$

if

for all $\epsilon > 0$ there exists $N \in \mathbb{N}$ such that $|x_n - x| < \epsilon$ for all $n > N$.

It is often convenient to express this using more symbols and fewer words in a compressed form such as

$$\forall \epsilon > 0 \ \exists N \in \mathbb{N} \ s.t. \ n > N \implies |x_n - x| < \epsilon.$$

To adapt this definition so that it will make sense in a general inner product space is very straightforward, because the only term we need to change is the term that represents the distance between x_n and its limit x, namely, $|x_n - x|$. No other part of the definition needs to change:

- '$\forall \epsilon > 0$' and '$\cdots < \epsilon$' involve the measuring of distance, and this always takes place in $\mathbb{R}$, because distances are (non-negative) real numbers; and
- '$\exists N \in \mathbb{N}$' and '$n > N \implies \cdots$' involve the values of the variable that indexes the sequence, not the values of the terms of the sequence, and the index variable always takes values in $\mathbb{N}$.

Finally, adapting the remaining term, $|x_n - x|$, is straightforward, because we have developed a notion of distance in arbitrary inner product spaces; all we need to do is to formally replace the absolute value signs by norm signs. Thus we are led to the following definition.

Definition 4.1 Let V be an inner product space. If $\{x_n\}$ is a sequence of vectors in V and x is a vector in V, we write

$$\lim_{n \to \infty} x_n = x \qquad \text{or} \qquad x_n \to x \text{ as } n \to \infty,$$

and we say **the limit of** x_n **is** x **as** n **tends to infinity**, if for all $\epsilon > 0$ there exists $N \in \mathbb{N}$ such that $\|x_n - x\| < \epsilon$ for all $n > N$.

4.1.2 Properties of Convergence

The following simple result allows us to reduce the question of the convergence of a sequence in a general inner product space to the question of the convergence of a related sequence in $\mathbb{R}$. This result will be used many times, but we will point out its use explicitly only on one or two occasions. The proof, which is left for Exercise 4.1, will reveal that the result is hardly more than a rephrasing of the definition of convergence.

Theorem 4.2 *Let V be an inner product space, and let $\{x_n\}$ be a sequence of vectors in V and x a vector in V. Then $x_n \to x$ in V if and only if $\|x_n - x\| \to 0$ in $\mathbb{R}$.*

Many minor and unsurprising results about the convergence of sequences of real numbers carry over without difficulty to general inner product spaces. Here are two such results.

Theorem 4.3 *Let V be an inner product space.*

(i) *A sequence $\{x_n\}$ in V can have at most one limit in V.*

(ii) *If $x_n \to x$ and $y_n \to y$ in V, then also $x_n + y_n \to x + y$. More generally, $\alpha x_n + \beta y_n \to \alpha x + \beta y$, for all scalars α and β.*

Proof We will give the proof that if $x_n \to x$ and $y_n \to y$ in V then also $x_n + y_n \to x + y$, leaving the other statements for Exercise 4.2. The proof will be written out in somewhat more detail than is required just for a logically complete argument, to assist the reader who is not familiar with proofs of statements such as these, with their use of ϵ's and N's.

Fix $\epsilon > 0$. Then the definition of convergence implies that

$$\exists N_1 \in \mathbb{N} \ \text{s.t.} \ n > N_1 \implies \|x_n - x\| < \tfrac{\epsilon}{2}$$

and

$$\exists N_2 \in \mathbb{N} \ \text{s.t.} \ n > N_2 \implies \|y_n - y\| < \tfrac{\epsilon}{2}.$$

We now need to show that the distance between the term $x_n + y_n$ and its supposed limit $x + y$, that is, $\|(x_n + y_n) - (x + y)\|$, can be made less than ϵ for sufficiently large n. But

$$\left\|(x_n + y_n) - (x + y)\right\| = \left\|(x_n - x) + (y_n - y)\right\| \le \|x_n - x\| + \|y_n - y\|,$$

by the triangle inequality, and we know that if $n > N_1$ then $\|x_n - x\| < \tfrac{\epsilon}{2}$ and if $n > N_2$ then $\|y_n - y\| < \tfrac{\epsilon}{2}$. Defining $N = \max\{N_1, N_2\}$, we observe that $n > N$ gives $N > N_1$ and $N > N_2$, and hence

$$\left\|(x_n + y_n) - (x + y)\right\| < \tfrac{\epsilon}{2} + \tfrac{\epsilon}{2} = \epsilon,$$

which is what we wanted to prove. Therefore, $x_n + y_n \to x + y$ as $n \to \infty$. □

▶ It is worth making two brief comments about the 'contrived' choice of $\tfrac{\epsilon}{2}$ above (such choices are common in arguments of this type). The first is that the choice is legitimate, because $\tfrac{\epsilon}{2}$ is a positive real number and the definition of convergence will supply us with a value of N for any positive real number we choose. The second is that this particular choice is a useful one, because we have looked ahead to see where the argument is heading, and have observed that a choice of $\tfrac{\epsilon}{2}$ at the start will result in an ϵ at the end, as required.

4.1.3 Closed Sets

Definition 4.4 Let X be a subset of an inner product space V. We say that X is **closed** in V if X contains the limits of all its convergent sequences; more precisely, whenever $\{x_n\}$ is a sequence of elements of X and $x_n \to x$ for some $x \in V$ as $n \to \infty$, we have $x \in X$.

The most important case of the definition for us will be when X is a subspace of V, rather than an arbitrary subset, but it is still useful to be able to apply the term more generally.

➤ The concept is also one that applies far more broadly than just to inner product spaces; it is important in topology and is fundamental in the theory of metric spaces. (Recall that we noted in Section 3.4 that inner product spaces can be regarded as metric spaces in a natural way.)

Let us note several simple examples, some details of which will be left for Exercise 4.3. Many other important examples will arise naturally in our later work.

Example 4.5   (i) Any closed interval X (in the usual sense of that phrase) is a closed set in $\mathbb{R}$ (in the sense of Definition 4.4). If $X = [a, b]$, say, and $\{x_n\}$ is a sequence in X, then $a \le x_n \le b$ for all n, and if the sequence converges to $x \in \mathbb{R}$ it is easy to check that $a \le x \le b$ as well, so that $x \in X$.

 (ii) The set $\mathbb{Z}$ of integers is closed in $\mathbb{R}$. It is easy to check that the only convergent sequences in $\mathbb{Z}$ are those sequences that are eventually constant (that is, are constant from some index on). The limit of such a sequence must then be the constant in question, and that is an element of $\mathbb{Z}$.

(iii) The set $\mathbb{Q}$ of rational numbers is not closed in $\mathbb{R}$. For this, it is enough to note that there is a sequence of rational numbers that converges to any given irrational number. (See Example 4.27 below for two concrete instances and further discussion, and see also Exercise 4.3.)

(iv) Any vector subspace X of $\mathbb{R}^n$ is closed in $\mathbb{R}^n$. (A proof of this could be attempted immediately, but it will be easier to deal with once we have progressed a bit further in this chapter; see Exercise 4.3.)

The next two definitions introduce two further concepts that are closely related to the idea of a closed set.

Definition 4.6   Let X be a subset of an inner product space V. The **closure** of X is the set of all limit points of all convergent sequences of elements of X.

Note that X is a subset of its closure, since each element x of X is trivially the limit of the sequence with constant value x. Moreover, if Y is any closed subset containing X, it is easy to check that Y contains the closure of X, and it follows that the closure of X can also be characterised either as *the smallest closed set containing X* or as *the intersection of all closed sets containing X*. (However, several points need to be checked in order to establish these claims rigorously; see Exercise 4.6 for further details.)

Let us again note a few examples.

Example 4.7   (i) The closure of a closed set X is X itself; this sounds
obvious, but should be checked.

(ii) The closure of the open interval (a, b) or either of the half-open intervals
$(a, b]$ and $[a, b)$ is the closed interval $[a, b]$.

(iii) The closure of $\mathbb{Q}$ is the whole of $\mathbb{R}$ – using the comment earlier that every
irrational number is the limit of a sequence of rationals.

Definition 4.8   A subset X of an inner product space V is said to be **dense**
in V if the closure of X is V.

Example 4.9   The set $\mathbb{Q}$ of rationals is dense in $\mathbb{R}$, by the previous example.

The following result gives a useful alternative characterisation of the closure
of a set.

Theorem 4.10   *Let X be a subset of an inner product space V. Then $v \in V$ is
in the closure of X if and only if for all $\epsilon > 0$ there exists $x \in X$ with $\|v - x\| < \epsilon$.*

Proof If v is in the closure of X, then there is a sequence $\{x_n\}$ in X such
that $x_n \to v$ as $n \to \infty$, and the desired condition follows directly from the
definition of convergence.

If the condition holds, then for every $n \in \mathbb{N}$ we can find $x_n \in X$ such that
$\|v - x\| < 1/n$. It follows that $x_n \to v$ as $n \to \infty$, so v is in the closure of X. □

4.2 Convergence in $\mathbb{R}^n$, ℓ^2 and the L^2 Spaces

We aim here to gain a reasonably detailed understanding of what the convergence of a sequence involves in the Euclidean spaces $\mathbb{R}^n$ and the sequence
space ℓ^2, and to study an example of convergence in an L^2 space.

To discuss the case of the Euclidean spaces, we will change notation from
our usual '$\mathbb{R}^n$' to '$\mathbb{R}^p$', simply because we will want the subscript 'n' to index
our sequences. We also introduce the following notation. If x is a vector in $\mathbb{R}^p$,
then we will use the notation $x^{(k)}$ for the kth entry of x, for $1 \le k \le p$. Thus, if
$x \in \mathbb{R}^p$, we can write

$$x = (x^{(1)}, x^{(2)}, \ldots, x^{(p)}).$$

Theorem 4.11   *Convergence in $\mathbb{R}^p$ is convergence entry by entry; that is,
$x_n \to x$ in $\mathbb{R}^p$ if and only if $x_n^{(k)} \to x^{(k)}$ in $\mathbb{R}$ for $1 \le k \le p$.*

Proof The theorem is in the form of an 'if and only if' statement, and we will
prove the 'if' and 'only if' parts by separate arguments.

First, assume that $x_n \to x$. Now for each k, we have

$$0 \leq \left| x_n^{(k)} - x^{(k)} \right| \leq \left(\sum_{\ell=1}^{p} \left| x_n^{(\ell)} - x^{(\ell)} \right|^2 \right)^{\frac{1}{2}} = \| x_n - x \|.$$

Therefore, for each k, application of Theorem 4.2 twice gives

$$x_n \to x \implies \| x_n - x \| \to 0$$
$$\implies \left| x_n^{(k)} - x^{(k)} \right| \to 0$$
$$\implies x_n^{(k)} \to x^{(k)}.$$

Second, assume that $x_n^{(k)} \to x^{(k)}$ for $1 \leq k \leq p$. Consider any fixed $\epsilon > 0$. It follows from our assumption that for each k in the range $1 \leq k \leq p$,

$$\exists N_k \in \mathbb{N} \text{ s.t. } n > N_k \implies \left| x_n^{(k)} - x^{(k)} \right| < \frac{\epsilon}{\sqrt{p}}.$$

Now set $N = \max\{N_1, N_2, \ldots, N_p\}$. Then $n > N$ implies $n > N_k$ for every k in the range $1 \leq k \leq p$, and hence

$$\| x_n - x \| = \left(\sum_{k=1}^{p} \left| x_n^{(k)} - x^{(k)} \right|^2 \right)^{\frac{1}{2}} < \left(\sum_{k=1}^{p} \left(\frac{\epsilon}{\sqrt{p}} \right)^2 \right)^{\frac{1}{2}} = \epsilon,$$

which shows that $x_n \to x$, as required. □

➤ Note that, as in the proof of Theorem 4.3, the odd-looking value $\epsilon/\sqrt{p}$ above was chosen precisely so that we would obtain ϵ at the end of the argument.

This is as simple a result as we could hope for: convergence in Euclidean spaces reduces completely to convergence in $\mathbb{R}$.

Turning to the case of ℓ^2, it is interesting to work out first how much of the previous argument remains valid. It is easy to see that the 'only if' part does carry over from $\mathbb{R}^p$ to ℓ^2: all we need to do is to replace the finite sum by an infinite one, and the argument then works. This means that we have the following theorem (as before, we write $x_n^{(k)}$ to represent the kth entry of $x \in \ell^2$, now for all $k \in \mathbb{N}$).

Theorem 4.12 *Convergence in ℓ^2 implies convergence entry by entry. More precisely, if $x_n \to x$ in ℓ^2, then $x_n^{(k)} \to x^{(k)}$ in $\mathbb{R}$ for all $k \in \mathbb{N}$.*

However, problems occur at two points if we try to carry over the 'if' part of the proof of Theorem 4.11 to ℓ^2. The first is where we made the choice $\epsilon/\sqrt{p}$ at the start of the argument; no such choice is possible in the ℓ^2 case, because p, which represented the number of entries in our vectors in the Euclidean case, is now infinite. The second is at the step where we set $N = \max\{N_1, N_2, \ldots, N_p\}$.

In the ℓ^2 case, we would need to set $N = \max\{N_1, N_2, \ldots\}$, and this would be the maximum of an infinite set of numbers, which in general is undefined.

Examples in Subsection 4.2.1 show that the converse of Theorem 4.12 is indeed false in general. Despite this, though, Theorem 4.12 is useful, because it tells us that if we form the entry-by-entry limit of a sequence in ℓ^2, then that entry-by-entry limit is *the only candidate* for the limit of the sequence in ℓ^2, though there may in fact be no limit. (We use this information in one of the examples.)

4.2.1 Three Examples in ℓ^2

We will consider three examples which illustrate Theorem 4.12 and give insight into the ways in which its converse fails. They show that convergence in ℓ^2 is not equivalent to entry-by-entry convergence and that the nature of convergence in ℓ^2 is intrinsically more complicated. (For further confirmation of the last point, compare the proofs of Theorems 4.29 and 4.31 below.)

Example 4.13   For any given $n \in \mathbb{N}$, consider the sequence

$$x_n = \left(\frac{1}{n}, \frac{1}{2n}, \frac{1}{3n}, \ldots \right).$$

We claim that $x_n \in \ell^2$, that is, that x_n is a square-summable sequence. Indeed, the required sum of squares is

$$\left(\frac{1}{n}\right)^2 + \left(\frac{1}{2n}\right)^2 + \left(\frac{1}{3n}\right)^2 + \cdots = \frac{1}{n^2} + \frac{1}{2^2n^2} + \frac{1}{3^2n^2} + \cdots$$

$$= \frac{1}{n^2}\left(\frac{1}{1^2} + \frac{1}{2^2} + \frac{1}{3^2} + \cdots\right),$$

and the parenthesised term in the last expression is a convergent series (a so-called 'p-series', with $p = 2$), establishing the claim.

Now the vector $x_n \in \ell^2$ is defined for each $n \in \mathbb{N}$, and this gives us a sequence of vectors $\{x_n\}$ in ℓ^2. (Note carefully that $\{x_n\}$ is a sequence of vectors each of which is itself a sequence – that is, *a sequence of sequences*.) We wish to determine whether the sequence $\{x_n\}$ converges in ℓ^2 and, if it does, what its limit is. In fact, the calculation above can be used to give us the answer. The sum of the p-series with $p = 2$ is known to be $\pi^2/6$ (though we do not need to know the specific value for this argument), so the calculation above shows that

$$\|x_n\| = \frac{\pi}{\sqrt{6}\,n}.$$

It now follows that the limit vector of the sequence $\{x_n\}$ in ℓ^2 is the zero vector $0 = (0, 0, 0, \ldots) \in \ell^2$, and this is so because

$$\|x_n - 0\| = \|x_n\| = \frac{\pi}{\sqrt{6}\, n} \to 0$$

as $n \to \infty$, from which Theorem 4.2 shows that $x_n \to 0$ in ℓ^2 as $n \to \infty$.

Let us look more closely at the example. Here we show each x_n, and the limit vector 0, displayed on a line by itself:

$$\left(\tfrac{1}{1\cdot 1}, \ \tfrac{1}{2\cdot 1}, \ \tfrac{1}{3\cdot 1}, \ \tfrac{1}{4\cdot 1}, \ \cdots\right)$$

$$\left(\tfrac{1}{1\cdot 2}, \ \tfrac{1}{2\cdot 2}, \ \tfrac{1}{3\cdot 2}, \ \tfrac{1}{4\cdot 2}, \ \cdots\right)$$

$$\left(\tfrac{1}{1\cdot 3}, \ \tfrac{1}{2\cdot 3}, \ \tfrac{1}{3\cdot 3}, \ \tfrac{1}{4\cdot 3}, \ \cdots\right)$$

$$\vdots \quad \vdots \quad \vdots \quad \vdots \quad \cdots$$

$$\downarrow \quad \downarrow \quad \downarrow \quad \downarrow \quad \cdots$$

$$\big(0, \quad 0, \quad 0, \quad 0, \quad \ldots\big).$$

Observe that we not only have $x_n \to 0$ in ℓ^2, as we just proved, but also that the kth entry of x_n converges in $\mathbb{R}$ to the kth entry of 0 for each $k \in \mathbb{N}$: as predicted by Theorem 4.12, the vectors x_n converge to their limit vector 0 entry by entry.

Example 4.14   Consider the familiar sequence

$$e_1 = (1, 0, 0, 0, \ldots), \quad e_2 = (0, 1, 0, 0, \ldots), \quad e_3 = (0, 0, 1, 0, \ldots), \ \ldots$$

in ℓ^2. What happens entry by entry? The picture this time is:

$$\big(1, \ 0, \ 0, \ 0, \ \ldots\big)$$

$$\big(0, \ 1, \ 0, \ 0, \ \ldots\big)$$

$$\big(0, \ 0, \ 1, \ 0, \ \ldots\big)$$

$$\vdots \quad \vdots \quad \vdots \quad \vdots \quad \cdots$$

$$\downarrow \quad \downarrow \quad \downarrow \quad \downarrow \quad \cdots$$

$$\big(0, \ 0, \ 0, \ 0, \ \ldots\big),$$

and we see that the limit formed entry by entry is the sequence $(0, 0, 0, 0, \ldots)$, the zero vector $0 \in \ell^2$, because the sequence of kth entries for each fixed k consists entirely of 0's apart from a solitary 1 in the kth position. However, we would be wrong to draw from this the conclusion that $e_n \to 0$ in ℓ^2. This is simply because

$$\|e_n - 0\| = \|e_n\| = 1,$$

from which Theorem 4.2 shows that e_n does not converge to 0 in ℓ^2. It follows that the sequence $\{e_n\}$ does not have a limit in ℓ^2 at all, because Theorem 4.12 implies that if it did have a limit, then that limit could only be the entry-by-entry limit 0, and we have just shown that this is not the limit.

Example 4.15   Consider the sequence

$$y_1 = (1, 0, 0, 0, \ldots), \quad y_2 = (1, 1, 0, 0, \ldots), \quad y_3 = (1, 1, 1, 0, \ldots), \ldots,$$

in which the first n entries of y_n are 1's and the remainder are 0's. It is clear that $y_n \in \ell^2$ for all n. Looking at the situation entry by entry, we have the picture:

$$\left(1, \ 0, \ 0, \ 0, \ \ldots\right)$$
$$\left(1, \ 1, \ 0, \ 0, \ \ldots\right)$$
$$\left(1, \ 1, \ 1, \ 0, \ \ldots\right)$$
$$\vdots \ \ \vdots \ \ \vdots \ \ \vdots \ \ \cdots$$
$$\downarrow \ \downarrow \ \downarrow \ \downarrow \ \cdots$$
$$\left(1, \ 1, \ 1, \ 1, \ \ldots\right).$$

This time, the limit formed entry by entry is $(1, 1, 1, 1, \ldots)$, because the sequence of kth entries for any fixed k consists entirely of 1's from the kth term onwards. But this 'limit vector' $(1, 1, 1, 1, \ldots)$ *is not in* ℓ^2: it is not square-summable, because $\sum_{n=1}^{\infty} 1^2$ does not converge. Hence, again by Theorem 4.12, the sequence $\{y_n\}$ has no limit in ℓ^2.

4.2.2 An Example in an L^2 Space

Consider the sequence of functions $\{f_n\}$ in $C([-1, 1])$, or $L^2([-1, 1])$, given by the formula

$$f_n(x) = \begin{cases} 0, & -1 \le x \le 0, \\ nx, & 0 < x \le 1/n, \\ 1, & 1/n < x \le 1 \end{cases}$$

for each $n \in \mathbb{N}$, and with graph of the typical form shown in Figure 4.1 (which is actually the graph for $n = 6$ specifically).

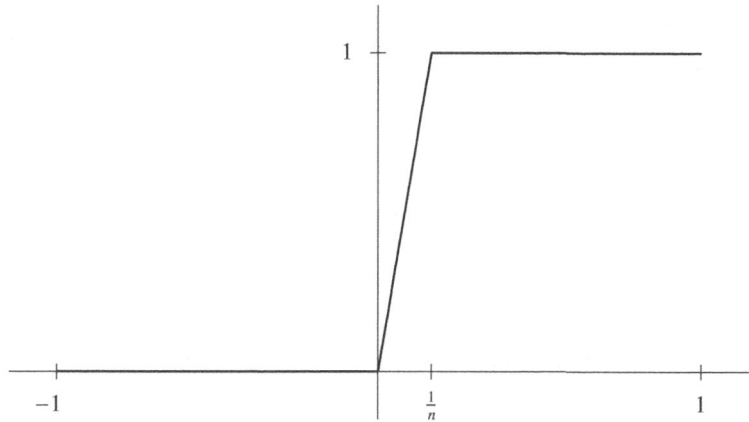

Figure 4.1 The graph of the function f_n

If we were to guess at the limit of the sequence $\{f_n\}$, our guess would probably be the function f given by

$$f(x) = \begin{cases} 0, & -1 \le x \le 0, \\ 1, & 0 < x \le 1. \end{cases}$$

Proving that this guess is correct is very simple: we calculate

$$\|f_n - f\| = \left(\int_{-1}^{1} (f_n(x) - f(x))^2 dx \right)^{\frac{1}{2}} = \left(\int_{0}^{\frac{1}{n}} (1 - nx)^2 \right)^{\frac{1}{2}} = \frac{1}{\sqrt{3n}},$$

and the last expression obviously tends to 0 as $n \to \infty$. Therefore, $f_n \to f$ in $L^2([-1, 1])$ as $n \to \infty$.

Note the following points.

- The functions f_n are all continuous and hence in $C([-1, 1])$ (and therefore also in $L^2([-1, 1])$), while f is not continuous and hence is not in $C([-1, 1])$, although it is in $L^2([-1, 1])$.
- There is more than one candidate for the limit function f; for example,

$$f(x) = \begin{cases} 0, & -1 \le x < 0, \\ 1, & 0 \le x \le 1 \end{cases}$$

would do equally well in the above argument. Note that this does not contradict part (i) of Theorem 4.3, which says that a sequence in an inner product space can have only one limit, because the definition of $L^2([-1, 1])$ treats functions as equal if the square of their difference has integral 0; this is the case for the two possible limit functions above, and for many others, and these functions actually represent *the same element of $L^2([-1, 1])$*.

- Although many different functions are limits, in the L^2 sense, for the sequence $\{f_n\}$, no *continuous* function f is a limit. (This is very reasonable intuitively, but nevertheless requires proof, and that is the content of Exercise 4.10.)

4.3 Series in Inner Product Spaces

4.3.1 The Definition of Convergence

The definition of the convergence of a sequence in an inner product space is a straightforward adaptation of the corresponding definition in $\mathbb{R}$. Similar comments apply to the convergence of infinite series in inner product spaces: we mimic the definition that is familiar in $\mathbb{R}$.

Definition 4.16   If $\{x_n\}$ is a sequence in an inner product space V, we say that **the infinite series $\sum_{n=1}^{\infty} x_n$ converges, with sum x,** and we write

$$\sum_{n=1}^{\infty} x_n = x,$$

if the sequence of partial sums $s_n = \sum_{k=1}^{n} x_k$ converges to x in V as $n \to \infty$. If the sequence of partial sums does not converge in V, then we say that the infinite series **diverges**.

▶ Notice that this definition is formally *identical* to that for convergence of an infinite series in $\mathbb{R}$. The dependence on the specific space we are working in is concealed in the condition requiring convergence of the sequence of partial sums, which, when written out in full, makes use of the norm in that space.

Since $\mathbb{R}$ is an inner product space, any series in $\mathbb{R}$ provides an example of a series in an inner product space, but we will soon encounter many examples of series in other spaces, and in particular in infinite-dimensional spaces.

4.3.2 Interchanging Sums with Inner Products and Norms

We are familiar (see, for example, Exercise 3.20) with the kinds of algebraic manipulations which show that if a finite set of vectors $e_1, e_2, \ldots, e_n$ in an inner product space is orthonormal, then

$$\left\langle \sum_{i=1}^{n} \alpha_i e_i, e_j \right\rangle = \alpha_j \quad \text{for } j = 1, 2, \ldots, n,$$

$$\left\langle \sum_{i=1}^{n} \alpha_i e_i, \sum_{i=1}^{n} \beta_i e_i \right\rangle = \sum_{i=1}^{n} \alpha_i \beta_i$$

and

$$\left\| \sum_{i=1}^{n} \alpha_i e_i \right\|^2 = \sum_{i=1}^{n} \alpha_i^2$$

for all scalars $\alpha_1, \alpha_2, \ldots, \alpha_n$ and $\beta_1, \beta_2, \ldots, \beta_n$.

It will be vital in our later work to know that analogous manipulations can be performed in the case of *infinite* sets of orthonormal vectors, and this subsection culminates in a theorem stating that this is indeed possible. To arrive at that point, however, we need a few preliminary results on the inner product and the norm and how they interact with convergence of sequences.

➤ All the preliminary results in this subsection could have been stated and proved immediately after Section 4.1 (and one of them even earlier), but their application to infinite series at the end of this subsection is our main reason for developing them.

We start with the following simple result.

Lemma 4.17   *For x and y in any inner product space V,*

$$\left| \|x\| - \|y\| \right| \leq \|x - y\|.$$

Proof   Using the triangle inequality in V, we have

$$\|x\| = \left\| y + (x - y) \right\| \leq \|y\| + \|x - y\|,$$

which gives

$$\|x\| - \|y\| \leq \|x - y\|.$$

A similar argument shows that

$$\|y\| - \|x\| \leq \|x - y\|,$$

and combination of the last two inequalities immediately yields the result.   □

Theorem 4.18   *If $x_n \to x$ is a convergent sequence in an inner product space V, then $\|x_n\| \to \|x\|$ in $\mathbb{R}$.*

Proof   By the lemma,

$$\left| \|x_n\| - \|x\| \right| \leq \|x_n - x\| \to 0,$$

since $x_n \to x$ in V (see Theorem 4.2). Therefore, $\|x_n\| \to \|x\|$ in $\mathbb{R}$.   □

Theorem 4.19   *If $x_n \to x$ is a convergent sequence in an inner product space V, then $\langle x_n, y \rangle \to \langle x, y \rangle$ for all $y \in V$. More generally, if $x_n \to x$ and $y_n \to y$ in V, then $\langle x_n, y_n \rangle \to \langle x, y \rangle$.*

Proof Using the triangle inequality, the Cauchy–Schwarz inequality and the previous theorem, we have

$$
\begin{aligned}
\left|\langle x_n, y_n \rangle - \langle x, y \rangle\right| &= \left|\langle x_n, y_n \rangle - \langle x_n, y \rangle + \langle x_n, y \rangle - \langle x, y \rangle\right| \\
&\leq \left|\langle x_n, y_n \rangle - \langle x_n, y \rangle\right| + \left|\langle x_n, y \rangle - \langle x, y \rangle\right| \\
&= \left|\langle x_n, y_n - y \rangle\right| + \left|\langle x_n - x, y \rangle\right| \\
&\leq \|x_n\| \cdot \|y_n - y\| + \|x_n - x\| \cdot \|y\| \\
&\to \lim_{n \to \infty} \|x_n\| \cdot \lim_{n \to \infty} \|y_n - y\| + \lim_{n \to \infty} \|x_n - x\| \cdot \|y\| \\
&= \|x\| \cdot 0 + 0 \cdot \|y\| \\
&= 0.
\end{aligned}
$$

Therefore, $\langle x_n, y_n \rangle \to \langle x, y \rangle$.   □

It is useful to restate the last two results in a way that avoids naming the various limits involved. In this form, the statements read: *if $\{x_n\}$ is a convergent sequence in V, then $\{\|x_n\|\}$ is a convergent sequence in $\mathbb{R}$ and*

$$
\lim_{n \to \infty} \|x_n\| = \left\| \lim_{n \to \infty} x_n \right\| ;
$$

and: *if $\{x_n\}$ and $\{y_n\}$ are convergent sequences in V, then $\{\|x_n\|\}$ and $\{\|y_n\|\}$ are convergent sequences in $\mathbb{R}$ and*

$$
\lim_{n \to \infty} \langle x_n, y_n \rangle = \left\langle \lim_{n \to \infty} x_n, \lim_{n \to \infty} y_n \right\rangle.
$$

An informal but useful way of summarising both statements is to say that *limits can be interchanged with norms and inner products.*

➤ Another, more abstract, way of reformulating the results, particularly if you are familiar with metric spaces, is to say that *the norm and the inner product are continuous functions.* It is a standard fact in analysis that a function $f \colon \mathbb{R} \to \mathbb{R}$ is continuous at $x \in \mathbb{R}$ if and only if for all sequences $x_n \to x$, we have $f(x_n) \to f(x)$. This result remains valid if we replace the domain $\mathbb{R}$ of f by an arbitrary metric space. Taking V as the metric space for Theorem 4.18 and $V \times V$ as the metric space for Theorem 4.19 shows that the reformulation in terms of continuity is correct.

Since an infinite sum is nothing more than the limit of a suitable sequence (namely, the sequence of partial sums), we can now, as promised earlier, generalise straightforwardly to infinite sums the three equations stated for finite sums at the start of this subsection. We will do this in two stages, first developing three more general equations from which the desired equations follow.

Theorem 4.20 *Suppose that $\sum_{n=1}^{\infty} \alpha_n x_n$ and $\sum_{n=1}^{\infty} \beta_n y_n$ are convergent series in an inner product space V. Then*

$$\left\langle \sum_{n=1}^{\infty} \alpha_n x_n, x \right\rangle = \sum_{n=1}^{\infty} \alpha_n \langle x_n, x \rangle \quad \text{for each } x \in V,$$

$$\left\langle \sum_{m=1}^{\infty} \alpha_m x_m, \sum_{n=1}^{\infty} \beta_n y_n \right\rangle = \sum_{m=1}^{\infty} \sum_{n=1}^{\infty} \alpha_m \beta_n \langle x_m, y_n \rangle$$

and

$$\left\| \sum_{n=1}^{\infty} \alpha_n x_n \right\|^2 = \sum_{m=1}^{\infty} \sum_{n=1}^{\infty} \alpha_m \alpha_n \langle x_m, x_n \rangle.$$

Note that each of these assertions requires that the series on the right-hand side, which is a series of real numbers, actually converges, and the assertion that it does converge has to be taken as part of the statement (a similar issue arose in our reformulations of Theorems 4.18 and 4.19). Thus a more precise formulation of the first part, for example, would read: *for each $x \in V$, the series $\sum_{n=1}^{\infty} \alpha_n \langle x_n, x \rangle$ converges, and $\langle \sum_{n=1}^{\infty} \alpha_n x_n, x \rangle = \sum_{n=1}^{\infty} \alpha_n \langle x_n, x \rangle$.*

Proof of Theorem 4.20 For the first assertion, the definition of the infinite sum gives

$$\left\langle \sum_{n=1}^{\infty} \alpha_n x_n, x \right\rangle = \left\langle \lim_{n \to \infty} \left(\sum_{i=1}^{n} \alpha_i x_i \right), x \right\rangle.$$

Also, Theorem 4.19 shows that the limit

$$\lim_{n \to \infty} \left\langle \sum_{i=1}^{n} \alpha_i x_i, x \right\rangle$$

exists and that

$$\left\langle \lim_{n \to \infty} \left(\sum_{i=1}^{n} \alpha_i x_i \right), x \right\rangle = \lim_{n \to \infty} \left\langle \sum_{i=1}^{n} \alpha_i x_i, x \right\rangle.$$

Finally,

$$\lim_{n \to \infty} \left\langle \sum_{i=1}^{n} \alpha_i x_i, x \right\rangle = \lim_{n \to \infty} \sum_{i=1}^{n} \alpha_i \langle x_i, x \rangle,$$

by routine manipulations.

The other two assertions follow easily from the first, and are left as exercises (see Exercise 4.12). □

It is now straightforward to deduce the following generalised versions of the three statements noted at the start of this subsection.

Theorem 4.21   *Suppose that $\{e_1, e_2, \ldots\}$ is an orthonormal sequence in an inner product space V and that $\sum_{n=1}^{\infty} \alpha_n e_n$ and $\sum_{n=1}^{\infty} \beta_n e_n$ are convergent series in V. Then*

$$\left\langle \sum_{n=1}^{\infty} \alpha_n e_n, e_m \right\rangle = \alpha_m \quad for \ m = 1, 2, \ldots,$$

$$\left\langle \sum_{n=1}^{\infty} \alpha_n e_n, \sum_{n=1}^{\infty} \beta_n e_n \right\rangle = \sum_{n=1}^{\infty} \alpha_n \beta_n$$

and

$$\left\| \sum_{n=1}^{\infty} \alpha_n e_n \right\|^2 = \sum_{n=1}^{\infty} \alpha_n^2.$$

4.4 Completeness in Inner Product Spaces

Now that we have developed the basics of the theory of sequences and series in inner product spaces, we can turn to examination of their deeper properties. We will introduce the key ideas by revising them in the case of $\mathbb{R}$, where they should be broadly familiar.

Suppose we have a convergent sequence $x_n \to x$ in $\mathbb{R}$. Since the terms x_n are getting closer and closer to x as $n \to \infty$, we should expect that they are also getting closer and closer to each other as $n \to \infty$. More formally, we should expect that the distance $|x_m - x_n|$ will become smaller and smaller as $m, n \to \infty$.

Recall that the definition of convergence says that $\lim_{n \to \infty} x_n = x$ if for all $\epsilon > 0$ there exists $N \in \mathbb{N}$ such that $|x_n - x| < \epsilon$ for all $n > N$ or, in condensed form,

$$\forall \epsilon > 0 \ \exists N \in \mathbb{N} \ \text{s.t.} \ n > N \implies |x_n - x| < \epsilon.$$

We can analogously phrase our expectation about $|x_m - x_n|$ by saying that for all $\epsilon > 0$ there exists $N \in \mathbb{N}$ such that $|x_m - x_n| < \epsilon$ for all $m, n > N$ or, in condensed form,

$$\forall \epsilon > 0 \ \exists N \in \mathbb{N} \ \text{s.t.} \ m, n > N \implies |x_m - x_n| < \epsilon.$$

We will prove shortly (see Theorem 4.23) that this expectation is justified.

First, however, note that we can rewrite the last condition in a way which applies not just in $\mathbb{R}$ but in a general inner product space, simply by replacing the absolute value signs by norm signs, just as we did with the definition of convergence of a sequence.

Definition 4.22 A sequence $\{x_n\}$ in an inner product space with the property that for all $\epsilon > 0$ there exists $N \in \mathbb{N}$ such that $\|x_m - x_n\| < \epsilon$ for all $m, n > N$, or

$$\forall \epsilon > 0 \ \exists N \in \mathbb{N} \ \text{s.t.} \ m, n > N \implies \|x_m - x_n\| < \epsilon,$$

is called a **Cauchy sequence**.

Theorem 4.23 *Every convergent sequence in an inner product space is a Cauchy sequence.*

Proof Suppose that $x_n \to x$, so that if we pick any $\epsilon > 0$, there exists $N \in \mathbb{N}$ such that

$$n > N \implies \|x_n - x\| < \tfrac{\epsilon}{2}.$$

Then if both m and n are greater than N, we have

$$\|x_m - x\| < \tfrac{\epsilon}{2} \qquad \text{and} \qquad \|x_n - x\| < \tfrac{\epsilon}{2},$$

and therefore, by the triangle inequality,

$$\|x_m - x_n\| \leq \|x_m - x\| + \|x_n - x\| < \tfrac{\epsilon}{2} + \tfrac{\epsilon}{2} = \epsilon,$$

which proves that $\{x_n\}$ is a Cauchy sequence. $\qquad\square$

It follows in particular that, as claimed above, convergent sequences in $\mathbb{R}$ are Cauchy sequences.

Note that the definition of a Cauchy sequence does not mention the existence or value of a limit – it talks purely about the terms of the given sequence. In a sense, Cauchy sequences are the sequences that 'ought to converge', and we might expect that the converse to the above theorem should hold: that a sequence which is Cauchy must have a limit. It is a fundamental fact about the real numbers that this is indeed the case in $\mathbb{R}$. Most of the deeper analytical facts about the real numbers – results such as the Monotone Convergence Theorem and the Intermediate Value Theorem, for example – depend on this.

Definition 4.24 A subset of an inner product space in which all Cauchy sequences converge is called **complete**; a complete inner product space is called a **Hilbert space**.

In this terminology, $\mathbb{R}$ is complete – a sequence in $\mathbb{R}$ converges if and only if it a Cauchy sequence – and is therefore a Hilbert space. In an informal sense, we can understand this as saying that $\mathbb{R}$ has no 'gaps': every sequence in $\mathbb{R}$ that 'ought to converge' actually has a limit to converge to.

Shortly, we will consider two examples of spaces that do have 'gaps', spaces in which some Cauchy sequences do not converge. First, we examine the relationship between closed subsets and complete subsets of a given inner product space.

Theorem 4.25   *Let X be a subset of an inner product space V.*

(i) *If X is complete then X is closed in V.*
(ii) *Suppose that V is complete. If X is closed in V then X is complete.*

Proof   Suppose first that X is complete, and let $\{x_n\}$ be a sequence in X that converges to some element $v \in V$. Since $\{x_n\}$ is convergent, it is a Cauchy sequence, by Theorem 4.23, so since X is complete, $\{x_n\}$ converges to some element $x \in X$. But limits are unique, by Theorem 4.3, so $x = v$, and hence $v \in X$. Therefore, X is closed.

Second, suppose V is complete and that X is closed in V, and let $\{x_n\}$ be a Cauchy sequence in X. Since V is complete, $\{x_n\}$ converges to some element $v \in V$; but then, since X closed, we must have $v \in X$. Therefore, X is complete.   □

Thus if we are working in the context of a complete space – which will usually be the case in the later parts of the book – the completeness and the closedness of a given subset are equivalent. It is still worthwhile, however, to keep the two concepts available individually. Note, for example, that they give different perspectives on the same property: the fact that a set X is complete is a purely *internal* property of X, while the fact that X is closed is an *external* property, since the full statement is that X is closed *in V*.

4.4.1 Two Examples in $\mathbb{R}$

Example 4.26   Consider the set $(0, 1] \subseteq \mathbb{R}$, and consider the sequence $\{1/n\}$ in $(0, 1]$. We claim that this is a Cauchy sequence. Let us prove this in two ways, one direct and one indirect.

First, note that

$$\left| \frac{1}{m} - \frac{1}{n} \right| = \begin{cases} \frac{1}{m} - \frac{1}{n}, & \text{if } m < n, \\ \frac{1}{n} - \frac{1}{m}, & \text{if } m \geq n. \end{cases}$$

When $m < n$, we have $0 < 1/m - 1/n < 1/m$, so if N is chosen to be at least $1/\epsilon$ and if $m, n > N$, then $\left| 1/m - 1/n \right| < \epsilon$. A similar argument applies in the case when $m \geq n$, so in either case, $m, n > N$ for any $N \geq 1/\epsilon$ implies $\left| 1/m - 1/n \right| < \epsilon$, and the sequence is therefore Cauchy, as claimed.

A simpler but indirect argument is to use the fact that $1/n \to 0$ in $\mathbb{R}$, and that by Theorem 4.23, $\{1/n\}$ is therefore a Cauchy sequence. (The fact that the convergence is to a limit that is actually outside of the original set $(0, 1]$ is immaterial, since the definition of a Cauchy sequence only requires a certain relationship between the terms of the sequence, which are in $(0, 1]$, and makes no reference to any possible limit.)

Thus $\{1/n\}$ is a Cauchy sequence in $(0, 1]$ that does not converge in $(0, 1]$: 0 is its only possible limit, but 0 is not in $(0, 1]$. Thus, $(0, 1]$ is not a complete set and is not closed in $\mathbb{R}$. (Indeed, as noted in Example 4.7, the closure of $(0, 1]$ is the strictly larger set $[0, 1]$.)

Example 4.27   A much more interesting and significant example is the set $\mathbb{Q} \subseteq \mathbb{R}$ of rational numbers (refer to Example 4.5). Now we know that $\sqrt{2}$ is *not* a rational number, and that it has a non-repeating decimal expansion $1.41421\ldots$. Consider the sequence whose nth term is the truncation of this expansion to n places. This sequence converges in $\mathbb{R}$ to $\sqrt{2}$, and each term is rational; the terms are

$$1.4, \quad 1.41, \quad 1.414, \quad 1.4142, \quad 1.41421, \ldots,$$

and these are expressible as the fractions

$$\frac{14}{10}, \quad \frac{141}{100}, \quad \frac{1414}{1000}, \quad \frac{14142}{10000}, \quad \frac{141421}{100000}, \ldots$$

Thus this sequence is Cauchy in $\mathbb{Q}$ (because it converges in $\mathbb{R}$), but has no limit in $\mathbb{Q}$: there is a 'gap' at $\sqrt{2}$ where its limit 'ought to be'. (Indeed, a straightforward generalisation of this argument, in which $\sqrt{2}$ is replaced by an arbitrary irrational number, provides a proof of the assertion about $\mathbb{Q}$ in Example 4.7.)

It may seem that the above argument is somehow circular, since it uses detailed information about $\sqrt{2}$, and yet depends precisely on the fact that $\sqrt{2}$ is missing from the set. An alternative would be to use an independently defined sequence such as

$$1.1, \quad 1.101, \quad 1.101001, \quad 1.1010010001, \quad 1.101001000100001, \ldots,$$

where the strings of 0's increase in length by one in each successive term. This sequence converges in $\mathbb{R}$ to the number

$$1.101001000100001000001000000100000001\ldots,$$

which is irrational because the pattern of 0's and 1's ensures that the decimal expansion is non-repeating. The remainder of the argument would be as for $\sqrt{2}$.

The only inner product space whose completeness we have discussed so far is $\mathbb{R}$; the fact that $\mathbb{R}$ is complete was noted following Definition 4.24. (The sets $(0, 1]$ and $\mathbb{Q}$ are incomplete subsets of the inner product space $\mathbb{R}$, but are not themselves inner product spaces.) In the following subsections, we will deal systematically with the completeness properties of our standard collection of inner product spaces: the Euclidean and sequence spaces and the function spaces.

➤ Although a direct analysis of the incompleteness of $(0, 1]$ and $\mathbb{Q}$ in terms of Cauchy sequences, as carried out above, is instructive, it should be noted that the most efficient proof of their incompleteness comes by combining some simple facts noted earlier. Specifically, Example 4.7 shows that $(0, 1]$ and $\mathbb{Q}$ are not closed in $\mathbb{R}$, and it follows immediately using Theorem 4.25 and the completeness of $\mathbb{R}$ that they are not complete.

4.4.2 The Completeness of $\mathbb{R}^n$

We first note the following technical result. (It is worth referring back to the statement and proof of Theorem 4.11 here; the structure of the two arguments is virtually identical.) Recall that $x_n^{(k)}$ denotes the kth entry of the vector x.

Theorem 4.28   *A sequence $\{x_n\}$ in $\mathbb{R}^p$ is a Cauchy sequence if and only if $\{x_n^{(k)}\}$ is a Cauchy sequence in $\mathbb{R}$ for $k = 1, \ldots, p$.*

Proof   Assume first that x_n is a Cauchy sequence. Now for each k, we have

$$0 \le \left| x_m^{(k)} - x_n^{(k)} \right| \le \left(\sum_{\ell=1}^{p} \left| x_m^{(\ell)} - x_n^{(\ell)} \right|^2 \right)^{\frac{1}{2}} = \| x_m - x_n \|.$$

Also, given $\epsilon > 0$, there exists, by the assumption, an $N \in \mathbb{N}$ such that

$$m, n > N \implies \| x_m - x_n \| < \epsilon.$$

Therefore, $m, n > N$ implies that $\left| x_m^{(k)} - x_n^{(k)} \right| < \epsilon$ for each k, which shows that each sequence $\{x_n^{(k)}\}$ is Cauchy.

Assume now that $\{x_n^{(k)}\}$ is Cauchy, for $1 \le k \le p$. Given $\epsilon > 0$, there exists, by the assumption, for each k in the range $1 \le k \le p$, an $N_k \in \mathbb{N}$ such that

$$m, n > N_k \implies \left| x_m^{(k)} - x_m^{(k)} \right| < \frac{\epsilon}{\sqrt{p}}.$$

If we set $N = \max\{N_1, N_2, \ldots, N_p\}$, then $m, n > N$ implies that $m, n > N_k$ for every k such that $1 \le k \le p$, and therefore

$$\| x_m - x_n \| = \left(\sum_{k=1}^{p} \left| x_m^{(k)} - x_n^{(k)} \right|^2 \right)^{\frac{1}{2}} < \left(\sum_{k=1}^{p} \left(\frac{\epsilon}{\sqrt{p}} \right)^2 \right)^{\frac{1}{2}} = \epsilon,$$

which shows that $\{x_n\}$ is Cauchy.   □

The completeness of $\mathbb{R}^p$ is now easy to prove, because we already have a result, Theorem 4.11, which tells us that convergence in $\mathbb{R}^p$ is convergence entry by entry. (See Exercise 4.14 for a more general result.)

Theorem 4.29 $\mathbb{R}^p$ *is complete, that is, is a Hilbert space.*

Proof   In somewhat compressed form, the argument is given by the following implications:

$$\{x_n\} \text{ Cauchy in } \mathbb{R}^p \implies \{x_n^{(k)}\} \text{ Cauchy in } \mathbb{R}, \text{ for } 1 \le k \le p$$

(by Theorem 4.28)

$$\implies \{x_n^{(k)}\} \text{ convergent in } \mathbb{R}, \text{ for } 1 \le k \le p$$

(by the completeness of $\mathbb{R}$)

$$\implies \{x_n\} \text{ convergent in } \mathbb{R}^p$$

(by Theorem 4.11).   □

4.4.3 The Completeness of ℓ^2

In ℓ^2, the analysis of completeness is considerably harder, because we do not have results as strong as Theorem 4.11 and Theorem 4.28 – compare Theorem 4.12 with Theorem 4.11, and compare Theorem 4.28 with the following statement, the proof of which (see Exercise 4.15) is given by a simple adaptation of the 'only if' part of the proof of Theorem 4.28.

Theorem 4.30   *If $\{x_n\}$ is a Cauchy sequence in ℓ^2, then $\{x_n^{(k)}\}$ is a Cauchy sequence in $\mathbb{R}$ for $k = 1, 2, \ldots$.*

The converse of Theorem 4.30 is false (compare this with Theorem 4.28), and the example of the sequence

$$e_1 = (1, 0, 0, 0, \ldots), \quad e_2 = (0, 1, 0, 0, \ldots), \quad e_3 = (0, 0, 1, 0, \ldots), \ldots$$

that we looked at in Subsection 4.2.1 shows this. Here, the sequence $\{e_n^{(k)}\}$ is a Cauchy sequence in $\mathbb{R}$ for each fixed k, since it consists eventually entirely of 0's (and therefore, as we noted earlier, trivially converges in $\mathbb{R}$). But the sequence $\{e_n\}$ is not a Cauchy sequence in ℓ^2, because $\|e_m - e_n\| = \sqrt{2}$ for all distinct m and n, and so we cannot make $\|e_m - e_n\|$ small merely by making m and n large.

Nevertheless, we have the following result, the proof of which is fairly intricate even though it requires no new mathematics.

Theorem 4.31   *The space ℓ^2 is complete, that is, is a Hilbert space.*

In the proof we need to work simultaneously with vectors in $\mathbb{R}^p$ and vectors in ℓ^2, so we introduce some notation and make some observations relevant to this. If $x = (x^{(1)}, x^{(2)}, \ldots)$ is any vector in ℓ^2, we will use the notation $x(p)$

to represent the 'truncated' vector $(x^{(1)}, x^{(2)}, \ldots, x^{(p)})$ in $\mathbb{R}^p$. Also, we will temporarily use subscripts on our norm signs to indicate the space in which the norm is defined, as in $\|x\|_{\ell^2}$, $\|x(p)\|_{\mathbb{R}^p}$ and so on.

We note two facts about the relationships between the various norms that are in play. First, if $x \in \mathbb{R}^\infty$, then

$$\|x(p)\|_{\mathbb{R}^p} \text{ is non-decreasing in } p, \tag{4.1}$$

since clearly

$$\left(\sum_{i=1}^{p} (x^{(i)})^2 \right)^{\frac{1}{2}} \leq \left(\sum_{i=1}^{q} (x^{(i)})^2 \right)^{\frac{1}{2}}$$

if $p \leq q$. Second, if $x \in \ell^2$, then

$$\|x(p)\|_{\mathbb{R}^p} \leq \|x\|_{\ell^2} \quad \text{for all } p \in \mathbb{N}, \tag{4.2}$$

since clearly

$$\left(\sum_{i=1}^{p} (x^{(i)})^2 \right)^{\frac{1}{2}} \leq \left(\sum_{i=1}^{\infty} (x^{(i)})^2 \right)^{\frac{1}{2}}.$$

Now to the proof itself.

Proof of Theorem 4.31   Let $\{x_n\}$ be a Cauchy sequence in ℓ^2. We are required to show that there is a vector $x \in \ell^2$ such that $x_n \to x$ as $n \to \infty$. Theorem 4.12 tells us how to find such an x, *if there is one*: it will be the entry-by-entry limit of the vectors x_n. Now since $\{x_n\}$ is a Cauchy sequence in ℓ^2, Theorem 4.30 shows that $\{x_n^{(k)}\}$ is a Cauchy sequence in $\mathbb{R}$ for $k = 1, 2, \ldots$, and so by the completeness of $\mathbb{R}$ there exists $x^{(k)} \in \mathbb{R}$ such that $x_n^{(k)} \to x^{(k)}$ as $n \to \infty$ for $k = 1, 2, \ldots$. We set

$$x = (x^{(1)}, x^{(2)}, \ldots),$$

and we claim that x is the vector we require. Thus, we have to prove that

$$x \in \ell^2 \quad \text{and} \quad x_n \to x \text{ as } n \to \infty.$$

(The second and third examples in Subsection 4.2.1 show that neither of these statements can be taken to be automatically true for a vector that is defined as an entry-by-entry limit. Note, however, that it is certainly true that $x(p) \in \mathbb{R}^p$ for all $p \in \mathbb{N}$.)

Consider a fixed $\epsilon > 0$. Because $\{x_n\}$ is a Cauchy sequence in ℓ^2, there is an $N \in \mathbb{N}$ such that

$$m, n > N \implies \|x_m - x_n\|_{\ell^2} < \epsilon. \tag{4.3}$$

Now consider any fixed $p \in \mathbb{N}$ and the truncated vectors $x_n(p) \in \mathbb{R}^p$. By the triangle inequality in $\mathbb{R}^p$ and the inequality (4.2), we have

$$\begin{aligned} \left\| x_n(p) - x(p) \right\|_{\mathbb{R}^p} &= \left\| (x_n(p) - x_m(p)) + (x_m(p) - x(p)) \right\|_{\mathbb{R}^p} \\ &\leq \left\| x_n(p) - x_m(p) \right\|_{\mathbb{R}^p} + \left\| x_m(p) - x(p) \right\|_{\mathbb{R}^p}, \\ &\leq \left\| x_n - x_m \right\|_{\ell^2} + \left\| x_m(p) - x(p) \right\|_{\mathbb{R}^p}, \end{aligned} \tag{4.4}$$

for all $m, n \in \mathbb{N}$. If we take $m, n > N$, then by (4.3), we have $\left\| x_n - x_m \right\|_{\ell^2} < \epsilon$, and so, from (4.4),

$$\left\| x_n(p) - x(p) \right\|_{\mathbb{R}^p} < \epsilon + \left\| x_m(p) - x(p) \right\|_{\mathbb{R}^p} \tag{4.5}$$

for $m, n > N$. Also, since $x_m^{(i)} \to x^{(i)}$ as $m \to \infty$ for $i = 1, 2, \ldots, p$ (in fact, for all $i \in \mathbb{N}$, but the first p values are all we need), Theorem 4.11 tells us that $x_m(p) \to x(p)$ in $\mathbb{R}^p$ as $m \to \infty$. Hence there exists $M \in \mathbb{N}$ such that

$$m > M \implies \left\| x_m(p) - x(p) \right\|_{\mathbb{R}^p} < \epsilon.$$

Therefore, applying (4.5) to any $m > \max\{M, N\}$, we have

$$\left\| x_n(p) - x(p) \right\|_{\mathbb{R}^p} < \epsilon + \epsilon = 2\epsilon,$$

for all $n > N$.

Now this argument holds for every $p \in \mathbb{N}$, and the value of N is independent of p, so the norms $\left\| x_n(p) - x(p) \right\|_{\mathbb{R}^p}$ are bounded above by 2ϵ for all $n > N$. Also, by (4.1), these norms are non-decreasing in p. Therefore, by the Monotone Convergence Theorem, they converge to some real number. But this means that the infinite sum defining $\| x_n - x \|_{\ell^2}$ exists, and hence for $n > N$ we have $x_n - x \in \ell^2$. But since $x_n \in \ell^2$ for all n, we have, using any chosen $n > N$,

$$x = x_n - (x_n - x) \in \ell^2,$$

which was the first of the two statements to be proved.

The second statement now follows: since the terms $\left\| x_n(p) - x(p) \right\|_{\mathbb{R}^p}$ are less than 2ϵ for $n > N$, and since $\| x_n - x \|_{\ell^2}$ is their limit as $p \to \infty$ for all $n > N$, we have

$$\| x_n - x \|_{\ell^2} \leq 2\epsilon \quad \text{for all } n > N.$$

Thus, we have shown that for our initial choice of ϵ there is an N such that

$$n > N \implies \| x_n - x \|_{\ell^2} \leq 2\epsilon,$$

and this implies that $x_n \to x$ as $n \to \infty$, as required, completing the proof. $\square$

➤ To make the above argument conform strictly with the definition of a limit as well as with our earlier arguments, we could 'tidy up' the proof so that the final inequality read '$\cdots < \epsilon$' rather than '$\cdots \leq 2\epsilon$'. For this, it would suffice to replace all the earlier occurrences of ϵ by $\epsilon/3$; then the last inequality would read '$\cdots \leq 2\epsilon/3$', to which we could append ' $< \epsilon$'. However, it would be unnecessarily pedantic to do this in practice, and its main effect would be to make a fairly complex argument look more complex than it really is.

4.4.4 The Completeness of the L^2 Spaces

In Subsection 4.2.2, we found a sequence of functions $f_n \in C([-1, 1])$ that was Cauchy but was not convergent to any continuous function – that is, had no limit in $C([-1, 1])$. Essentially the same argument applies to $C(I)$ for any non-trivial interval I, and we conclude that $C(I)$ is not complete for any non-trivial interval. (The fact that the sequence $\{f_n\}$ does not converge to any continuous function was noted explicitly in Subsection 4.2.2, and the fact that $\{f_n\}$ is Cauchy, though not noted there explicitly, is immediate from a fact which was noted, namely, that $\{f_n\}$ is convergent in the larger space $L^2([-1, 1])$. See Exercise 4.16 for a subtle issue raised by this discussion.)

We are therefore prompted to ask whether $L^2([-1, 1])$ in particular, or $L^2(I)$ in general, is large enough to contain the limits of all of its Cauchy sequences. The following theorem tells us that this is indeed the case. (Not surprisingly, detailed information from the Lebesgue theory is necessary for the proof of the theorem, which we therefore omit.)

Theorem 4.32   *The space $L^2(I)$ is complete for every interval I, that is, $L^2(I)$ is a Hilbert space.*

The incomplete space $C(I)$ is thus a subspace of the complete space $L^2(I)$, and it can be further shown that $L^2(I)$ is, in a sense that can be made precise without too much difficulty, the *smallest* complete space containing $C(I)$, and $L^2(I)$ is referred to as the **completion** of $C(I)$.

There is in fact a general result stating that every inner product space has a completion. Intuitively, the completion of a given space is obtained by adding just enough new vectors to provide limits for all the Cauchy sequences in the space (which implies that if the original space is already complete, then it is its own completion). Any inner product space will be a subspace of many complete spaces, but as in the case of $C(I)$ and $L^2(I)$, the completion is the 'smallest' of them: it contains just enough extra vectors to provide limits for all Cauchy sequences, but no more. We will not formalise these ideas further here, or prove the result just mentioned; doing so is not particularly difficult, but the issues are not central to our main concerns and we do not need to return to them.

➤ Note that although the general result just mentioned tells us that $C(I)$ has a completion, the result does not tell us that the completion can be identified specifically as the space $L^2(I)$. A separate argument is needed to establish this fact, and for this the Lebesgue theory must once again be invoked.

We can view the relation between $C(I)$ and $L^2(I)$ (or between any space and its completion) as paralleling the relation between $\mathbb{Q}$ and $\mathbb{R}$. It is clear that $\mathbb{Q}$ has many 'gaps', which we can 'fill in' by including the 'missing' irrational numbers. In a similar though less immediately obvious way, $C(I)$ also has many 'gaps' – we have merely identified one of them in the case of the interval $I = [-1, 1]$ – and $L^2(I)$ 'fills in the gaps' in $C(I)$.

In Subsection 1.2.2, we spoke about trying to find the 'right' space of functions on I in which to work; we said that the space $F(I)$ of all functions was 'too large', while the space $C(I)$ was 'not large enough'. In $L^2(I)$, we claim we have finally identified the right space. It is the completeness of $L^2(I)$ that ultimately provides the justification of this claim, because it is on the basis of completeness that the theory in the remainder of this chapter unfolds, and on that, in turn, the wavelet theory of the later chapters depends.

4.4.5 Riemann Integration and Lebesgue Integration

The theory of Lebesgue integration approaches these issues from another direction entirely. Here, the theory of integration is developed for its own sake, principally with a view to rectifying some of the deficiencies of the usual theory of Riemann integration (alluded to in Subsection 2.3.3, for example). But the Lebesgue theory, once fully developed, happens to provide us with the inner product space $L^2(I)$ of (equivalence classes of) square-integrable functions on I that forms the completion of $C(I)$.

Although we are not able to expand upon the Lebesgue theory in this book, it is worthwhile listing some of the deficiencies of Riemann integration and giving an indication of how and to what extent the Lebesgue theory rectifies them (the entries in the list are not entirely independent).

(i) Every Riemann-integrable function is bounded, but there are also many unbounded functions that 'ought' to be integrable. For example, the function

$$f(x) = \begin{cases} \dfrac{1}{\sqrt{x}}, & 0 < x \le 1, \\ 0, & x = 0 \end{cases}$$

is not Riemann-integrable on $[0, 1]$, but is Lebesgue-integrable. Of course, we can evaluate the integral $\int_0^1 f$ as an 'improper Riemann integral', but this already requires an extension of the Riemann theory.

(ii) 'Not enough' discontinuous functions are Riemann-integrable. The integrand in (i) is an example, but while that function is unbounded, there are also bounded examples. Consider, for example, the function

$$f(x) = \begin{cases} 1, & x \text{ rational}, \\ 0, & x \text{ irrational}, \end{cases}$$

which has arisen in earlier discussion. This function is not Riemann-integrable, but there are good theoretical reasons for wanting it to be integrable, even if such a function would rarely arise in applications. In the Lebesgue theory, as noted previously, this function is indeed integrable, and its integral over $[0, 1]$ is 0.

(iii) Many arguments in analysis require the interchange of limits and integrals, as in the equation

$$\int_a^b \left(\lim_{n \to \infty} f_n \right) = \lim_{n \to \infty} \int_a^b f_n.$$

If we are using Riemann integration, this interchange is allowed only under fairly stringent conditions, such as when the convergence in $\lim_{n \to \infty} f_n$ is uniform. Lebesgue integration allows the interchange much more generally.

For example, suppose we list the rational numbers in $[0, 1]$ in some order as

$$r_1, r_2, r_3, \ldots$$

(which we can do because they form a countable set), and define

$$f_n(x) = \begin{cases} 1, & x = r_1 \text{ or } r_2 \text{ or } \ldots \text{ or } r_n, \\ 0, & \text{otherwise}. \end{cases}$$

Each f_n is Riemann-integrable, with integral 0, since it differs from the zero function at only a finite number of points. But the pointwise limit of the sequence $\{f_n\}$ is the function f considered in (ii), which is not Riemann-integrable, and so the interchange of limit and integral for these functions fails in the Riemann theory. The interchange succeeds, however, in the Lebesgue theory because f is Lebesgue-integrable and has integral 0.

This example shows one way in which the interchange of a limit and an integral can fail when we are using the Riemann integral: the limit function may simply fail to be Riemann-integrable. But there are also cases where the limit function is Riemann-integrable, or even just Lebesgue-integrable, but where the limit of the integral and the integral of the limit, while both properly defined, are still not equal (see Exercise 4.17 for an example). In these circumstances, a more powerful integration theory cannot help.

(iv) Closely related to (iii), and taking us back to our discussion of completeness, $C(I)$ is not complete (and neither is the space of Riemann-integrable functions on I).

Lebesgue's theory overcomes these difficulties in something like an optimal fashion. As we have already remarked, however, we cannot go more deeply into the Lebesgue theory here. We will, when we need to (and as we have already done), make use of standard and unsurprising facts from the Lebesgue theory.

4.5 Orthonormal Bases in Hilbert Spaces

We used the idea of an orthonormal basis informally at several places in Chapter 1. We are now at the point where we must introduce the definition formally.

Definition 4.33 A set B in a Hilbert space V is called an **orthonormal basis**, or a **complete orthonormal set**, if it is a maximal orthonormal set in V.

The last phrase in the definition needs explanation. To say that B is a maximal orthonormal set means two things:

(i) B is an orthonormal subset of V, and
(ii) no subset of V that contains B but is strictly larger than B is orthonormal.

Equivalently to (ii) (in the presence of (i)):

(ii)′ if $x \in V$ is any non-zero vector not in B, then $\langle x, b \rangle \neq 0$ for some $b \in B$;

or again:

(ii)″ if $x \in V$ is such that $\langle x, b \rangle = 0$ for all $b \in B$, then $x = 0$.

(It is important to understand clearly why these equivalences hold; see Exercise 4.18.)

The use of the word 'basis' in the phrase 'orthonormal basis' is standard, as noted in Chapter 1, but is perhaps slightly unfortunate. The set B is certainly orthonormal, and therefore linearly independent (Theorem 3.15), but it need not be a spanning set for V, and therefore need not be a basis in the ordinary sense of linear algebra. (Theorem 4.39 below, on Fourier series in Hilbert spaces, will explain in exactly what sense B is a basis.)

Theorem 4.34 *Every Hilbert space has an orthonormal basis.*

We will not give the proof of this result, even though it is neither long nor difficult. This is because the proof depends on a fact from set theory known as Zorn's lemma, and a proper discussion of Zorn's lemma would take us too far afield.

➤ You may be familiar with the Axiom of Choice in set theory; Zorn's lemma is equivalent to the Axiom of Choice, given the rest of set theory, even though the two statements appear quite different.

It is in fact the case that *every* inner product space has a maximal orthonormal set – the proof does not require completeness. However, without completeness, none of the important parts of the later theory work usefully. In particular, the properties of maximal orthonormal sets that make it reasonable to call them orthonormal bases do not hold without completeness. Consequently, we will never refer to such sets in other than complete spaces.

We turn immediately to an examination of the standard orthonormal bases in our standard Hilbert spaces.

4.5.1 The Euclidean Space $\mathbb{R}^n$

In these finite-dimensional spaces, the standard bases (using the term in its ordinary sense) are also orthonormal bases: the vectors

$$e_1 = (1, 0, 0, \ldots, 0),$$
$$e_2 = (0, 1, 0, \ldots, 0),$$
$$\vdots$$
$$e_n = (0, 0, 0, \ldots, 1)$$

certainly form an orthonormal set, and this set is also maximal, since a vector x that is orthogonal to $e_1, e_2, \ldots, e_n$ must clearly be 0, which proves condition (ii)″.

4.5.2 The Sequence Space ℓ^2

Here the infinite sequence

$$e_1 = (1, 0, 0, \ldots),$$
$$e_2 = (0, 1, 0, \ldots),$$
$$e_3 = (0, 0, 1, \ldots),$$
$$\vdots$$

forms an orthonormal basis: the argument used in $\mathbb{R}^n$ applies essentially unchanged. (Note, however, that this orthonormal basis is *not* a basis in the ordinary sense, as we observed in Chapter 2.)

4.5.3 The L^2 Spaces

Theorem 4.35 *In the space $L^2([-\pi, \pi])$, the standard collection of orthonormalised trigonometric functions*

$$\frac{1}{\sqrt{2\pi}} \quad and \quad \frac{1}{\sqrt{\pi}} \cos nx \quad and \quad \frac{1}{\sqrt{\pi}} \sin nx \ for\ n \in \mathbb{N},$$

forms an orthonormal basis.

In the L^2 spaces, in contrast to the Euclidean spaces and ℓ^2, it is often difficult to prove that the sets we claim to be orthonormal bases actually are: the proof of orthonormality is generally routine, but the proof of maximality is generally hard.

In the case of the trigonometric functions, the proof of orthonormality is a straightforward integration exercise, as we noted in Subsection 1.2.2, but there is no easy proof of maximality. Our lack of a developed theory of Lebesgue integration is an obvious difficulty, but even given the Lebesgue theory, a fairly substantial mathematical argument is needed.

Theorem 4.36 *In the space $L^2([-1, 1])$, the sequence*

$$Q_0(x), Q_1(x), Q_2(x), \ldots$$

of orthonormalised Legendre polynomials forms an orthonormal basis.

Orthonormality is of course automatic for the Legendre polynomials, since they are constructed using the Gram–Schmidt process, but as with the trigonometric functions, there is no elementary proof of maximality. (However, if we assume two standard results, one from the Lebesgue theory and the other from classical analysis, a proof is fairly straightforward, and the argument is outlined in Exercise 4.19.)

Theorem 4.37 *In the space $L^2(\mathbb{R})$, the doubly infinite set consisting of the scaled, dilated, translated copies*

$$H_{j,k}(x) = 2^{j/2} H(2^j x - k) \quad for\ j, k \in \mathbb{Z}$$

of the Haar wavelet H forms an orthonormal basis.

Orthonormality is fairly simple, and has been noted (though so far without proof) several times, but maximality is more difficult. (Exercises 4.20, 4.21, 4.22 and 4.23 suggest proofs for both facts.)

4.6 Fourier Series in Hilbert Spaces

We can at last state and prove the major theorem on the existence of Fourier series in Hilbert spaces of (in general) infinite dimension.

In fact, we will prove two theorems here that together generalise all parts of Theorem 3.17 from the finite-dimensional to the infinite-dimensional case. Thus in the first theorem, Theorem 4.38, we deal with orthogonal projections into subspaces of (in general) infinite dimension, and in the second, Theorem 4.39, we deal with Fourier series. It may be helpful to refer back to Theorem 3.17 before going on, to be clear about exactly what is being generalised.

The extra ingredient needed for a successful generalisation from the finite-dimensional case to the general case is to impose the condition that the relevant space or subspace is complete. This completeness condition was already present implicitly in the finite-dimensional case of Theorem 3.17, because every finite-dimensional inner product space is automatically complete (see Exercise 4.14).

Since the finite-dimensional case is already covered fully by Theorem 3.17, we will deal in the two theorems that follow with the infinite-dimensional case only; this will allow us to simplify our notation somewhat, since we will not need to introduce indexing that represents both cases.

➤ To be more precise, we will deal here with Hilbert spaces that have *countably infinite* orthonormal bases – bases that can be listed as *sequences*, such as $\{e_1, e_2, e_3, \ldots\}$. All Hilbert spaces have (finite or infinite) orthonormal bases, by Theorem 4.34, but an infinite orthonormal basis need not necessarily be countable. Generalisation beyond the countable case is not difficult, but we will not pursue it here. In any case, the Hilbert spaces most often encountered, and certainly the ones of interest in this text, have countable orthonormal bases.

The first of our two theorems deals with projection.

Theorem 4.38 *Let W be a complete subspace of an inner product space V and suppose that W has an orthonormal basis $\{e_1, e_2, \ldots\}$. Let x be any vector in V. Then*

$$\sum_{n=1}^{\infty} \langle x, e_n \rangle \, e_n \ converges, \tag{4.6}$$

$$\sum_{n=1}^{\infty} \langle x, e_n \rangle^2 \leq \|x\|^2, \tag{4.7}$$

and

$$x - \sum_{n=1}^{\infty} \langle x, e_n \rangle \, e_n \perp e_m \tag{4.8}$$

for $m = 1, 2, \ldots$.

As in the finite-dimensional case of Chapter 3:

- the inequality $\sum_{n=1}^{\infty} \langle x, e_n \rangle^2 \leq \|x\|^2$ is called **Bessel's inequality** (compare Theorem 3.17); and
- the vector $\sum_{n=1}^{\infty} \langle x, e_n \rangle e_n$ is the **orthogonal projection**, or just the **projection**, of x into the subspace W (compare Definition 3.18, and see Section 4.7 below).

Proof of Theorem 4.38 Let s_n be the nth partial sum $s_n = \sum_{i=1}^{n} \langle x, e_i \rangle e_i$ of the series in (4.6) for each n, and note that $s_n \in W$. We will show first that $\{s_n\}$ is a Cauchy sequence. Consider $\|s_n - s_m\|$, and suppose for concreteness that $n \geq m$. Then, using standard manipulations involving orthonormal sets, we have

$$\|s_n - s_m\|^2 = \left\| \sum_{i=1}^{n} \langle x, e_i \rangle e_i - \sum_{i=1}^{m} \langle x, e_i \rangle e_i \right\|^2$$

$$= \left\| \sum_{i=m+1}^{n} \langle x, e_i \rangle e_i \right\|^2$$

$$= \sum_{i=m+1}^{n} \langle x, e_i \rangle^2.$$

Consider the infinite series $\sum_{n=1}^{\infty} \langle x, e_n \rangle^2$ in $\mathbb{R}$. By Theorem 3.17, all its partial sums $\sum_{i=1}^{n} \langle x, e_i \rangle^2$ are bounded above by $\|x\|^2$, and the partial sums are clearly non-decreasing with n. Hence, by the Monotone Convergence Theorem, the series $\sum_{n=1}^{\infty} \langle x, e_n \rangle^2$ converges. But this implies that $\sum_{i=m+1}^{n} \langle x, e_i \rangle^2$ can be made as small as we wish by making m (and hence also n) sufficiently large, and it follows that $\{s_n\}$ is a Cauchy sequence, as claimed.

Therefore, because W is complete, $\sum_{n=1}^{\infty} \langle x, e_n \rangle e_n$ is convergent, which establishes (4.6). Also, we showed in the previous paragraph that $\sum_{i=1}^{n} \langle x, e_i \rangle^2 \leq \|x\|^2$ for all n, and it is immediate from this that

$$\sum_{n=1}^{\infty} \langle x, e_n \rangle^2 \leq \|x\|^2,$$

which is (4.7).

Finally, using Theorem 4.21, we find that

$$\left\langle x - \sum_{n=1}^{\infty} \langle x, e_n \rangle e_n, e_m \right\rangle = \langle x, e_m \rangle - \left\langle \sum_{n=1}^{\infty} \langle x, e_n \rangle e_n, e_m \right\rangle$$

$$= \langle x, e_m \rangle - \sum_{n=1}^{\infty} \langle x, e_n \rangle \langle e_n, e_m \rangle$$

$$= \langle x, e_m \rangle - \langle x, e_m \rangle$$
$$= 0$$

for each m, which gives (4.8).   □

We now turn to our second theorem, dealing with Fourier series.

Theorem 4.39 *Suppose that V is a Hilbert space with orthonormal basis $\{e_1, e_2, \ldots\}$, and let x be any vector in V. Then*

$$x = \sum_{n=1}^{\infty} \langle x, e_n \rangle e_n \tag{4.9}$$

and

$$\|x\|^2 = \sum_{n=1}^{\infty} \langle x, e_n \rangle^2. \tag{4.10}$$

The standard terminology to describe this situation is borrowed from the familiar case when the space is $L^2([-\pi, \pi])$ and the orthonormal basis consists of the usual trigonometric functions (but refer also to Theorem 3.17 and the discussion that follows it):

- $\sum_{n=1}^{\infty} \langle x, e_n \rangle e_n$ is the **Fourier series for** x with respect to the orthonormal basis $\{e_1, e_2, \ldots\}$;
- the terms $\langle x, e_n \rangle$ are the **Fourier coefficients of** x with respect to $\{e_1, e_2, \ldots\}$; and
- the identity $\|x\|^2 = \sum_{n=1}^{\infty} \langle x, e_n \rangle^2$ is called **Parseval's relation**.

Proof of Theorem 4.39   By the previous result, the series $\sum_{n=1}^{\infty} \langle x, e_n \rangle e_n$ converges, and

$$x - \sum_{n=1}^{\infty} \langle x, e_n \rangle e_n \perp e_n$$

for all n. Hence, by the definition of an orthonormal basis (see especially condition (ii)″ following Definition 4.33), we have $x - \sum_{n=1}^{\infty} \langle x, e_n \rangle e_n = 0$, that is,

$$x = \sum_{n=1}^{\infty} \langle x, e_n \rangle e_n,$$

which is Equation (4.9). Equation (4.10) now follows by application of Theorem 4.21 to Equation (4.9).   □

This theorem of course applies to all of our standard Hilbert spaces with their standard orthonormal bases, as follows.

(i) For $x = (x_1, x_2, \ldots) \in \ell^2$, we have

$$x = \sum_{n=1}^{\infty} \langle x, e_n \rangle e_n ;$$

and since

$$\langle x, e_n \rangle = x_n$$

for all $n \in \mathbb{N}$, this expansion simply reads

$$x = (x_1, x_2, \ldots) = \sum_{n=1}^{\infty} x_n e_n$$

– which hardly seems momentous enough to require a major theorem for its justification.

(ii) For a function f in the space $L^2([-\pi, \pi])$, we have

$$f(x) = a_0 \frac{1}{\sqrt{2\pi}} + \sum_{n=1}^{\infty} a_n \frac{\cos nx}{\sqrt{\pi}} + \sum_{n=1}^{\infty} b_n \frac{\sin nx}{\sqrt{\pi}},$$

where the Fourier coefficients a_n for $n \geq 0$ and b_n for $n \geq 1$ are defined by

$$a_0 = \left\langle f, \frac{1}{\sqrt{2\pi}} \right\rangle = \frac{1}{\sqrt{2\pi}} \int_{-\pi}^{\pi} f(x) \, dx,$$

and, for all $n \geq 1$,

$$a_n = \left\langle f, \frac{\cos nx}{\sqrt{\pi}} \right\rangle = \frac{1}{\sqrt{\pi}} \int_{-\pi}^{\pi} f(x) \cos nx \, dx$$

and

$$b_n = \left\langle f, \frac{\sin nx}{\sqrt{\pi}} \right\rangle = \frac{1}{\sqrt{\pi}} \int_{-\pi}^{\pi} f(x) \sin nx \, dx.$$

(If necessary, refer here and in (iv) below to the notational convention established in Subsection 1.2.2.)

(iii) For a function f in the space $L^2([-1, 1])$, we have

$$f = \sum_{n=0}^{\infty} \langle f, Q_n \rangle Q_n,$$

where $Q_0(x), Q_1(x), Q_2(x), \ldots$ are the orthonormalised Legendre polynomials.

(iv) For a function f in the space $L^2(\mathbb{R})$, we have

$$f = \sum_{j=-\infty}^{\infty} \sum_{k=-\infty}^{\infty} w_{j,k} H_{j,k},$$

where the $H_{j,k}$, for $j, k \in \mathbb{Z}$, are the usual scaled, dilated, translated copies of the Haar wavelet H, and the Haar wavelet series coefficients $w_{j,k}$ are given by

$$w_{j,k} = \langle f, H_{j,k} \rangle = \int_{-\infty}^{\infty} f H_{j,k},$$

for all $j, k \in \mathbb{Z}$.

➤ For an infinite series of real numbers, rearrangement of the order of summation can affect whether or not the series converges, and can affect the value of the sum when it does converge (see Exercise 4.36 for a striking illustration of this). It is only for the so-called *absolutely convergent* series that rearrangement is guaranteed to have no effect.

It is therefore reasonable to ask whether the convergence or the sum of a Fourier series in the general sense of Theorem 4.39 can be affected if the order of the terms in the sum is changed. Fortunately, it can be shown (though the proof is not completely straightforward) that for these series rearrangements cannot affect the sum – Fourier series behave like absolutely convergent series in $\mathbb{R}$. This has the useful implication that we do not need to bother to impose a 'standard order' on the elements of an orthonormal basis when it does not come equipped with a natural order. (It might seem reasonable to say that there is a natural order for the Legendre polynomials in $L^2([-1, 1])$, since they are indexed by the natural numbers, but there is no obvious order for the trigonometric basis of $L^2([-\pi, \pi])$, since three different classes of functions are involved.)

4.6.1 Pointwise Convergence

It is important to understand precisely what the above expansions of functions as Fourier series with respect to various orthonormal bases of functions mean – and also what they do not mean.

Consider the classical case of trigonometric Fourier series. Just above, we used Theorem 4.39 to justify writing

$$f(x) = a_0 \frac{1}{\sqrt{2\pi}} + \sum_{n=1}^{\infty} a_n \frac{\cos nx}{\sqrt{\pi}} + \sum_{n=1}^{\infty} b_n \frac{\sin nx}{\sqrt{\pi}}, \qquad (4.11)$$

with the Fourier coefficients a_n and b_n as earlier. However, the equality given by the theorem between a function and its Fourier series is equality in the usual L^2 sense, which is equality almost everywhere, and does not in general entail pointwise equality, that is, equality for every value of x.

It may be tempting to interpret equation (4.11) as implying pointwise convergence because of the form of the equation: it reads '$f(x) = \cdots$', and thus appears to tell us the *value* $f(x)$ for all x. The temptation might be lessened if we defined, say,

$$A_0(x) = \frac{1}{\sqrt{2\pi}}, \quad A_n(x) = \frac{\cos nx}{\sqrt{\pi}} \quad \text{and} \quad B_n(x) = \frac{\sin nx}{\sqrt{\pi}}$$

for $x \in [-\pi, \pi]$ and $n \in \mathbb{N}$, and then wrote the equation in the form

$$f = a_0 A_0 + \sum_{n=1}^{\infty} a_n A_n + \sum_{n=1}^{\infty} b_n B_n,$$

without mention of the argument to any function, though we still need to remain aware of the special meaning of the equality symbol. But normally we do not go to the trouble of giving names to the functions like this, and we are therefore forced to give the functions explicit arguments. (Compare these comments to those made near the end of Subsection 1.2.2.)

An important topic in any detailed study of trigonometric Fourier series is to find conditions under which pointwise convergence does occur. (Some of the issues here, however, are very difficult, an indication of which is given by the fact that a completely satisfactory analysis of pointwise convergence in $L^2([-\pi, \pi])$ was only given as recently as 1966.) The following theorem is often sufficient for applications.

Theorem 4.40 *Let f be piecewise smooth on $[-\pi, \pi]$ and let $x_0 \in (-\pi, \pi)$.*

 (i) *If f is continuous at x_0, then the Fourier series of f converges at x_0 to $f(x_0)$.*
 (ii) *If f is discontinuous at x_0, then the Fourier series of f converges at x_0 to*

$$\tfrac{1}{2}(f(x_0^-) + f(x_0^+)),$$

the average of the left- and right-hand limits of f at x_0.

➤ A function is **piecewise continuous** on $[-\pi, \pi]$ if it is continuous at all but a finite number of points in $(-\pi, \pi)$, its left- and right-hand limits exist at every point in $(-\pi, \pi)$, and its right- and left-hand limits exist at $-\pi$ and π, respectively. A function is **piecewise smooth** on $[-\pi, \pi]$ if both the function and its derivative are piecewise continuous.

Part (ii) of the theorem can also be applied at $x_0 = \pm\pi$, provided that we replace the limit value in (ii) by the average of the right-hand limit of f at $-\pi$ and the left-hand limit at π.

One can enquire about the pointwise convergence of the Fourier series of functions that are not in $L^2([-\pi, \pi])$. Here, things can go badly wrong: there is a famous example due to Kolmogorov of a function that is Lebesgue-integrable but not in $L^2([-\pi, \pi])$, and whose Fourier series diverges *at every point*.

Similar comments to those above apply to non-trigonometric Fourier expansions in L^2 spaces, such as wavelet series and expansions with respect to Legendre polynomials: pointwise convergence cannot be inferred automatically from the L^2 convergence given by Theorem 4.39.

In this light, we can see that our discussion of Haar wavelet series in Chapter 1 was seriously oversimplified. There we claimed something to the effect that the Haar wavelet coefficients $\{w_{j,k} : j, k \in \mathbb{Z}\}$ for a function $f \in L^2(\mathbb{R})$

contain enough information to reconstruct f. We see now that such a statement is too strong: we cannot guarantee that f is indeed reconstructed at each point of the real line by its wavelet series, but only that the function defined by the wavelet series is indistinguishable from f *according to the measure used in $L^2(\mathbb{R})$*: that the L^2 distance between f and the function defined by the wavelet series is 0.

4.7 Orthogonal Projections in Hilbert Spaces

We will note here some facts about projections that are implicit in or follow from Theorem 4.38. We assume, as in that theorem, that we are dealing with a complete subspace W of an inner product space V and that W has an orthonormal basis $\{e_1, e_2, \ldots\}$. Everything said will apply whether the orthonormal basis of W is finite or infinite, but for notational convenience we will deal, as earlier, with the infinite case only. (Also as earlier, the finite-dimensional case is covered in Chapter 3.)

Although it is the subspace W that is required to be complete rather than the enclosing space V, in most of our applications the enclosing space will in fact be complete – it will typically be an L^2-space – and given this, Theorem 4.25 shows that the hypothesis that W is complete is equivalent to the hypothesis that W is closed.

We begin by reformulating some parts of Theorem 4.38.

Theorem 4.41 *Let W be a complete subspace of an inner product space V. Denote the projection from V to W by P, and let x be any vector in V.*

  (i) *Px is in W, Px is the closest element to x in W, and Px is the unique such element of W.*
 (ii) *$x - Px \perp W$, in the sense that $x - Px \perp w$ for all $w \in W$.*
(iii) *$\|Px\| \leq \|x\|$.*
 (iv) *If $x \in W$, then $Px = x$.*
  (v) *$PPx = Px$.*

Proof The proof of (i) is almost identical to the proof of Theorem 3.17, using Theorem 4.21 to justify the interchange of infinite sums and inner products. Property (ii) is a straightforward consequence of the third statement of Theorem 4.38. Inequality (iii) is just a reinterpretation of Bessel's inequality from Theorem 4.38, while (iv) is immediate from Theorem 4.39. Finally, since $Px \in W$, (iv) yields (v). □

There is a significant inference to be drawn from the fact that for every element x of V there exists a unique element of W that is closest to x.

The theorem tells us, of course, that this element is Px, so we can compute Px in terms of a given orthonormal basis for W by using the definition of the projection. But because of uniqueness, the projection is independent of the orthonormal basis chosen for W, and it can therefore be computed by the use of any orthonormal basis, or indeed by any other indirect method that may be available in a particular case: any argument that gives us the unique element of W that is closest to x must have given us Px.

Consider the situation where X is a complete subspace of W and W is a complete subspace of V. Denote the projection of V to W by P_W, the projection of V to X by P_X and the projection of W to X by P_{WX}. Our intuition from cases of low dimension suggests that $P_X = P_{WX} \circ P_W$, that is, that performing two consecutive orthogonal projections is equivalent to performing just one. (For example, consider the case when V is $\mathbb{R}^3$, when W is a 2-dimensional subspace of V, that is, a plane through the origin, and when X is a 1-dimensional subspace of W, that is, a line through the origin that lies in the plane.) This assertion holds in the general case.

Theorem 4.42 *In the circumstances just described, we have $P_X = P_{WX} \circ P_W$.*

The proof is short, but is left for Exercise 4.40. We record one simple consequence of this result.

Theorem 4.43 *Let X be a complete subspace of W, and W a complete subspace of V. Then using the notation above, we have $\|P_X v\| \le \|P_W v\|$ for all $v \in V$.*

Proof From $P_X = P_{WX} \circ P_W$, we have $P_X v = P_{WX}(P_W v)$, and applying part (iii) of Theorem 4.41 gives

$$\|P_X v\| = \|P_{WX}(P_W v)\| \le \|P_W v\|. \qquad \square$$

4.7.1 Orthogonal Complements and Direct Sums

The ideas of orthogonal complements and direct sums arise naturally in the context of orthogonal projections.

For the discussion, we introduce two straightforward extensions of the way in which the orthogonality symbol '$\perp$' is used. First, if V is an inner product space and $v \in V$ and $W \subseteq V$, then we write $v \perp W$ to mean that $v \perp w$ for all $w \in W$ (this notation was already used in part (ii) of Theorem 4.41). Second, if $W, X \subseteq V$, then we write $W \perp X$ to mean that $w \perp x$ for all $w \in W$ and $x \in X$.

Suppose that W is a closed subspace of a complete space V. Then we define the **orthogonal complement** of W in V to be the set

$$W^\perp = \{v \in V : v \perp W\}.$$

➤ We know that requiring that W be closed is equivalent to requiring that it be complete, given the completeness of V. The orthogonal complement can be defined for subspaces that are not closed, but the case of a closed subspace is enough for our purposes.

Note that the definition permits the situation – a common one in applications – where W is a subspace of V and V is in turn a subspace of a larger enclosing complete space. The notion of the orthogonal complement of a subspace is a *relative* one: relative to whichever enclosing space is under consideration. Thus W has an orthogonal complement relative to V and also an orthogonal complement relative to any larger enclosing space (and these two complements are not equal).

The notation '$W^\perp$' is often read as 'W-perp'.

We record a few basic facts about orthogonal complements in the following result.

Theorem 4.44 *Let W be a closed subspace of a complete space V, and denote the projection of V to W by P.*

(i) *W and $W^\perp$ are mutually orthogonal; that is, $W \perp W^\perp$.*
(ii) *$W^\perp$ is a closed subspace of V.*
(iii) *$W^\perp = \{v \in V : Pv = 0\}$.*
(iv) *$W^\perp = \{v - Pv : v \in V\}$.*
(v) *$W^{\perp\perp} = W$.*
(vi) *Each $v \in V$ has a unique representation as a sum $v = w + x$, where $w \in W$ and $x \in W^\perp$. Moreover, $w = Pv$, and if $P^\perp$ denotes the projection of V to $W^\perp$, then $x = P^\perp v$.*

Observe that since part (ii) of the theorem says that $W^\perp$ is a closed subspace of V, the space $(W^\perp)^\perp$, the orthogonal complement of $W^\perp$, is defined; $(W^\perp)^\perp$ is normally denoted by $W^{\perp\perp}$, as in part (v) of the theorem. With perhaps the exception of one step in the proof of part (v), the proofs are straightforward, and are left for Exercise 4.41.

If a space V has mutually orthogonal closed subspaces W and X such that each $v \in V$ has a unique representation as a sum $v = w + x$, where $w \in W$ and $x \in W$, then V is referred to as the **direct sum** of W and X, and we write $V = W \oplus X$. Therefore, in the circumstances of the theorem above, and in view of parts (i), (ii) and (v), we have $V = W \oplus W^\perp$.

We will meet these ideas again briefly in Chapter 5, and at more length and in a more general form in Chapter 6.

Exercises

4.1   Prove Theorem 4.2 directly from the definition of convergence in V
       and $\mathbb{R}$. (A suggested first step if you do not have much experience in
       arguing about limits directly from the ϵ, N-definition is to write out in full
       what it means to say that $x_n \to x$ in V and $\|x_n - x\| \to 0$ in $\mathbb{R}$; in fact, not
       much more should be needed.)

4.2   (a) Use an ϵ, N argument to prove part (i) of Theorem 4.3. (Suppose that
           x and y are two distinct limits of $\{x_n\}$, and consider the special value
           $\epsilon = \frac{1}{2}\|x - y\|$ in the definition of convergence.)
       (b) Give a full proof of part (ii) of Theorem 4.3.

4.3   Prove the claims made without proof in Example 4.5. (For the case of
       the rational numbers $\mathbb{Q}$, show that every irrational number is the limit
       of a sequence of rational numbers – or indeed of infinitely many such
       sequences. For the case of a subspace X of $\mathbb{R}^n$, pick an orthonormal basis
       for X and then argue as in Theorem 4.11. See also Exercise 4.14.)

4.4   Show that every rational number is the limit of a sequence of irrational
       numbers (or of infinitely many such sequences), and deduce that the set
       of irrationals is dense in $\mathbb{R}$.

4.5   Show that there are infinitely many irrational numbers between any
       two given rational numbers and that there are infinitely many rational
       numbers between any two given irrational numbers.

4.6   Let V be an inner product space.

       (a) Prove that the intersection of any collection of closed subsets of V is
           closed.
       (b) Prove that V is closed in V.
       (c) Deduce that the characterisations of the closure of a set X in V
           following Definition 4.6 as the smallest closed set containing X and
           as the intersection of all closed sets containing X are correct.

4.7   Let V be an inner product space.

       (a) Show that the union of a finite family of closed sets of V is closed.
       (b) Give an example to show that the union of an infinite family of
           closed sets need not be closed. (Since $\mathbb{R}$ is an inner product space,
           an example in $\mathbb{R}$ will suffice.)

4.8   Informally, to form the closure of X, we 'add to X all of its limit points'.
       Prove that the set so formed is closed. Specifically, show that if each of
       $x_1, x_2, x_3, \ldots$ is the limit of a sequence of elements of X, and if $x_n \to x$
       as $n \to \infty$ for some x, then x is the limit of a sequence of elements of X.
       (This tells us that in forming the closure we do not need to add the limit

points, then look for new limit points of the enlarged set and add them, and so on, some indefinite number of times.)

4.9   Prove the claims made without proof in Example 4.7.

4.10 Prove the claim at the end of Subsection 4.2.2 that no continuous function f is the limit of the sequence $\{f_n\}$ defined in that subsection. (Assume that some continuous function f is the limit; show that f must be 0 on $[-1, 0)$ and 1 on $(0, 1]$, which gives a contradiction immediately.)

4.11 Refer to Theorem 4.19 in the main text. As an exercise in understanding and applying the inner product space axioms, rewrite the proof of the theorem, using the same steps, but adding after each line a short comment that precisely justifies the manipulation just performed. (One line of justification should be enough in each case – something like 'by inner product space axiom IP3' might be typical.)

4.12 Give proofs of the unproved parts of Theorem 4.20, and apply Theorem 4.20 to deduce Theorem 4.21.

4.13 Prove formally that the infinite decimal introduced in Example 4.27 does not eventually repeat (which implies that it represents an irrational number), and confirm that the given sequence of terminating decimals converges to it.

4.14 Show that every finite-dimensional inner product space V is complete. (Choose an orthonormal basis for V, say $e_1, e_2, \ldots, e_m$, for some m. Each vector $x \in V$ is expressible as usual in the form $x = \sum_{k=1}^{m} x^{(k)} e_k$, where $x^{(k)} = \langle x, e_k \rangle$ for each k. Show that if $\{x_n\}$ is a Cauchy sequence in V, then for each k the sequence of 'coordinates' $\{x_n^{(k)}\}$ is a Cauchy sequence in $\mathbb{R}$. Now use the completeness of $\mathbb{R}$, and argue as in Theorem 4.11.)

4.15 Prove Theorem 4.30.

4.16 Fix a bounded interval I. Now $C(I)$ is defined as a space whose elements are functions, but we also regard $C(I)$ as a subspace of $L^2(I)$, and the elements of $L^2(I)$ are not functions but equivalence classes of functions. We therefore have two different characterisations of $C(I)$. The aim of this exercise is to reconcile the two characterisations. (We can consider this done if we show that any equivalence class in $L^2(I)$ that contains a continuous function can contain no other continuous function, so that the equivalence class identifies the continuous function uniquely. Thus we need to show that any two continuous functions that are equivalent in the L^2 sense are equal as functions. For this, see Exercise 3.4.)

4.17 For each $n \in \mathbb{N}$, let f_n be the function on $[0, 1]$ that has value 0 at 0, value n on $(0, 1/n]$ and value 0 on $(1/n, 1]$. Find the pointwise limit f of the sequence $\{f_n\}$. Show that f and all of the f_n are Riemann-integrable (and hence Lebesgue-integrable) on $[0, 1]$ and write down their integrals.

Show that the interchange of limit and integral, in the sense discussed in Subsection 4.4.5, fails for these functions.

4.18 Prove the equivalence of the conditions (ii), (ii)' and (ii)" in Definition 4.33 and the discussion following.

4.19 This exercise is designed to show why the Legendre polynomials form an orthonormal basis in $L^2([-1, 1])$. The proof depends on the following two important results (which we cannot prove in this book).

- The set $C([a, b])$ of continuous functions on the interval $[a, b]$ is dense in $L^2([a, b])$. (In more detail, for any $f \in L^2([a, b])$ and any $\epsilon > 0$, there is a continuous function g on $[a, b]$ such that $\|f - g\| < \epsilon$.)
- Weierstrass's Approximation Theorem: For any continuous function f on an interval $[a, b]$ and for any $\epsilon > 0$, there exists a polynomial p such that $|f(x) - p(x)| < \epsilon$ for all $x \in [a, b]$. (Informally, continuous functions can be uniformly approximated arbitrarily closely by polynomials.)

Now prove the result: show that if a function $f \in L^2([-1, 1])$ is orthogonal to all the Legendre polynomials then it is orthogonal to all polynomials, and use the above results to show that $f = 0$.

4.20 This exercise asks for a detailed proof of the orthonormality property for the Haar wavelet H, a property that we have discussed and used frequently, but not proved. Consider the scaled, dilated, translated copies $H_{j,k}$ of H, for $j, k \in \mathbb{Z}$.

(a) Suppose that $j < j'$ and that k and k' are arbitrary. Show that the support of $H_{j',k'}$ is either disjoint from the support of $H_{j,k}$ or lies entirely within the left-hand half or the right-hand half of the support of $H_{j,k}$.

A possible argument might run as follows. The support of $H_{j,k}$ is the interval $[k/2^j, (k + 1)/2^j)$. For fixed j and k, the three points

$$\frac{k}{2^j}, \quad \frac{k + \frac{1}{2}}{2^j}, \quad \frac{k + 1}{2^j}$$

divide the real line into four intervals. Consider four cases, corresponding to the interval in which $k'/2^{j'}$ lies.

(b) Hence derive the orthonormality relation $\langle H_{j,k}, H_{j',k'} \rangle = \delta_{j,j'} \delta_{k,k'}$ for all $j, k, j', k' \in \mathbb{Z}$.

4.21 We know that the statement that the scaled, dilated, translated copies $H_{j,k}$ of the Haar wavelet H form a *maximal* orthonormal set in $L^2(\mathbb{R})$ is equivalent to the claim that the only function in $L^2(\mathbb{R})$ that is orthogonal to every function $H_{j,k}$ is the zero function. The aim of this exercise is to prove this claim relative to the *continuous* functions in $L^2(\mathbb{R})$; that is, to show that if $f \in L^2(\mathbb{R})$ is continuous and

$$\langle f, H_{j,k} \rangle = \int_{-\infty}^{\infty} f H_{j,k} = 0 \quad \text{for all } j, k \in \mathbb{Z},$$

then $f = 0$. The steps below suggest a possible line of argument (step (a) is the only one that presents much difficulty).

Let f be a continuous function such that all the inner products $\langle f, H_{j,k} \rangle$ are 0, and consider the function F defined by the formula

$$F(x) = \int_0^x f(t)\, dt$$

for $x \in \mathbb{R}$. Note that by the Fundamental Theorem of Calculus, F is continuous and differentiable, and satisfies $F'(x) = f(x)$, for all x.

(a) Show that for each fixed $j \in \mathbb{Z}$, the integrals $\int_{k/2^j}^{(k+1)/2^j} f$ are equal for all $k \in \mathbb{Z}$.

(b) Show that for each fixed $j \in \mathbb{Z}$, we have $F(k/2^j) = kF(1/2^j)$ for all $k \in \mathbb{Z}$.

(c) Deduce that $F(k/2^j) = (k/2^j)F(1)$ for all $j, k \in \mathbb{Z}$.

(d) Deduce that $F(x) = xF(1)$ for all $x \in \mathbb{R}$.

(e) Deduce that f must be a constant function, and hence must be the zero function.

4.22 If we impose a stronger condition than continuity on the function f in the previous exercise, it is possible to give a considerably simpler argument leading to the same conclusion. Let $f : \mathbb{R} \to R$ be continuously differentiable.

(a) Show that if f is not constant, then there is a non-empty open interval (a, b) on which f is either strictly increasing or strictly decreasing. (Use the Mean Value Theorem.)

(b) Deduce that if $\langle f, H_{j,k} \rangle = 0$ for all $j, k \in \mathbb{Z}$, then $f = 0$.

4.23 With the help of appropriate results from the Lebesgue theory, we can use either Exercise 4.21 or Exercise 4.22 to deduce that the collection of functions $H_{j,k}$ is a maximal orthonormal set in $L^2(\mathbb{R})$.

(a) To use Exercise 4.21, the result we require is just the first of the two results quoted in Exercise 4.19, but reformulated for $L^2(\mathbb{R})$: the set of continuous functions in $L^2(\mathbb{R})$ with bounded support is dense in $L^2(\mathbb{R})$. Given this result, deduce the maximality.

(b) To use Exercise 4.22, we require the stronger result that the set of continuously differentiable functions in $L^2(\mathbb{R})$ with bounded support is dense in $L^2(\mathbb{R})$. Again, confirm that maximality follows.

4.24 (a) Show that the functions defined by the expressions

$$\sqrt{2}\cos(n\pi(2x-1)) \quad \text{and} \quad \sqrt{2}\sin(n\pi(2x-1)),$$

for $n \in \mathbb{N}$, together with the function with constant value 1, form an orthonormal basis for the Hilbert space $L^2([0,1])$.

(b) Generalise (a), writing down an orthonormal basis of trigonometric functions for the general closed, bounded interval $[a, b]$. (Hint: Start by writing down the formula for the linear function from $[-\pi, \pi]$ onto $[a, b]$ that maps $-\pi$ to a and π to b.)

4.25 Suppose that $\{e_1, e_2, \ldots\}$ is an orthonormal set in an inner product space V and that $\alpha = \{\alpha_n\}$ is a sequence of scalars for which the series $\sum_{n=1}^{\infty} \alpha_n e_n$ converges in V. Write $x = \sum_{n=1}^{\infty} \alpha_n e_n$.

(a) Show that $\alpha_n = \langle x, e_n \rangle$ for all n.

(b) Show that the sequence $\alpha = \{\alpha_n\}$ belongs to ℓ^2 and that $\|x\| = \|\alpha\|$, where the norm on the left is in V and the norm on the right is in ℓ^2.

(c) Suppose that the series $\sum_{n=1}^{\infty} \beta_n e_n$ converges in V, and write $\beta = \{\beta_n\}$ and $y = \sum_{n=1}^{\infty} \beta_n e_n$. Show that $\langle x, y \rangle = \langle \alpha, \beta \rangle$, where the inner product on the left is that of V and the inner product on the right is that of ℓ^2.

4.26 Let $\{e_1, e_2, \ldots\}$ be an orthonormal set in a Hilbert space V. Show that for any given sequence $\{\alpha_n\}$ of scalars, the series $\sum_{n=1}^{\infty} \alpha_n e_n$ converges in V if and only if $\sum_{n=1}^{\infty} \alpha_n^2$ converges in $\mathbb{R}$.

4.27 Let $\{e_1, e_2, \ldots\}$ be any sequence of vectors in an inner product space V. For the purposes of this exercise, call a sequence of scalars $\alpha = \{\alpha_n\}$ *sparse* if only a finite number of its terms are non-zero. (Note if $\alpha = \{\alpha_n\}$ is sparse then, trivially, α belongs to ℓ^2 and $\sum_{n=1}^{\infty} \alpha_n e_n$ converges.) Show that if $\left\|\sum_{n=1}^{\infty} \alpha_n e_n\right\| = \|\alpha\|$ for every sparse sequence $\alpha = \{\alpha_n\}$, then $\{e_1, e_2, \ldots\}$ is an orthonormal set.

4.28 Let $\{e_1, e_2, \ldots\}$ be an orthonormal set in a Hilbert space. Define

$$V = \left\{ \sum_{n=1}^{\infty} \alpha_n e_n : \sum_{n=1}^{\infty} \alpha_n^2 \text{ converges} \right\}.$$

(a) Show that V is a complete (and hence closed) subspace of the given space and has $\{e_1, e_2, \ldots\}$ as an orthonormal basis. (Note that the proof of the completeness of V will require use of the fact that ℓ^2 is complete.)

(b) Show that V is the closure of the linear span of the set $\{e_1, e_2, \ldots\}$ (the linear span of a set S is the collection of *finite* linear combinations of elements of S).

4.29 Suppose that V is a Hilbert space with orthonormal basis $\{e_1, e_2, \ldots\}$. Show that $V = \left\{ \sum_{n=1}^{\infty} \alpha_n e_n : \sum_{n=1}^{\infty} \alpha_n^2 \text{ converges} \right\}$.

4.30 Deduce from Exercise 4.25 that the Fourier coefficients of an element of a Hilbert space with respect to a given orthonormal basis are unique. More precisely, show that if the element is expressed as a convergent series with respect to the basis elements, then the coefficients in the series must be the respective Fourier coefficients of the element.

4.31 Consider the simple function f defined in Subsection 1.4.3, and its Haar wavelet coefficients.

(a) Writing $\hat{f}$ to represent the sequence of wavelet coefficients, show that $\hat{f} \in \ell^2$.
(b) Show that $\|f\| = \|\hat{f}\|$, where the norm on the left is that of $L^2(\mathbb{R})$ and the norm on the right is that of ℓ^2.

4.32 Let f be the function on $[-\pi, \pi]$ defined by $f(x) = \begin{cases} \frac{1}{2}x + \pi, & \text{if } x < 0, \\ \frac{1}{2}x, & \text{if } x \geq 0. \end{cases}$

(a) Compute the Fourier coefficients of f with respect to the orthonormalised trigonometric functions, and show that the Fourier series of f is the function defined by the formula

$$\frac{\pi}{2} - \sum_{n=1}^{\infty} \frac{\sin nx}{n}.$$

(Why can you predict from the graph of the function that the coefficients of the cosine terms will be 0?)
(b) Writing $\hat{f}$ to represent the sequence of Fourier coefficients, show that $\hat{f} \in \ell^2$.
(c) Show that $\|f\| = \|\hat{f}\|$, where the norm on the left is that of $L^2([-\pi, \pi])$ and the norm on the right is that of ℓ^2.

4.33 Relate your findings in Exercises 4.31 and 4.32 to the results of Exercise 4.25.

4.34 Provide full details of the rather informal argument at the end of the first paragraph of the proof of Theorem 4.38 that the sequence $\{s_n\}$ is a Cauchy sequence.

4.35 Confirm the final claim in the proof of Theorem 4.39.

4.36 In the context of the comments just preceding Subsection 4.6.1, this exercise uses a well-known example to show concretely how rearrangement of the order of summation of a series of real numbers can change the value of the sum. Consider the alternating harmonic series

$$\sum_{n=1}^{\infty} \frac{(-1)^{n-1}}{n} = 1 - \frac{1}{2} + \frac{1}{3} - \frac{1}{4} + \frac{1}{5} - \frac{1}{6} + \frac{1}{7} - \frac{1}{8} + \frac{1}{9} - \frac{1}{10} + \cdots.$$

(a) Use the alternating series test to show that the series converges, with sum S, say, and use the error estimate associated with the test to show that $S \neq 0$. (In fact, $S = \ln 2$, though we do not need that information here.) Deduce that the series is conditionally convergent.

(b) Rearrange the order of the series, always taking one positive term from the original series followed by two negative terms, giving the series

$$1 - \frac{1}{2} - \frac{1}{4} + \frac{1}{3} - \frac{1}{6} - \frac{1}{8} + \frac{1}{5} - \frac{1}{10} - \frac{1}{12} + \cdots.$$

Confirm that this is genuinely a rearrangement, in the sense that the new series contains each term from the original series once and once only.

(c) By grouping the terms of the new series suitably (but without changing their order again), show that the series has sum $\frac{1}{2}S$.

4.37 In the notation of Theorem 4.41, use that result to show that

$$\|x - Px\|^2 + \|Px\|^2 = \|x\|^2.$$

(See also Exercise 3.40 of Chapter 3.) Hence tighten the statement of part (iii) of the theorem by deducing that if $\|Px\| = \|x\|$ then $Px = x$.

4.38 (a) Use the example in Subsection 4.2.2 to show that if a subspace W of a space V is not complete (even though V itself may be complete), then for $x \in V$, there may not exist a closest point to x in W. (Hint: Referring to the function f defined in the example, show that there there is no function in $C([-1, 1])$ that is of distance 0 from f, but that for every $\epsilon > 0$ there is a function in $C([-1, 1])$ whose distance from f is less than ϵ.)

(b) Use the same example to show that under the same conditions there may not exist an element $w \in W$ such that $x - w \perp W$.

4.39 Let W be a subspace of an inner product space V and let $x \in V$.

(a) Show there cannot exist more than one element of W at minimal distance from x. (This, with the previous exercise, shows that the problem of best approximation in an incomplete subspace of an inner

product space may not have a solution, but that if a solution does
exist then it is unique.)

(b) Show that there cannot exist more than one element $w \in W$ such that
$x - w \perp W$. (Hint: Use Exercise 3.6.)

4.40 Prove Theorem 4.42. (For $v \in V$, show that both $v - P_X v \perp X$ and
$v - P_{WX} P_W v \perp X$, and then apply part (b) of Exercise 4.39.)

4.41 Prove Theorem 4.44. (Parts (i) and (ii) are very simple. For (iii), use
results from Theorem 4.41 to prove separately that $x \perp W$ implies
$Px = 0$ and that $Px = 0$ implies $x \perp W$. Part (iv) follows easily from
part (iii). For part (v) it is essential first to derive a clear criterion for
the membership of a vector x in $W^{\perp\perp}$. The inclusion $W \subseteq W^{\perp\perp}$ is then
straightforward. For the converse, assume that $x \notin W$; deduce using
Theorem 4.41 that $y = x - Px \neq 0$; then prove that $y \in W^{\perp}$ but that
$\langle y, x \rangle \neq 0$; finally, conclude that $x \notin W^{\perp\perp}$. The first statement of part (vi)
again requires use of Theorem 4.41. For $x = P^{\perp} v$, show that x is the
closest element to v in $W^{\perp}$, and apply part (i) of Theorem 4.41.)

4.42 Let I be a non-trivial interval on the real line. For any $p \geq 1$, we can
define a vector space $L^p(I)$ of functions on I by specifying that $f \in L^p(I)$
if $\int_I |f|^p$ is defined and finite. There is a norm $\| \cdot \|_p$ on $L^p(I)$ defined by

$$\|f\|_p = \left(\int_I |f|^p \right)^{1/p},$$

and it can be shown that this norm has the properties given in Theorem 3.8 for the norm on $L^2(I)$. Exactly as in Section 3.4, the norm
on $L^p(I)$ can be used to define a notion of distance, and the convergence
of sequences and series can then be discussed as in the present chapter.
Cauchy sequences always converge in $L^p(I)$, so the space is complete. (A
vector space that has a norm, whether or not it has an inner product, is
called a **normed space**, and a complete normed space is called a **Banach
space**; thus $L^p(I)$ is a Banach space for all $p \geq 1$.) It is clear that the value
$p = 2$ gives us the familiar space $L^2(I)$ (observe that the absolute value
signs in the definition of the space and the norm are not needed when
$p = 2$), and that the norm $\| \cdot \|_2$ is the usual norm $\| \cdot \|$ on $L^2(I)$.

(a) Show that the norm $\| \cdot \|_p$ does not arise from an inner product unless
$p = 2$. (A norm that arises from an inner product satisfies the parallelogram law, by Theorem 3.9. Find $f, g \in L^p(I)$ for which the law
fails when $p \neq 2$; it may be helpful to consider the case $I = [0, 1]$
first, and then generalise.)

(b) Show that neither of $L^1(\mathbb{R})$ and $L^2(\mathbb{R})$ is contained in the other.

(c) Let I be a bounded interval and let $f \in L^2(I)$. Use the Cauchy–Schwarz inequality to show that $\|f\|_1 \leq \ell(I)^{1/2}\|f\|_2$, where $\ell(I)$ is the length of I.

(d) Show that if I is a bounded interval, then $L^1(I)$ is not contained in $L^2(I)$, but that $L^2(I)$ is contained in $L^1(I)$.

(e) Find generalisations of parts (b), (c) and (d) from L^1 and L^2 to the case of L^p and L^q for arbitrary $p, q \geq 1$. (This is more difficult.)

5

The Haar Wavelet

5.1 Introduction

In the previous chapters, we have developed the general ideas necessary to be able to speak with precision about wavelets and wavelet series. The key concepts were those of Hilbert spaces and of orthonormal bases and Fourier series in Hilbert spaces, and we have seen that Haar wavelet series are examples of such series.

But although we have developed a substantial body of relevant background material, we have so far developed no general theory of wavelets. Indeed, the simple and rather primitive Haar wavelet is the only wavelet we have encountered, and we have not even hinted at ways of constructing other wavelets, perhaps with 'better' properties than the Haar wavelet, or at the desirability of being able to do so. These are the issues that we turn to in the final chapters of the book.

Specifically, our major remaining goals are as follows:

- to develop a general framework, called a *multiresolution analysis*, for the construction of wavelets;
- to develop within this framework as much of the general theory of wavelets as our available methods allow; and
- to use that theory to show how other wavelets can be constructed with certain specified properties, culminating in the construction of the infinite family of wavelets known as the *Daubechies wavelets*, of which the Haar wavelet is just the first member.

The main aim of the present short chapter is to introduce the idea of a multiresolution analysis by defining and studying the multiresolution analysis that corresponds to the Haar wavelet.

126

5.2 The Haar Wavelet Multiresolution Analysis

The concept of a multiresolution analysis was developed in about 1986 by the French mathematicians Stéphane Mallat and Yves Meyer, providing a framework for the systematic creation of wavelets, a role which it still retains. Mirroring all of our discussion of wavelets up to this point, we will introduce the idea of a multiresolution analysis by first examining it carefully in the specific case of the Haar wavelet. As we have found in our previous discussion, the Haar wavelet is simple enough for us to see in a very concrete fashion how the various constructions and arguments work.

5.2.1 The Wavelet

Let us begin by briefly revising the main relevant facts from earlier chapters. First, recall that the definition of the Haar wavelet H is

$$H(x) = \begin{cases} 1, & 0 \le x < \frac{1}{2}, \\ -1, & \frac{1}{2} \le x < 1, \\ 0, & \text{otherwise}, \end{cases}$$

and that its graph is as shown in Figure 5.1.

Second, recall that, given the Haar wavelet, we work with its scaled binary dilations and dyadic translations $H_{j,k}$, defined by

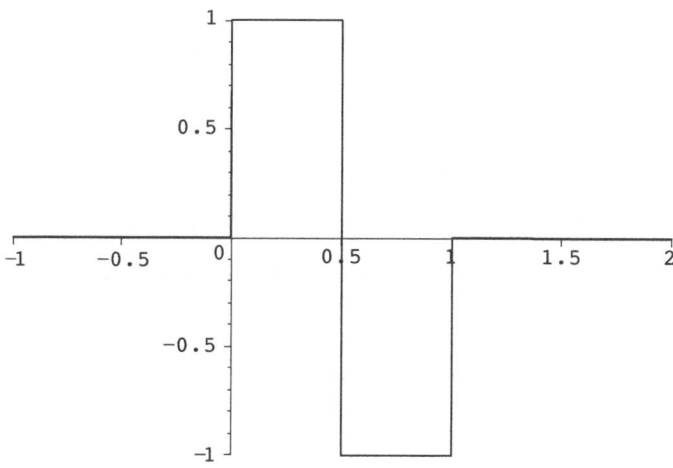

Figure 5.1 The graph of the Haar wavelet

$$H_{j,k}(x) = 2^{j/2}H(2^j x - k) \quad \text{for all } j, k \in \mathbb{Z}.$$

From Sections 4.5 and 4.6, we know that the collection $\{H_{j,k} : j, k \in \mathbb{Z}\}$ forms an orthonormal basis for $L^2(\mathbb{R})$, and that every function $f \in L^2(\mathbb{R})$ can therefore be expanded as a Haar wavelet series (that is, as a Fourier series in $L^2(\mathbb{R})$ with respect to the functions $\{H_{j,k}\}$) in the form

$$f = \sum_{j=-\infty}^{\infty} \sum_{k=-\infty}^{\infty} w_{j,k} H_{j,k},$$

where the Haar wavelet coefficients $\{w_{j,k} : j, k \in \mathbb{Z}\}$ are given by

$$w_{j,k} = \langle f, H_{j,k} \rangle = \int_{-\infty}^{\infty} f H_{j,k} \quad \text{for all } j, k \in \mathbb{Z}.$$

5.2.2 The Multiresolution Analysis

We now begin the specification of the multiresolution analysis associated with the Haar wavelet. The multiresolution analysis consists of a certain collection of subspaces $\{V_j : j \in \mathbb{Z}\}$ of $L^2(\mathbb{R})$ and a certain function $\phi \in L^2(\mathbb{R})$ which together satisfy six properties that we will denote by (MRA1)–(MRA6).

The subspaces are as follows. For every $j \in \mathbb{Z}$, we define the subspace V_j of $L^2(\mathbb{R})$ by setting

$$V_j = \left\{ f \in L^2(\mathbb{R}) : f \text{ is constant on } \left[\frac{k}{2^j}, \frac{k+1}{2^j}\right) \text{ for all } k \in \mathbb{Z} \right\}.$$

Just as it is important to be clear in detail on how the factors j and k scale, dilate and translate H to produce $H_{j,k}$ (see Subsection 1.4.2), it is important also to be clear on the significance of the value of j in the formation of V_j. For example,

$$V_{-1} = \left\{ f \in L^2(\mathbb{R}) : f \text{ is constant on } [2k, 2k + 2) \text{ for all } k \in \mathbb{Z} \right\},$$

$$V_0 = \left\{ f \in L^2(\mathbb{R}) : f \text{ is constant on } [k, k + 1) \text{ for all } k \in \mathbb{Z} \right\},$$

$$V_1 = \left\{ f \in L^2(\mathbb{R}) : f \text{ is constant on } \left[\frac{k}{2}, \frac{k+1}{2}\right) \text{ for all } k \in \mathbb{Z} \right\} :$$

as j increases, the length of the interval on which the functions in V_j are constant decreases.

An important preliminary observation is that *the subspaces V_j are all closed* or, equivalently, *complete* (see Section 4.4, and especially Theorem 4.25). (The proof of this is left for Exercise 5.1.)

We now proceed to record and prove the properties (MRA1)–(MRA6), leaving detailed checking of the arguments for Exercise 5.2. (Note that we have not yet introduced the function ϕ mentioned above; this will be defined in the context of property (MRA6), the last of the six multiresolution analysis properties.)

(i) The first property expresses a nesting or inclusion relation among the subspaces V_j:

$$V_j \subset V_{j+1} \qquad \text{(MRA1)}$$

for all $j \in \mathbb{Z}$; or, in more detail, $\cdots \subset V_{-2} \subset V_{-1} \subset V_0 \subset V_1 \subset V_2 \subset \cdots$. (A doubly infinite sequence of nested spaces like this is sometimes referred to in the literature as a 'ladder of spaces'.)

Proof   Observe first that for any j and k, we have

$$\left[\frac{k}{2^j}, \frac{k+1}{2^j} \right) = \left[\frac{2k}{2^{j+1}}, \frac{2k+2}{2^{j+1}} \right) = \left[\frac{2k}{2^{j+1}}, \frac{2k+1}{2^{j+1}} \right) \cup \left[\frac{2k+1}{2^{j+1}}, \frac{2k+2}{2^{j+1}} \right).$$

It follows easily that for any $j \in \mathbb{Z}$,

$$f \in V_j \implies f \text{ constant on } \left[\frac{k}{2^j}, \frac{k+1}{2^j} \right) \text{ for all } k \in \mathbb{Z}$$

$$\implies f \text{ constant on } \left[\frac{k}{2^{j+1}}, \frac{k+1}{2^{j+1}} \right) \text{ for all } k \in \mathbb{Z}$$

$$\implies f \in V_{j+1}. \qquad \qquad \square$$

(ii) Every function $f \in L^2(\mathbb{R})$ is the limit of a sequence of functions in $\bigcup_{j \in \mathbb{Z}} V_j$; that is:

$$\text{the set } \bigcup_{j \in \mathbb{Z}} V_j \text{ is dense in } L^2(\mathbb{R}). \qquad \text{(MRA2)}$$

Proof   The fact that f can be expanded as a wavelet series

$$\sum_{j=-\infty}^{\infty} \sum_{k=-\infty}^{\infty} w_{j,k} H_{j,k}$$

implies that f is the limit of the sequence of partial sums

$$\sum_{j=-N}^{N} \sum_{k=-\infty}^{\infty} w_{j,k} H_{j,k}$$

of the wavelet series as $N \to \infty$, and these partial sums are all in $\bigcup_{j \in \mathbb{Z}} V_j$. $\qquad \square$

(iii) The only function lying in all the spaces V_j is the zero function; that is:

$$\bigcap_{j \in \mathbb{Z}} V_j = \{0\}. \qquad \text{(MRA3)}$$

Proof If $f \in V_j$ for all $j \in \mathbb{Z}$, then f is constant on

$$\left[\frac{k}{2^j}, \frac{k+1}{2^j} \right),$$

for all $j, k \in \mathbb{Z}$. Therefore, f is constant on $[0, \infty)$, since any two numbers $r, s \in [0, \infty)$ lie in some interval

$$\left[0, \frac{1}{2^j} \right),$$

when j is sufficiently large and negative. By a similar argument, f is constant on $(-\infty, 0)$. Thus, f must be given by

$$f(x) = \begin{cases} a, & x < 0, \\ b, & x \geq 0, \end{cases}$$

for some constants a and b. But the only function of this form *that belongs to $L^2(\mathbb{R})$* is the function that has $a = b = 0$, namely, the function that is identically 0. □

(iv) This property expresses the relationship between a given function in V_0 and its binary dilations:

$$f(x) \in V_0 \text{ if and only if } f(2^j x) \in V_j \qquad (MRA4)$$

for all $j \in \mathbb{Z}$.

Proof For any $j \in \mathbb{Z}$,

$$f(2^j x) \in V_j \iff f(2^j x) \text{ is constant on } \left[\frac{k}{2^j}, \frac{k+1}{2^j} \right) \text{ for all } k \in \mathbb{Z}$$

$$\iff f \text{ is constant on } [k, k+1) \text{ for all } k \in \mathbb{Z}$$

$$\iff f(x) \in V_0. \qquad □$$

(v) This property expresses the relationship between a given function in V_0 and its integer translations:

$$f(x) \in V_0 \text{ if and only if } f(x-k) \in V_0 \qquad (MRA5)$$

for all $k \in \mathbb{Z}$.

Proof For any $k \in \mathbb{Z}$,

$$f(x) \in V_0 \iff f(x) \text{ is constant on } [n, n+1) \text{ for all } n \in \mathbb{Z}$$

$$\iff f(x-k) \text{ is constant on } [n, n+1) \text{ for all } n \in \mathbb{Z}$$

$$\iff f(x-k) \in V_0. \qquad □$$

(vi) We now introduce the special function ϕ that we noted earlier is a component of a multiresolution analysis; property (MRA6) then tells us how ϕ is
related to the spaces V_j. This property is thus different in character from
the preceding five properties, even though it is nearly as easy to check.
It may not be clear at first sight why ϕ and (MRA6) are significant – the
function ϕ does not appear to be directly related to the Haar wavelet, for
example, as we might have expected it to be. However, the discussion in
the remainder of the chapter should make the central role of ϕ clear.
    We define

$$\phi(x) = \begin{cases} 1, & 0 \le x < 1, \\ 0, & \text{otherwise.} \end{cases}$$

We claim that $\phi \in V_0$ and that $\{\phi(x - k) : k \in \mathbb{Z}\}$ is an orthonormal basis
for V_0.

Proof   Clearly, $\phi \in V_0$, since ϕ is constant on each interval $[k, k + 1)$.
Now $\{\phi(x - k) : k \in \mathbb{Z}\}$ is an orthonormal set, since

$$\langle \phi(x - k), \phi(x - l) \rangle = \int_{-\infty}^{\infty} \phi(x - k)\phi(x - l)\,dx = \begin{cases} 1, & k = l, \\ 0, & k \ne l. \end{cases}$$

To see that $\{\phi(x - k) : k \in \mathbb{Z}\}$ is a maximal orthonormal set in V_0, note
that for any $f \in V_0$,

$$\langle f(x), \phi(x - k) \rangle = \int_{-\infty}^{\infty} f(x)\phi(x - k)\,dx = \int_{k}^{k+1} f(x)\,dx.$$

If f is orthogonal to $\phi(x - k)$ for all k, we therefore have $\int_{k}^{k+1} f(x)\,dx = 0$
for all k. But the fact that $f \in V_0$ means that f is constant on each interval
$[k, k + 1)$, and since its integral on $[k, k + 1)$ is 0, its value there must
be identically 0. Since this holds for all k, it follows that f is the zero
function.   □

    Thus we have established that:

there exists $\phi \in V_0$ such that the set $\{\phi(x - k) : k \in \mathbb{Z}\}$
is an orthonormal basis for V_0.   (MRA6)

The function ϕ is known as the *scaling function* of the multiresolution
analysis.

    We have now introduced subspaces V_j for $j \in \mathbb{Z}$ and a function ϕ that together
satisfy properties (MRA1)–(MRA6), and this completes the specification of
the multiresolution analysis for the Haar wavelet.

➤ Observe that in properties (MRA4), (MRA5) and (MRA6) we have made use of the notational convention of Subsection 1.2.2 in several ways.

In each case, however, we could have expressed the property without invoking the convention by making use of our double-subscript notation for scaling, dilation and translation. Specifically, property (MRA4) could be written in the equivalent form

$$f \in V_0 \text{ if and only if } f_{j,0} \in V_j \text{ for all } j \in \mathbb{Z},$$

property (MRA5) in the form

$$f \in V_0 \text{ if and only if } f_{0,k} \in V_0 \text{ for all } k \in \mathbb{Z},$$

and the orthonormality condition from property (MRA6) in the form

$$\{\phi_{0,k} : k \in \mathbb{Z}\} \text{ is an orthonormal basis for } V_0$$

(see Exercise 5.3).

5.2.3 Some Observations

The specification of the multiresolution analysis (including the scaling function) did not in any way depend on, or even mention, the Haar wavelet itself, so our next task is to understand its connection to the wavelet.

Now V_0 consists of the functions that are constant on all the integer-length intervals $[k, k + 1)$. The Haar wavelet H is not constant on these intervals, and is therefore not in V_0. However, H is constant on the *half-integer* intervals $[k/2, (k + 1)/2)$, and by property (MRA5) we therefore have $H \in V_1$. The integer translates $H_{0,k}$ of H, for $k \in \mathbb{Z}$, likewise lie in V_1. Observe (see Exercise 5.4) that each $H_{0,k}$ is orthogonal to every function in V_0, a fact that we can express compactly using the notation of Subsection 4.7.1 by writing $H_{0,k} \perp V_0$.

We record a few properties of ϕ and H and their relation to the ladder of spaces V_j. These help to explain the relation between the wavelet and the multiresolution analysis, and will also be of use below.

Theorem 5.1 *Let the subspaces $\{V_j : j \in \mathbb{Z}\}$ and the function ϕ be as above. Then for each $j \in \mathbb{Z}$,*

(i) *the collection of functions $\{\phi_{j,k} : k \in \mathbb{Z}\}$ forms an orthonormal basis for V_j,*

(ii) *the collection of functions $\{\phi_{j,k} : k \in \mathbb{Z}\} \cup \{H_{j,k} : k \in \mathbb{Z}\}$ forms an orthonormal basis for V_{j+1}, and*

(iii) *the collection of functions $\{H_{j',k} : j' < j \text{ and } k \in \mathbb{Z}\}$ forms an orthonormal basis for V_j.*

Two comments are worthwhile.

- Part (i) is given explicitly by the property (MRA6) in the case $j = 0$, so (i) is telling us simply that the analogue of (MRA6) holds at *every* level in the hierarchy of spaces V_j (or at each 'rung' of the ladder of spaces).

- Part (i) provides some of the information needed for part (ii), but the extra information required is that the collection $\{\phi_{j,k} : k \in \mathbb{Z}\}$, which is an orthonormal subset of V_j, and hence of V_{j+1}, can be *extended* by the addition of the collection $\{H_{j,k} : k \in \mathbb{Z}\}$ to give an orthonormal basis for V_{j+1}.

We will prove part (iii) of the theorem only, and leave the proofs of parts (i) and (ii) for Exercise 5.5.

Proof of part (iii) of Theorem 5.1   We will consider the case $j = 0$ and show that $\{H_{j,k} : j < 0 \text{ and } k \in \mathbb{Z}\}$ is an orthonormal basis for V_0. Note first that these functions are indeed members of V_0, so that what we have to show are the orthonormality and maximality of the collection.

Orthonormality requires no argument, since we already know that the entire collection of scaled, dilated, translated functions $\{H_{j,k} : j, k \in \mathbb{Z}\}$ is orthonormal. (See Exercise 4.20 and discussion at several earlier points; the collection is moreover an orthonormal basis for $L^2(\mathbb{R})$ by Exercises 4.21, 4.22 and 4.23.)

Maximality is nearly as easy. Suppose that f is a function in V_0 that is orthogonal to each member of the collection $\{H_{j,k} : j < 0 \text{ and } k \in \mathbb{Z}\}$. Now the fact that $f \in V_0$ means that f is constant on all the integer-length intervals $[k, k+1)$, and this implies that f is orthogonal to each member of the collection $\{H_{j,k} : j \geq 0 \text{ and } k \in \mathbb{Z}\}$ (consideration of the relevant graphs should make this clear), and so we find that f is in fact orthogonal to $H_{j,k}$ *for all* $j, k \in \mathbb{Z}$. Hence, applying the fact that the collection $\{H_{j,k} : j, k \in \mathbb{Z}\}$ is an orthonormal basis for $L^2(\mathbb{R})$, we conclude that f is the zero function, as required.

The argument for general j is similar. (Alternatively, we can use the technical simplifications outlined in Section 5.4 below to deduce the general case from the case for $j = 0$.) $\square$

5.3 Approximation Spaces and Detail Spaces

The spaces V_j in the multiresolution analysis are sometimes called **approximation spaces**. The reason for this name, and for the name of the associated **detail spaces**, will become clear if we analyse how the spaces V_j are related to the Haar wavelet series of a function $f \in L^2(\mathbb{R})$. Theorem 5.1 provides the technical information we need.

Consider a fixed $j \in \mathbb{Z}$. Now V_j is closed, and hence complete, as we observed in Section 5.2, and by part (i) of Theorem 5.1 we know that V_j has the collection $\{\phi_{j,k} : k \in \mathbb{Z}\}$ as an orthonormal basis. Therefore, the projection of $f \in L^2(\mathbb{R})$ into V_j is given by the series

$$\sum_{k=-\infty}^{\infty} v_{j,k} \phi_{j,k}, \qquad (5.1)$$

where $v_{j,k} = \langle f, \phi_{j,k} \rangle$ for all k (see Section 4.6).

We could write down a similar expression for the projection of f into V_{j+1}, but we will use the alternative provided by part (ii) of Theorem 5.1: since, by that result, the collection $\{\phi_{j,k} : k \in \mathbb{Z}\} \cup \{H_{j,k} : k \in \mathbb{Z}\}$ is an orthonormal basis for V_{j+1}, the series

$$\sum_{k=-\infty}^{\infty} v_{j,k} \phi_{j,k} + \sum_{k=-\infty}^{\infty} w_{j,k} H_{j,k}, \qquad (5.2)$$

where $w_{j,k} = \langle f, H_{j,k} \rangle$ for all k, is the projection of f into V_{j+1}. That is, we can obtain the projection of f into V_{j+1} by adding to its projection into V_j the terms

$$\sum_{k=-\infty}^{\infty} w_{j,k} H_{j,k}. \qquad (5.3)$$

Recall from Theorem 4.41 that the projection of a vector into a closed subspace is the unique vector in the subspace that is closest to the original vector – we could refer to the projection, using the terminology of Chapter 3, as its best approximation in the subspace. In particular, the projection given by (5.1) of f into V_j is the best approximation to f in V_j, and similarly for the projection given by (5.2) into V_{j+1}. Hence we can interpret the analysis above as saying that, starting with the best level-j approximation to f, we have added just enough extra detail about f at level $j + 1$ to obtain its best level-$(j + 1)$ approximation. Also, the 'extra detail' was the series (5.3), and this is also a projection of f, this time into the closed subspace, say W_j, that has the collection $\{H_{j,k} : k \in \mathbb{Z}\}$ as orthonormal basis (see Exercise 4.28).

It is therefore reasonable to call the spaces V_j the approximation spaces and the spaces W_j the detail spaces of the multiresolution analysis.

One further observation is of interest in the light of the discussion of orthogonal complements and direct sums in Subsection 4.7.1. The spaces V_j and W_j are closed subspaces of V_{j+1} and are mutually orthogonal (that is, $V_j \perp W_j$), and every element of V_{j+1} can be expressed in a unique way as the sum of an element of V_j and an element of W_j. From this we can say, according to Subsection 4.7.1, that W_j is the orthogonal complement of V_j in V_{j+1}, that is, $W_j = V_j^\perp$, and we can write $V_{j+1} = V_j \oplus W_j$; further, $V_j = W_j^\perp$. (Detailed confirmation of all the claims here is left for Exercise 5.6.)

Let us work through an example in which it is simple enough for us to calculate the projections into both the approximation spaces and the detail spaces explicitly. Consider the function

$$f(x) = \begin{cases} x, & 0 \leq x < 1, \\ 0, & \text{otherwise.} \end{cases}$$

The projections of f into the approximation space V_j and the detail space W_j are given by the series

$$\sum_{k=-\infty}^{\infty} v_{j,k} \phi_{j,k} \quad \text{and} \quad \sum_{k=-\infty}^{\infty} w_{j,k} H_{j,k},$$

respectively, where $v_{j,k} = \langle f, \phi_{j,k} \rangle$ and $w_{j,k} = \langle f, H_{j,k} \rangle$ for all k. It is straightforward to calculate the coefficients $v_{j,k}$ and $w_{j,k}$ (see Exercise 5.7), and we find that for $j \geq 0$

$$v_{j,k} = \begin{cases} 2^{-(3j+2)/2}(2k+1), & 0 \leq k \leq 2^j - 1, \\ 0, & k < 0 \text{ or } k > 2^j - 1 \end{cases}$$

and

$$w_{j,k} = \begin{cases} -2^{-(3j+4)/2}, & 0 \leq k \leq 2^j - 1, \\ 0, & k < 0 \text{ or } k > 2^j - 1. \end{cases}$$

Using this and the corresponding information for $j < 0$ (see Exercise 5.7), we plot in Figure 5.2 the graphs of the projections of f into V_j and W_j for $j = 0, 1, 2, 3, 4, 5$, represented, respectively, by the upper line and the lower line in each plot. (It is easy to see by examining the coefficients of the two projections that all of these functions, like f itself, have support lying in $[0, 1]$, so they are plotted only over that interval.) It is clear graphically, as we know must be the case analytically, that in each diagram the sum of the two functions plotted is the function whose plot is uppermost in the next diagram.

The trend in the sequence is easy to see. The projection of f into the approximation space V_j is the closest function to f in V_j, that is, *the closest function to f that is constant on all the intervals of the form $[k/2^j, (k+1)/2^j)$ for $k \in \mathbb{Z}$*. We can say, repeating a description that we used in Chapter 1, that this projection constitutes a coarse-grained image of f: as accurate as possible to a resolution of $1/2^j$, but containing no information at any finer resolution.

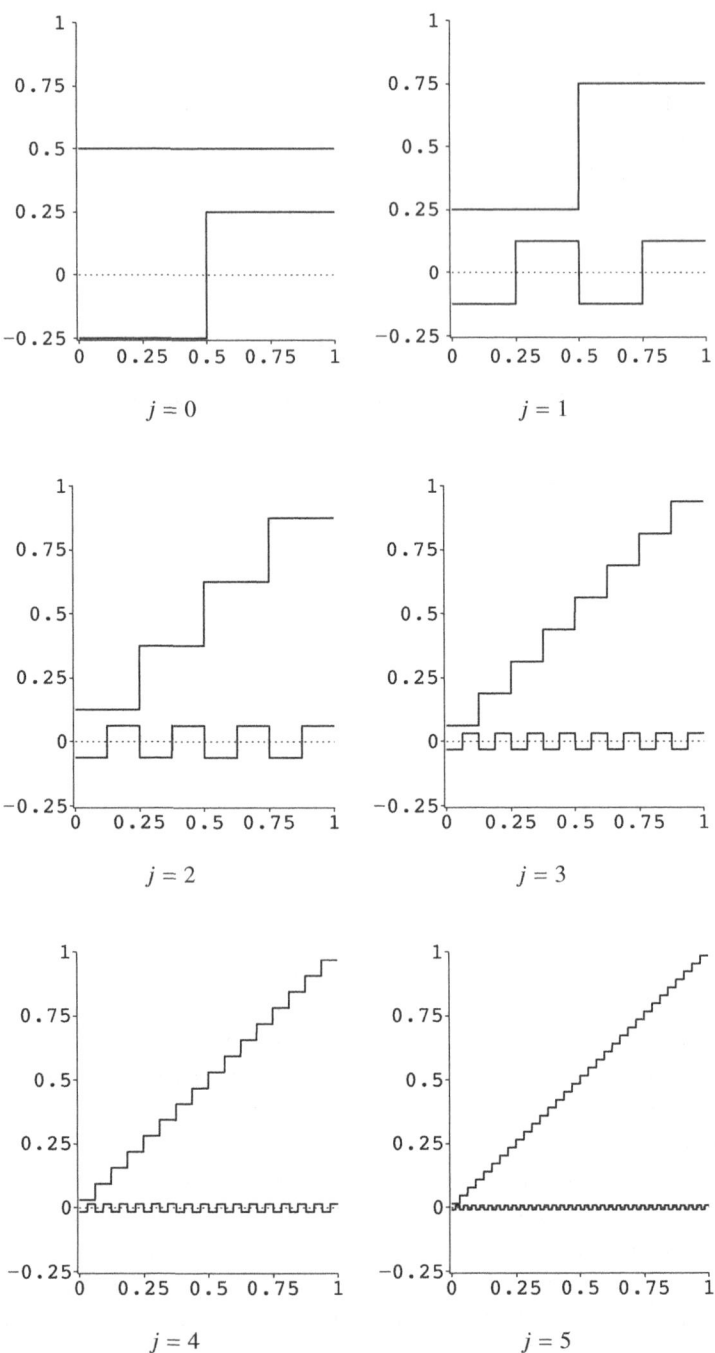

Figure 5.2 Graphs of the projections of f into the approximation spaces V_j (the upper line) and the detail spaces W_j (the lower line) for $j = 0, 1, 2, 3, 4, 5$

The projection into the corresponding detail space W_j then contains exactly the information needed to refine the best approximation at a resolution of $1/2^j$ to give the best approximation at a resolution of $1/2^{j+1}$.

5.4 Some Technical Simplifications

In this section, we record a few observations that allow simplification of some of the arguments in both this chapter and the next. These are arguments that involve the manipulation of sets of scaled, dilated, translated functions (which may or may not be wavelets). The observations allow us to reduce certain arguments to special cases without loss of generality. The two statements of the following theorem are general enough to deal with all our applications.

Theorem 5.2 *Suppose that $f \in L^2(\mathbb{R})$, and let $i, j, k, \ell \in \mathbb{Z}$. Then*

(i) $\langle f_{i,k}, f_{j,\ell} \rangle = \langle f_{0,k}, f_{j-i,\ell} \rangle$ *and*
(ii) $\langle f_{i,k}, f_{j,\ell} \rangle = \langle f_{i,0}, f_{j,\ell-2^{j-i}k} \rangle$.

The proofs are left as an exercise (see Exercise 5.9); the arguments involve nothing more than suitable changes of variable in the integrals that define the inner products involved.

We note a few typical illustrations of the way in which these observations can be useful.

- Suppose we need to prove that whenever i and j are distinct integers, all members of the set $\{f_{i,k} : k \in \mathbb{Z}\}$ are orthogonal to all members of the set $\{f_{j,\ell} : \ell \in \mathbb{Z}\}$. This means we must prove that $\langle f_{i,k}, f_{j,\ell} \rangle = 0$ for $i \neq j$ and for all k and ℓ. Our simplifying observation is that it is enough to prove just that $\langle f_{0,k}, f_{j,\ell} \rangle = 0$ for all $j \neq 0$ and for all k and ℓ. This is because if the last statement is proved, then for $i \neq j$ we have $j - i \neq 0$ and hence $\langle f_{0,k}, f_{j-i,\ell} \rangle = 0$ for all j and k, and so part (i) of the theorem gives

$$\langle f_{i,k}, f_{j,\ell} \rangle = \langle f_{0,k}, f_{j-i,\ell} \rangle = 0$$

for all j and k, as required.
- Suppose we need to prove that for some function f the set of translates $\{f_{0,k} : k \in \mathbb{Z}\}$ is orthonormal. This requires us to show that $\langle f_{0,k}, f_{0,\ell} \rangle = \delta_{k,\ell}$ for all k and ℓ. We claim that it is enough for us prove the simpler statement that $\langle f_{0,0}, f_{0,k} \rangle = \delta_{0,k}$ for all k. Indeed, if this is proved, then by part (ii) of the theorem we have

$$\langle f_{0,k}, f_{0,\ell} \rangle = \langle f_{0,0}, f_{0,\ell-k} \rangle = \delta_{0,\ell-k} = \delta_{k,\ell},$$

for all k and ℓ, as required.

- Suppose we need to prove that the set $\{f_{j,k} : k \in \mathbb{Z}\}$ is orthonormal for each $j \in \mathbb{Z}$. Here it suffices, using part (i) of the theorem, to prove just that the single set $\{f_{0,k} : k \in \mathbb{Z}\}$ is orthonormal; the details are left for Exercise 5.10.

Exercises

5.1 Prove that the subspaces V_j introduced in Subsection 5.2.2 are complete (and hence closed), perhaps as follows. Work for concreteness with V_0, and let $\{f_n\}$ be a Cauchy sequence in V_0. Since $L^2(\mathbb{R})$ is complete, there exists $f \in L^2(\mathbb{R})$ such that $f_n \to f$ as $n \to \infty$, and we need to show that $f \in V_0$, that is, that f is constant on the integer-length intervals $[k, k + 1)$ for all $k \in \mathbb{Z}$. For each k, denote the restrictions of f_n and f to $[k, k + 1)$ by $f_n^{(k)}$ and $f^{(k)}$, respectively, and suppose that $f_n^{(k)}$ has the constant value $c_n^{(k)}$.

   (a) Show that $f_n^{(k)} \to f^{(k)}$ in $L^2([k, k + 1))$ as $n \to \infty$.
   (b) Show that $\{c_n^{(k)}\}$ is a Cauchy sequence of real numbers for each k, and therefore converges to some $c^{(k)} \in \mathbb{R}$.
   (c) Deduce that in $L^2([k, k + 1))$ the sequence $\{f_n^{(k)}\}$ converges to the constant function with value $c^{(k)}$, and hence that $f^{(k)}$ is constant on $[k, k + 1)$ with value $c^{(k)}$.
   (d) Deduce that $f \in V_0$.
   (e) Check finally that an essentially identical argument carries through in V_j for arbitrary $j \in \mathbb{Z}$.

5.2 Confirm all the details of the outline proofs of the multiresolution analysis properties (MRA1)–(MRA6) in Subsection 5.2.2. (For example, the proof of (MRA2) requires the information that the spaces V_j are closed; this is not mentioned explicitly in the proof given, but a careful check of the proof will reveal where and why it is needed.)

5.3 Confirm the correctness of the reformulations of properties (MRA4), (MRA5) and (MRA6) given in the remarks at the end of Subsection 5.2.2. (Note that for (MRA4) in particular it is necessary to make use of the fact that V_j is a *subspace* of $L^2(\mathbb{R})$.)

5.4 Confirm the claim made in the second paragraph of Subsection 5.2.3 that $H_{0,k} \perp V_0$. (Inspection of the relevant graphs should almost be enough.)

5.5 Prove parts (i) and (ii) of Theorem 5.1, complete the proof of part (iii), and confirm the two claims following the statement of the theorem. (Use the ideas of Section 5.4 wherever possible.)

5.6   Confirm the claim made in the discussion of approximation and detail spaces in Section 5.3 that $V_{j+1} = V_j \oplus W_j$.

5.7   (a) Carry out the calculations of the coefficients $\{v_{j,k}\}$ and $\{w_{j,k}\}$ for the example in Section 5.3, making sure to cover the case when $j < 0$, for which the coefficients are not given in the main text.

       (b) Draw by hand the first two or three graphs of Figure 5.2.

5.8   In relation to the example in Section 5.3, observe that parts (i) and (iii) of Theorem 5.1 give two different orthonormal bases for V_j, and we could in principle compute the projection of the function f into V_j with respect to either of them. The calculations shown use the basis given by part (i) of the theorem. Write down the Fourier series expression for the projection in terms of the basis given by part (iii) of the theorem. Confirm directly that although the two series expressions are quite different, the projections themselves are nevertheless the same (as is guaranteed by not only by Theorem 5.1 itself but also by the comments following Theorem 4.41).

5.9   Prove Theorem 5.2. (Note that our use of the double-subscript notation for scaling, dilation and translation has always been restricted to the case where both subscripts have integer values, while in part (ii) there are clearly values of i, j, k and ℓ for which the value of the subscript term $\ell - 2^{j-i}k$ is not an integer. Confirm, by inspection of the argument for part (ii), that the result is nevertheless valid exactly as stated.)

5.10   Give a proof of the third illustration in Section 5.4.

5.11   Suppose that in investigating the Haar wavelet H, we have proved the partial orthonormality relation

$$\langle H_{0,0}, H_{j,k} \rangle = \delta_{0,j}\delta_{0,k} \text{ for all } j, k \in \mathbb{Z}.$$

       Use Theorem 5.2 to deduce from this the general orthonormality relation

$$\langle H_{j,k}, H_{j',k'} \rangle = \delta_{j,j'}\delta_{k,k'} \text{ for all } j, k, j', k' \in \mathbb{Z}.$$

5.12   For $f \in L^2(\mathbb{R})$, we have $f = \sum_{j\in\mathbb{Z}} \sum_{k\in\mathbb{Z}} w_{j,k} H_{j,k}$, where the right-hand side is the usual Haar wavelet series for f. Clearly, $\int_{-\infty}^{\infty} H_{j,k} = \int_{-\infty}^{\infty} H = 0$ for all j and k, but it is also clearly not the case that $\int_{-\infty}^{\infty} f = 0$ for all $f \in L^2(\mathbb{R})$. (For a specific example, refer to Subsection 1.4.3.) Explain the apparent paradox.

5.13   On a strict view, the elements of $L^2(\mathbb{R})$ are equivalence classes of square-integrable functions under the relation of equality almost everywhere, even if we adopt, as often as possible, the more relaxed view that the elements are simply the square-integrable functions themselves (see Sections 3.2 and 3.3). This exercise suggests the experiment of thinking through carefully how the concept of a multiresolution analysis, as presented in Subsection 5.2.2, should be formulated when the strict view is taken: in what ways should the formulations of the statements and proofs of the multiresolution analysis conditions be modified?

6

Wavelets in General

6.1 Introduction

Anticipating the formal definition below, we note that a wavelet is a function $\psi \in L^2(\mathbb{R})$ whose scaled, dilated, translated copies $\psi_{j,k}$ for $j, k \in \mathbb{Z}$ form an orthonormal basis for $L^2(\mathbb{R})$. Given such a function, the results of Section 4.6 tell us that any function in $L^2(\mathbb{R})$ can be expressed as a Fourier series with respect to the functions $\psi_{j,k}$, though we will usually call such a series a *wavelet series* and the corresponding Fourier coefficients the *wavelet coefficients*. All of this, of course, corresponds exactly to what we know from earlier chapters in the specific case of the Haar wavelet H.

Recall from Section 2.3 that for $k = 0, 1, 2, \ldots$ we denote by $C^k(I)$ the space of functions on an interval I which have continuous derivatives up to and including order k. Note that the case $k = 0$ is included: $C^0(I)$ is just $C(I)$, the space of continuous functions on I. Further, recall that we define $C^\infty(I)$ to be the space of infinitely differentiable functions on I. Terms such as *smooth* and *smoothness* are often used in an informal way to describe the location of a function in the hierarchy of spaces $C^k(\mathbb{R})$; a function in $C^{k+1}(\mathbb{R})$ might be said to be *smoother* than one in $C^k(\mathbb{R})$, for example.

Applications of wavelets usually call for the use of wavelets that satisfy specific conditions of one kind or another, and the condition that a wavelet have some degree of smoothness is almost universal. The minimal requirement for a wavelet in practice is therefore usually continuity (membership of $C^0(\mathbb{R})$), though a higher order of smoothness might well be preferred (membership of $C^1(\mathbb{R})$ or $C^2(\mathbb{R})$, for example). This means that we need to have methods for finding wavelets that are smoother than the Haar wavelet, which is not continuous and so is not in $C^k(\mathbb{R})$ for any $k \geq 0$.

For example, consider the use of wavelets for data compression. (This will often involve data in several dimensions – an image file is an obvious

141

two-dimensional example – while we are only considering wavelets in one dimension, in $L^2(\mathbb{R})$, but the principle is essentially the same.) The input data is regarded as a function (of one variable in our case) and our aim, in principle, is to compute and store the wavelet coefficients of that function rather than the values of the function itself, and in doing so to achieve an overall saving in storage – that is, to achieve compression of the input data.

Now the Fourier series theory from Chapter 4 tells us that the wavelet coefficients of the function contain enough information to reconstruct the function exactly, as measured by the usual definition of equality in $L^2(\mathbb{R})$. In practice, however, there will generally be an infinite number of non-zero wavelet coefficients, which we certainly cannot store, and even if there are only finitely many non-zero coefficients, there may still be far too many to allow any compression to be achieved.

What we must usually do in practice is therefore to ignore many small coefficient values by treating them as though they are exactly rather than just approximately zero. Then, when we try to reconstruct the original data from its wavelet coefficients, we will hope that the small coefficient values that we decided to ignore will not affect the accuracy of the reconstruction too seriously.

There are two competing requirements here. On the one hand, we want good compression, meaning that we do not want to have to store too many coefficients (and of course we cannot store infinitely many); but on the other hand, we want reasonably faithful reproduction of the original data. These requirements conflict, since better reproduction requires more coefficients to be stored, and therefore results in worse compression. Thus the scheme is satisfactory if it permits a reasonable balance between the quantity of the compressed data and the quality of the reconstructed data.

It turns out in general that the smoother the wavelet, the better the compression will be for a given quality of reproduction, so that, as noted above, the Haar wavelet is normally unsatisfactory, and one looks for smoother wavelets. (Just how smooth raises other issues: increasing smoothness imposes its own costs in the compression process.)

Aside from its lack of smoothness, another notable feature of the Haar wavelet is that it has bounded support. (Recall that the support of a function f is $\{x : f(x) \neq 0\}$, or the smallest closed set containing this – the distinction is not important here.) The support of the Haar wavelet is the interval $[0, 1)$ (or $[0, 1]$ on the alternative definition). There are applications where wavelets with unbounded support are of interest, but it is easy to see that wavelets with bounded support are likely to be most useful when the function whose wavelet series is to be computed has bounded support itself. For example, data for compression will often be supported over a bounded region (consider an image file again). If a wavelet has bounded support, then many of the wavelet coefficients corresponding to such data will be exactly zero (rather than just approximately zero), since they will correspond to inner products

$$\langle f, g \rangle = \int_{-\infty}^{\infty} fg$$

in which the supports of f and g simply do not overlap. If the wavelet had unbounded support, then it might well be the case that all of the wavelet coefficients were non-zero.

Thus the combination of bounded support and good smoothness properties is generally a desirable one to aim for. In the earliest days of the development of wavelet theory, it was not clear that wavelets with this combination of properties existed at all, and the first construction of such wavelets by Ingrid Daubechies in 1988 was therefore a major breakthrough. Daubechies in fact constructed an infinite sequence of wavelets, which are now known as the Daubechies wavelets. The first member of the sequence is the Haar wavelet, and the smoothness and support length of the wavelets increase in step along the sequence. It is also known that there does not exist an infinitely differentiable wavelet with bounded support, suggesting that if we are seeking to combine smoothness with bounded support, it is not possible to improve very much on the results of Daubechies.

6.1.1 Our Program

Our program over this chapter and the next is first to develop some of the general theory of wavelets and second to apply that theory to show how to construct the Daubechies wavelets. In the first half of the present chapter, we develop the general theory as far as we can without the imposition of any restrictive assumptions. In the second half of the chapter, it becomes necessary to impose an assumption if we are to take the theory much further using the methods available to us. The assumption needed is one of the conditions discussed above: that our wavelets have bounded support. In the next chapter, certain further aspects of the general theory in the case of bounded support are discussed, but the focus is predominantly on the construction and properties of the Daubechies wavelets.

There are two points to note about the restriction to the case of wavelets of bounded support in the later parts of the present chapter and throughout Chapter 7. The first point, as argued at the start of this section, is that these are the wavelets that are likely to be most useful in applications, and that the

Daubechies wavelets in particular have bounded support. This implies that a development of the theory for the case of bounded support is enough for our specific purposes in this book.

The second point, given above as the initial reason for imposing the restriction halfway through this chapter, relates to the mathematical techniques that we are assuming to be available: significant simplifications are possible in some of the arguments given the assumption of bounded support. A common step in our arguments is the interchange of sums and integrals (recall the brief discussion of this issue in Subsection 4.4.5). When the sums involved are infinite, which is generally the case if we do not assume bounded support, then these interchanges require analytical justification, a problem that typically needs to be handled either by using results from Lebesgue integration theory or by using other techniques altogether (such as those mentioned in Subsection 6.1.2 below). However, if we assume bounded support and the sums consequently become finite sums, then the interchanges become trivial, because the integral of a finite sum is always the sum of the integrals.

Most of the results that we develop on the assumption of bounded support exist in forms that are more general than ours, but our proof techniques are often no longer applicable in the general case, or are applicable only with the aid of more powerful analytical machinery.

6.1.2 Analytical Tools

We will expand a little here on the comments above about the analytical tools that we use and those that we prefer to avoid.

First, as we just noted, and have also noted at various earlier points, the theory of Lebesgue integration is one major analytical tool whose use at any deep level we are taking care to avoid. For any result whose proof requires serious use of the Lebesgue theory, we will either deal with a restricted form of the result for which the theory is not needed, as we noted we would do by restricting to the case of wavelets of bounded support later in this chapter, or we will simply state the result without proof. (In Chapter 7, especially, several important results will be stated without proof.)

If the Lebesgue theory is available, then access to other analytical machinery is opened up, above all to Fourier analysis, that is, the use of the Fourier integral transform, and all of the general literature on wavelets makes free use of the techniques of Fourier analysis. The Fourier transform allows us in principle to convert a given analytical problem into an equivalent but different analytical problem in the 'Fourier domain', which usually has a quite different character and is amenable to treatment by different mathematical techniques. A minor illustration of this is the fact that there are few wavelets for which closed analytic formulae can be written down (the Haar wavelet is the most notable exception), but that closed analytic expressions (albeit usually somewhat complicated ones) do exist for some of these wavelets in the Fourier domain, with the consequence that it can be considerably more convenient to define and manipulate them there.

➤ It is important to note that the lack of closed formulae for wavelets is no obstacle to their use in applications. What is important for applications is that there should exist good *algorithms* for computing with wavelets, and these do indeed exist, whether or not a given wavelet can be expressed by a simple formula. In particular, there are good algorithms for producing numerical approximations to wavelets – so that their graphs can be plotted, for example. (We will briefly discuss such algorithms in Chapter 7.)

The types of arguments we favour in this text are ones that are often referred to informally as 'elementary'. This term should not be taken to mean that the arguments are necessarily easy (a few of the arguments in this chapter and the next are at least moderately difficult), but simply that they avoid the use of more advanced mathematical theories such as those of Lebesgue integration and Fourier analysis.

Although our arguments in this chapter and the next are elementary in the above sense, we do make a brief excursion into the Fourier-analytic approach to wavelet theory in the final chapter, Chapter 8.

6.2 Two Fundamental Definitions

In the previous chapter, we started with the Haar wavelet and derived a multiresolution analysis from it. In this chapter, we will proceed in reverse, giving the general definition of a multiresolution analysis and then showing how a multiresolution analysis can be used to construct a wavelet. This will allow us, at least in principle, to produce wavelets at will, and we will see over the course of this chapter and the next that the Haar wavelet is just the first in the infinite sequence of progressively smoother Daubechies wavelets.

The two definitions referred to in the heading of the section are the definitions of a multiresolution analysis and of a wavelet; they are truly fundamental, and they form the basis for the rest of our work.

6.2.1 The Definition of a Multiresolution Analysis

Definition 6.1   Any doubly infinite sequence of closed subspaces $\{V_j\}$ of $L^2(\mathbb{R})$, along with a function $\phi \in V_0$, that together satisfy conditions

(MRA1)–(MRA6) from Chapter 5 is called a **multiresolution analysis**. The function ϕ is called the **scaling function** of the multiresolution analysis.

In detail, a multiresolution analysis consists of a collection $\{V_j : j \in \mathbb{Z}\}$ of closed subspaces of $L^2(\mathbb{R})$ and a function $\phi \in V_0$ with the following properties.

(MRA1)   The spaces $\{V_j\}$ are 'nested', or form a 'ladder', in the sense that $\cdots \subset V_{-2} \subset V_{-1} \subset V_0 \subset V_1 \subset V_2 \subset \cdots$; that is, $V_j \subset V_{j+1}$ for all $j \in \mathbb{Z}$.

(MRA2)   The set $\bigcup_{j \in \mathbb{Z}} V_j$ is dense in $L^2(\mathbb{R})$; that is, every function $f \in L^2(\mathbb{R})$ is the limit of a sequence of functions in $\bigcup_{j \in \mathbb{Z}} V_j$.

(MRA3)   The only function lying in all the spaces V_j is the zero function; that is, $\bigcap_{j \in \mathbb{Z}} V_j = \{0\}$.

(MRA4)   For all $j \in \mathbb{Z}$, $f(x) \in V_0$ if and only if $f(2^j x) \in V_j$.

(MRA5)   For all $k \in \mathbb{Z}$, $f(x) \in V_0$ and only if $f(x - k) \in V_0$.

(MRA6)   The collection $\{\phi(x - k) : k \in \mathbb{Z}\}$ is an orthonormal basis for V_0.

As we noted at the end of Subsection 5.2.2 and in Exercise 5.3 in the special case of the Haar multiresolution analysis and scaling function, the last three properties can be restated in equivalent forms, as follows, by making use of our standard double-subscript notation for scaling, dilation and translation.

(MRA4)′   For all $j \in \mathbb{Z}$, $f \in V_0$ if and only if $f_{j,0} \in V_j$.

(MRA5)′   For all $k \in \mathbb{Z}$, $f \in V_0$ and only if $f_{0,k} \in V_0$.

(MRA6)′   There exists $\phi \in V_0$ such that $\{\phi_{0,k} : k \in \mathbb{Z}\}$ is an orthonormal basis for V_0.

6.2.2 Consequences of the Definition

In this subsection we will explore, in a preliminary way, some of the consequences of the multiresolution analysis conditions. We will in fact use here only part of the information provided to us by the conditions: *we will use only the fact that the V_j are closed subspaces of $L^2(\mathbb{R})$ such that conditions (MRA1), (MRA2) and (MRA3) hold*, and no use will be made of conditions (MRA4), (MRA5) and (MRA6), or of the scaling function ϕ. The full power of the complete definition will be used only in later sections.

We can gain a useful perspective on the significance of the first three multiresolution analysis conditions by bringing into play the ideas of orthogonal projections, orthogonal complements and direct sums from Section 4.7.

For each $j \in \mathbb{Z}$, denote the projection of $L^2(\mathbb{R})$ into the closed subspace V_j by P_j. Let f be any fixed element of $L^2(\mathbb{R})$. Then from the nesting relations among the subspaces V_j given in condition (MRA1), Theorem 4.43 yields the inequality

$$\|P_j f\| \le \|P_{j+1} f\|$$

for all $j \in \mathbb{Z}$, and applying Theorem 4.41 as well, we have

$$0 \le \cdots \le \|P_{-2} f\| \le \|P_{-1} f\| \le \|P_0 f\| \le \|P_1 f\| \le \|P_2 f\| \le \cdots \le \|f\|.$$

The fact that the 0 at the left and the $\|f\|$ at the right here are the best possible bounds is immediate from the following result.

Theorem 6.2 *Suppose that the spaces $\{V_j : j \in \mathbb{Z}\}$ form a multiresolution analysis with scaling function $\phi \in V_0$. Let $f \in L^2(\mathbb{R})$. Then*

(i) $\lim\limits_{j \to \infty} P_j f = f$ *and*

(ii) $\lim\limits_{j \to -\infty} P_j f = 0.$

Proof For part (i), let $\epsilon > 0$, and use the density condition (MRA2) to pick $g \in \bigcup_{j \in \mathbb{Z}} V_j$ such that $\|f - g\| < \epsilon$. Suppose specifically that $g \in V_{j_0}$, and observe that it follows immediately by the nesting condition (MRA1) that $g \in V_j$ for all $j \ge j_0$. But by part (i) of Theorem 4.41, $P_j f$ is the closest element to f in V_j for every j, so for $j \ge j_0$ we have $\|f - P_j f\| \le \|f - g\| < \epsilon$. Since $\epsilon > 0$ was arbitrary, it follows that $\lim_{j \to \infty} \|f - P_j f\| = 0$, as required for (i).

For part (ii), it suffices to prove that $\lim_{j \to -\infty} P_j f = 0$ when $f \in \bigcup_{j \in \mathbb{Z}} V_j$. Indeed, suppose that this statement has been proved. Then for an arbitrary $f \in L^2(\mathbb{R})$ we have $P_0 f \in V_0 \subseteq \bigcup_{j \in \mathbb{Z}} V_j$, and so $\lim_{j \to -\infty} P_j P_0 f = 0$. But by Theorem 4.42, $P_j P_0 f = P_j f$ for all $j \le 0$, and the fact that $\lim_{j \to -\infty} P_j f = 0$ follows immediately.

For each $j \in \mathbb{Z}$, we write $W_j = V_j^\perp$, the orthogonal complement of V_j relative to V_{j+1} (see Subsection 4.7.1). We observe that $W_j \perp W_{j'}$ whenever $j \ne j'$: if, say, $j < j'$, then $W_j \subset V_{j+1} \subseteq V_{j'}$, while $V_{j'} \perp W_{j'}$.

We now fix $f \in \bigcup_{j \in \mathbb{Z}} V_j$ and show that $\lim_{j \to -\infty} P_j f = 0$. Almost the entire argument is devoted just to establishing that $\lim_{j \to -\infty} P_j f$ *exists*; once we know it exists, only a few lines are required to confirm that its limit is 0. Towards the end of the argument, we make essential use of the completeness of $L^2(\mathbb{R})$, in a way that parallels the completeness argument used in the proof of Theorem 4.38.

Suppose specifically that $f \in V_{i+1}$ for some $i \in \mathbb{Z}$ (our use of $i + 1$ rather than i as the index value is just for notational convenience). By part (vi) of Theorem 4.44, f can be uniquely represented as a sum $f = a_i + b_i$ with $a_i \in V_i$ and $b_i \in W_i$. The same result tells us specifically that $a_i = P_i f$, so part (iii) of Theorem 4.41 gives $\|a_i\| = \|P_i f\| \le \|f\|$.

Since $a_i \in V_i$, we can repeat this argument with a_i in place of f, and we obtain a unique representation of a_i as a sum $a_i = a_{i-1} + b_{i-1}$ with $a_{i-1} \in V_{i-1}$ and $b_{i-1} \in W_{i-1}$, and $\|a_{i-1}\| = \|P_{i-1}a_i\| \le \|a_i\| \le \|f\|$. By combining the results of this and the previous paragraph, we also have $f = a_i + b_i = a_{i-1} + b_{i-1} + b_i$.

We now continue this argument (strictly, by induction) 'down the ladder' of spaces V_j, expressing f, for each $r \ge 0$, as the sum

$$f = a_{i-r} + b_{i-r} + \cdots + b_{i-1} + b_i, \tag{6.1}$$

where

$$a_{i-r} = P_{i-r}f \in V_{i-r}, \quad b_{i-r} \in W_{i-r} \quad \text{and} \quad \|a_{i-r}\| \le \|f\|. \tag{6.2}$$

Considering a fixed r for the moment, we note that since $b_{i-r}, \ldots, b_{i-1}, b_i$ belong to $W_{i-r}, \ldots, W_{i-1}, W_i$, respectively, and since the subspaces in this sequence are mutually orthogonal, it follows that the vectors $b_{i-r}, \ldots, b_{i-1}, b_i$ are mutually orthogonal. From orthogonality, we have

$$\|b_{i-r} + \cdots + b_{i-1} + b_i\|^2 = \|b_{i-r}\|^2 + \cdots + \|b_{i-1}\|^2 + \|b_i\|^2$$

(see Exercise 3.19), and since $f = a_{i-r} + b_{i-r} + \cdots + b_{i-1} + b_i$ by (6.1), we therefore have

$$\begin{aligned}
\|b_{i-r}\|^2 + \cdots + \|b_{i-1}\|^2 + \|b_i\|^2 &= \|f - a_{i-r}\|^2 \\
&\le (\|f\| + \|a_{i-r}\|)^2 \\
&\quad \text{(by the triangle inequality)} \\
&\le (\|f\| + \|f\|)^2 \\
&\quad \text{(by (6.2))} \\
&= 4\|f\|^2.
\end{aligned}$$

But this holds for all $r \ge 0$, so the sums $\|b_{i-r}\|^2 + \cdots + \|b_{i-1}\|^2 + \|b_i\|^2$ are bounded above for $r \ge 0$, and since they are also non-decreasing, the Monotone Convergence Theorem tells us that they converge to a limit; that is, the series $\sum_{r=0}^{\infty} \|b_{i-r}\|^2$ converges.

This implies that the sequence of 'tail-ends' of this series, that is, the sequence of sums $\sum_{r=R}^{\infty} \|b_{i-r}\|^2$, converges to 0 as $R \to \infty$. But by orthogonality again, we have

$$\left\| \sum_{r=R}^{\infty} b_{i-r} \right\|^2 = \sum_{r=R}^{\infty} \|b_{i-r}\|^2.$$

(We need here a statement analogous to that of Exercise 3.19, but for infinite sums; for this, see the second equation of Theorem 4.20.) It follows that $\left\| \sum_{r=R}^{\infty} b_{i-r} \right\|^2$ converges to 0 as $R \to \infty$, which in turn implies that the sequence of partial sums of the series $\sum_{r=0}^{\infty} b_{i-r}$ is a Cauchy sequence in $L^2(\mathbb{R})$. Since $L^2(\mathbb{R})$ is complete, we conclude that the series $\sum_{r=0}^{\infty} b_{i-r}$ converges.

It now follows immediately from (6.1) that $\lim_{r\to\infty} a_{i-r}$ exists. But by (6.2), $a_{i-r} = P_{i-r}f$, and so $\lim_{r\to\infty} P_{i-r}f$ exists, which is equivalent to the assertion that $\lim_{j\to-\infty} P_j f$ exists. Finally, since, for each $k \in \mathbb{Z}$ we have $P_j f \in V_k$ for all $j \le k$, and V_k is closed, we have $\lim_{j\to-\infty} P_j f \in V_k$, and condition (MRA3) therefore gives $\lim_{j\to-\infty} P_j f = 0$, completing the proof of part (ii). $\qquad\square$

➤ It is worth noting that conditions (i) and (ii) of the theorem are in fact *equivalent* to the multiresolution analysis conditions (MRA2) and (MRA3), respectively, given the remaining conditions (see Exercise 6.7).

If a complete space V has mutually orthogonal closed subspaces $W_1, W_2, \ldots$ with the property that every $v \in V$ is expressible uniquely as a convergent series $v = w_1 + w_2 + \cdots$, with $w_j \in W_j$ for $j = 1, 2, \ldots$, then we say that V is the **direct sum** of $W_1, W_2, \ldots$, and we write

$$V = W_1 \oplus W_2 \oplus \cdots = \bigoplus_{j=1}^{\infty} W_j.$$

This definition is of course a natural generalisation of the definition of the direct sum of two closed subspaces given in Subsection 4.7.1. Note that although the mutually orthogonal closed subspaces above are indexed specifically by the natural numbers, the definition works in a similar way for indexing over any countable set – an observation important for the next theorem.

The generalised definition allows the following convenient reformulation and extension of the findings from our discussion above.

Theorem 6.3 *Suppose that the spaces $\{V_j : j \in \mathbb{Z}\}$ form a multiresolution analysis with scaling function $\phi \in V_0$. Denote by W_j the orthogonal complement of V_j relative to V_{j+1} for each $j \in \mathbb{Z}$. Then for each $i \in \mathbb{Z}$,*

$$V_{i+1} = W_i \oplus W_{i-1} \oplus \cdots = \bigoplus_{j=-\infty}^{i} W_j.$$

Further,

$$L^2(\mathbb{R}) = \bigoplus_{j=-\infty}^{\infty} W_j.$$

Proof   We continue using the notation from the proof of Theorem 6.2. From that proof, we know that $f \in V_{i+1}$ can be expressed for each $r \ge 0$ in the form $f = a_{i-r} + b_{i-r} + \cdots + b_{i-1} + b_i$, where $b_{i-r} \in W_{i-r}$ for all r and

$$\lim_{r\to\infty} a_{i-r} = \lim_{r\to\infty} P_{i-r}f = \lim_{j\to-\infty} P_j f = 0.$$

From this, it follows immediately that f has the (convergent) infinite series expansion $f = b_i + b_{i-1} + \cdots$. Further, f has only one such expansion. Indeed, if also $f = b'_i + b'_{i-1} + \cdots$, where $b'_{i-r} \in W_{i-r}$ for all $r \geq 0$, then we have $(b_i - b'_i) + (b_{i-1} - b'_{i-1}) + \cdots = 0$. But now, using an orthogonality argument of exactly the type used twice in the proof of Theorem 6.2, we have

$$\left\| (b_i - b'_i) + (b_{i-1} - b'_{i-1}) + \cdots \right\|^2 = \|b_i - b'_i\|^2 + \|b_{i-1} - b'_{i-1}\|^2 + \cdots = 0,$$

and so $\|b_{i-r} - b'_{i-r}\| = 0$ and then $b_{i-r} = b'_{i-r}$ for all $r \geq 0$, proving the claimed uniqueness. This proves the first statement of the theorem.

The essentials of the argument for the second statement of the theorem are as follows, but the details are left for verification in Exercise 6.8. We first observe that for any fixed $f \in L^2(\mathbb{R})$ there exist vectors $b_j \in W_j$ for all $j \in \mathbb{Z}$ such that $P_{j+1}f = b_j + b_{j-1} + \cdots$ for each $j \in \mathbb{Z}$. In fact, it suffices to define $b_j = P_{j+1}f - P_j f$ for each j; the fact that $b_j \in W_j$ follows from the properties of projections, and the convergence of the series for $P_{j+1}f$ follows by part (ii) of Theorem 6.2. Now, if we let j tend to ∞, part (i) of Theorem 6.2 shows that $f = \sum_{j=-\infty}^{\infty} b_j$, giving a (convergent) series representation of f of the kind required. Finally, the series for f is unique by an orthogonality argument of the type used for the first statement of the theorem, and this completes the proof. $\square$

6.2.3 The Definition of a Wavelet

We now introduce the second of our two fundamental definitions.

Definition 6.4   A **wavelet** is a function $\psi \in L^2(\mathbb{R})$ with the property that the family of functions

$$\psi_{j,k}(x) = 2^{j/2}\psi(2^j x - k) \quad \text{for } j, k \in \mathbb{Z}$$

is an orthonormal basis for $L^2(\mathbb{R})$.

Note that now, according to Theorem 4.37, we can say that H, the function we have been referring to as the Haar wavelet, really *is* a wavelet!

6.3 Constructing a Wavelet

The two fundamental definitions of the previous section may at first sight appear unrelated, but there is in fact a deep connection between them. We explored this connection in detail in Chapter 5 in the special case of the

Haar wavelet, and understanding it in the general case is our main task in this section.

Specifically, we will link the two definitions by showing in quite explicit terms how a multiresolution analysis can be used to construct a wavelet. We will see that the scaling function and the associated condition (MRA6), which are perhaps the least obvious parts of the definition of a multiresolution analysis, play a central role. In contrast to the previous section, we now need to bring into play the full power of the multiresolution analysis conditions.

We will construct the wavelet in two main steps. The first step is taken in Subsection 6.3.1, the main theorem of which gives a sufficient condition for a function to be a wavelet. The theorem of Subsection 6.3.4 then shows how to manufacture a function satisfying that condition, which is the required second step. An important new technical tool needed for the second step, the *scaling equation* of a scaling function, is meanwhile introduced in Subsection 6.3.3; this idea is used not only here but at many later places in this chapter and the next.

In the short Subsections 6.3.2 and 6.3.5, we introduce approximation and detail spaces (closely following the account in the Haar case of Chapter 5) and we observe how the important problems of controlling the smoothness and bounding the support of a wavelet can be reduced to the corresponding problems for the scaling function.

6.3.1 A Sufficient Condition for a Wavelet

The main result of this subsection, Theorem 6.6, states that a certain quite simple condition on a function in relation to a given scaling function (or, strictly, to the multiresolution analysis of which it is a part) is sufficient to ensure that the function is a wavelet. The importance and depth of the result are perhaps belied by its rather short proof, which relies heavily on the work of the preceding section.

For use in the proof of the main result, we first note the following straightforward, general statement about orthonormal bases in direct sums; the proof is left for Exercise 6.9.

Theorem 6.5 *Suppose that the Hilbert space V can be expressed as a direct sum $\bigoplus_{j=-\infty}^{\infty} V_j$. If B_j is an orthonormal basis for V_j for all j, then the collection $B = \bigcup_{j=-\infty}^{\infty} B_j$ is an orthonormal basis for V.*

Theorem 6.6 *Suppose that the spaces $\{V_j : j \in \mathbb{Z}\}$ form a multiresolution analysis with scaling function $\phi \in V_0$. If $\psi \in V_1$ is such that the family of functions*

$$\{\phi_{0,k} : k \in \mathbb{Z}\} \ \cup \ \{\psi_{0,k} : k \in \mathbb{Z}\}$$

is an orthonormal basis for V_1, then ψ is a wavelet.

Recall that multiresolution analysis condition (MRA6) tells us that $\phi \in V_0$ and that $\{\phi_{0,k} : k \in \mathbb{Z}\} = \{\phi(x-k) : k \in \mathbb{Z}\}$ is an orthonormal basis for V_0. Since $V_0 \subset V_1$, it is automatic that $\{\phi_{0,k} : k \in \mathbb{Z}\}$ is an orthonormal subset of V_1. The extra information in the hypothesis of the theorem is that this orthonormal set *can be extended, by inclusion of the set $\{\psi_{0,k} : k \in \mathbb{Z}\} = \{\psi(x-k) : k \in \mathbb{Z}\}$, to an orthonormal basis for V_1.*

Proof of Theorem 6.6   As just noted, the set $\{\phi_{0,k}\}$ is an orthonormal basis for the closed subspace V_0 of V_1, and by Exercise 4.28, the set $\{\psi_{0,k}\}$ is an orthonormal basis for a closed subspace W_0 of V_1. We claim that W_0 is $V_0^\perp$, the orthogonal complement of V_0 in V_1. Let $f \in V_1$. If $f \in W_0$, then f can be expressed as a series in the functions $\{\psi_{0,k}\}$, which are orthogonal to V_0 by the hypothesis of the theorem, and it follows that $f \perp V_0$. Conversely, since f is an element of V_1, we can express f as a series in the functions $\{\phi_{0,k}\} \cup \{\psi_{0,k}\}$, and if $f \perp V_0$ then the coefficients of f in the series corresponding to the functions $\{\phi_{0,k}\}$ must all be 0, and so $f \in W_0$. Therefore, by definition of the orthogonal complement, $W_0 = V_0^\perp$.

Using the simplifying observations of Section 5.4, it is easy to check that the following assertions, exactly analogous to those of the previous paragraph, hold for all $j \in \mathbb{Z}$: $\{\phi_{j,k}\}$ is an orthonormal basis for the closed subspace V_j of V_{j+1}; $\{\phi_{j,k}\} \cup \{\psi_{j,k}\}$ is an orthonormal basis for V_{j+1}; $\{\psi_{j,k}\}$ is an orthonormal basis for a closed subspace W_j of V_{j+1}; and W_j is $V_j^\perp$, the orthogonal complement of V_j in V_{j+1}.

But by Theorem 6.3, we have $L^2(\mathbb{R}) = \bigoplus_{j \in \mathbb{Z}} W_j$, and so by Theorem 6.5 it follows that $\bigcup_{j \in \mathbb{Z}} \{\psi_{j,k} : k \in \mathbb{Z}\} = \{\psi_{j,k} : j, k \in \mathbb{Z}\}$ is an orthonormal basis for $L^2(\mathbb{R})$, which by Definition 6.4 establishes that ψ is a wavelet.   □

6.3.2 Approximation Spaces and Detail Spaces

Our discussion of approximation and detail spaces in this chapter can be very brief, since the ideas closely follow those of Section 5.3, which was for the specific case of the Haar wavelet multiresolution analysis. Moreover, these spaces, in all but name, have played an important role in this chapter already, so all that is really needed here is to formally identify them.

Under the hypotheses of Theorem 6.6, the space V_j is closed and hence complete for each j, and V_j has the collection $\{\phi_{j,k} : k \in \mathbb{Z}\}$ as an orthonormal basis. For any fixed $f \in L^2(\mathbb{R})$, the series

$$\sum_{k=-\infty}^{\infty} v_{j,k}\phi_{j,k},$$

where $v_{j,k} = \langle f, \phi_{j,k} \rangle$ for all k, therefore represents the projection of f into V_j or, equivalently, the best approximation to f in V_j. The spaces V_j in the multiresolution analysis can therefore be referred to as **approximation spaces**. Further, as we saw in the proof of Theorem 6.6, for each j the collection $\{\psi_{j,k} : k \in \mathbb{Z}\}$ is an orthonormal basis for the closed subspace $W_j = V_j^{\perp}$ of V_{j+1}, and so the series

$$\sum_{k=-\infty}^{\infty} w_{j,k}\psi_{j,k},$$

where $w_{j,k} = \langle f, \psi_{j,k} \rangle$ for all k, provides the extra detail which when added to the best approximation of f in V_j will produce its best approximation in V_{j+1}. The spaces W_j can therefore be referred to as **detail spaces**.

6.3.3 The Scaling Equation and Scaling Coefficients

Consider a fixed a multiresolution analysis $\{V_j : j \in \mathbb{Z}\}$ with scaling function ϕ. Now $\phi \in V_0$ and $V_0 \subset V_1$, so we have $\phi \in V_1$, and it follows by condition (MRA4) that

$$\phi\left(\frac{x}{2}\right) \in V_0.$$

Therefore, since the collection $\{\phi(x - k)\}$ forms an orthonormal basis for V_0, we can express $\phi(x/2)$ as a Fourier series

$$\phi\left(\frac{x}{2}\right) = \sum_{k \in \mathbb{Z}} a_k \phi(x - k), \tag{6.3}$$

where the Fourier coefficients are

$$a_k = \left\langle \phi\left(\frac{x}{2}\right), \phi(x - k) \right\rangle = \int_{-\infty}^{\infty} \phi\left(\frac{x}{2}\right)\phi(x - k)\,dx$$

for all $k \in \mathbb{Z}$. Replacing x by $2x$ in the Fourier series, and similarly changing variable in the integral defining a_k, we have the alternative and somewhat more notationally convenient forms

$$\phi(x) = \sum_{k \in \mathbb{Z}} a_k \phi(2x - k) \tag{6.4}$$

and

$$a_k = 2 \int_{-\infty}^{\infty} \phi(x)\phi(2x - k)\,dx \quad \text{for } k \in \mathbb{Z}. \tag{6.5}$$

Definition 6.7   The equation (6.4) for $\phi(x)$ is called the **scaling equation** of the multiresolution analysis, and we refer to the coefficients a_k defined by equation (6.5) for all $k \in \mathbb{Z}$ as the **scaling coefficients**.

➤ The terms we have introduced here are far from standard, and several others are in use. In the literature, the scaling equation is also referred to as the *dilation equation*, the *refinement equation* and the *two-scale equation*, and the scaling coefficients are referred to as the *scaling filter coefficients*, the *scaling parameters*, the *structure constants* and the *two-scale sequence*. (There are no doubt further variants.)

Note also that since we derived the scaling equation $\phi(x) = \sum_{k \in \mathbb{Z}} a_k \phi(2x - k)$ from a Fourier series in V_0, the equals sign in the equation denotes not equality for all x but rather the usual notion of equality in $L^2(\mathbb{R})$, namely, equality almost everywhere. We will generally not add the 'almost everywhere' side-condition to the scaling equation explicitly, but it should be borne in mind that it is always present implicitly. (In this connection, see the scaling function for the Haar multiresolution analysis below, and Exercise 6.11.)

Note carefully how the scaling equation *expresses ϕ as a series of scaled, dilated, translated copies of itself* (the dilation in this case being uniformly a *contraction* by a factor of 2).

In the case of the Haar multiresolution analysis, where the scaling function is

$$\phi(x) = \begin{cases} 1, & 0 \le x < 1, \\ 0, & \text{otherwise,} \end{cases}$$

we compute easily that

$$a_k = 2 \int_{-\infty}^{\infty} \phi(x)\phi(2x - k)\,dx = \begin{cases} 1, & k = 0, 1, \\ 0, & \text{otherwise.} \end{cases}$$

The scaling equation for the Haar multiresolution analysis is therefore

$$\phi(x) = \phi(2x) + \phi(2x - 1).$$

Given this equation, it takes only a moment to confirm its correctness directly (see Exercises 6.10 and 6.11).

We now record a family of relations among the scaling coefficients that will play an important role throughout this chapter and the next.

Theorem 6.8   *With scaling coefficients a_k for $k \in \mathbb{Z}$ defined as above,*

$$\sum_{k \in \mathbb{Z}} a_k a_{k-2\ell} = 2\delta_{0,\ell} \quad \text{for all } \ell \in \mathbb{Z}.$$

Proof We compute as follows:

$$\delta_{0,\ell} = \langle \phi(x), \phi(x-\ell) \rangle$$

$$= \left\langle \sum_{k\in\mathbb{Z}} a_k \phi(2x-k), \phi(x-\ell) \right\rangle$$

(using the scaling equation)

$$= \sum_{k\in\mathbb{Z}} a_k \langle \phi(2x-k), \phi(x-\ell) \rangle$$

(by Theorem 4.19)

$$= \sum_{k\in\mathbb{Z}} a_k \int_{-\infty}^{\infty} \phi(2x-k)\phi(x-\ell)\,dx$$

$$= \sum_{k\in\mathbb{Z}} a_k \int_{-\infty}^{\infty} \phi(2x+2\ell-k)\phi(x)\,dx$$

(by a change of variable)

$$= \sum_{k\in\mathbb{Z}} a_k \int_{-\infty}^{\infty} \phi(2x-(k-2\ell))\phi(x)\,dx$$

$$= \frac{1}{2} \sum_{k\in\mathbb{Z}} a_k a_{k-2\ell}$$

(by definition of the scaling coefficients),

which gives the conclusion.   □

We observe for future reference that since $\delta_{0,\ell} = \delta_{0,-\ell}$ for all ℓ, we can replace ℓ by $-\ell$ in the theorem to yield the equivalent form

$$\sum_{k\in\mathbb{Z}} a_k a_{k+2\ell} = 2\delta_{0,\ell} \quad \text{for all } \ell \in \mathbb{Z}. \tag{6.6}$$

6.3.4 From Scaling Function to Wavelet

We now take the second and final step in the construction of a wavelet from a multiresolution analysis, by showing how to construct a function ψ that satisfies the sufficient condition of Theorem 6.6. The function is got by reversing the order of the coefficients in the scaling equation and incorporating alternating signs.

Theorem 6.9   *If the scaling equation for the multiresolution analysis is*

$$\phi(x) = \sum_{k \in \mathbb{Z}} a_k \phi(2x - k),$$

then the function ψ defined by

$$\psi(x) = \sum_{k \in \mathbb{Z}} (-1)^k a_{1-k} \phi(2x - k) \tag{6.7}$$

satisfies the condition in Theorem 6.6, and is therefore a wavelet.

Definition 6.10   We refer to the equation (6.7) for $\psi(x)$ as the **wavelet equation**.

Given the scaling coefficients in the Haar case, which we calculated above, the theorem gives as the corresponding wavelet the function ψ defined by the wavelet equation

$$\begin{aligned}
\psi(x) &= \sum_{k \in \mathbb{Z}} (-1)^k a_{1-k} \phi(2x - k) \\
&= (-1)^0 a_1 \phi(2x) + (-1)^1 a_0 \phi(2x - 1) \\
&= \phi(2x) - \phi(2x - 1),
\end{aligned}$$

and a moment's inspection shows that this ψ is precisely the Haar wavelet H (see Exercise 6.10).

Proof of Theorem 6.9   According to Theorem 6.6, it is enough if we prove that $\psi \in V_1$ (so that $\psi \in L^2(\mathbb{R})$ in particular) and that

$$\{\phi(x - k) : k \in \mathbb{Z}\} \cup \{\psi(x - k) : k \in \mathbb{Z}\}$$

is an orthonormal basis for V_1.

For the fact that the series that defines ψ converges to an element of $L^2(\mathbb{R})$, see Exercise 6.12. The fact that $\psi \in V_1$ is then immediate, because the series expresses ψ (after minor manipulation) in terms of the functions $\phi_{1,k}$ for $k \in \mathbb{Z}$, which form an orthonormal basis for V_1.

For orthonormality, there are three cases to consider: we have to show that

$$\langle \phi(x - k), \phi(x - \ell) \rangle = \delta_{k,\ell},$$

that

$$\langle \psi(x - k), \psi(x - \ell) \rangle = \delta_{k,\ell}$$

and that

$$\langle \phi(x - k), \psi(x - \ell) \rangle = 0,$$

for all k and ℓ. Now the first case is given immediately by condition (MRA6), which says that $\{\phi(x-k) : k \in \mathbb{Z}\}$ is an orthonormal basis for V_0. For the second case, it is enough if we prove that

$$\langle \psi(x), \psi(x - k) \rangle = \delta_{0,k},$$

since the general case can be obtained from this by a translation, using the observations in Section 5.4. We then have

$$\langle \psi(x), \psi(x - k) \rangle$$

$$= \left\langle \sum_{n \in \mathbb{Z}} (-1)^n a_{1-n} \phi(2x - n), \sum_{m \in \mathbb{Z}} (-1)^m a_{1-m} \phi(2(x - k) - m) \right\rangle$$

$$= \sum_{n \in \mathbb{Z}} \sum_{m \in \mathbb{Z}} (-1)^{n+m} a_{1-n} a_{1-m} \left\langle \phi(2x - n), \phi(2x - (2k + m)) \right\rangle$$

(by Theorem 4.19)

$$= \frac{1}{2} \sum_{n \in \mathbb{Z}} \sum_{m \in \mathbb{Z}} (-1)^{n+m} a_{1-n} a_{1-m} \langle \phi_{1,n}, \phi_{1,2k+m} \rangle$$

$$= \frac{1}{2} \sum_{m \in \mathbb{Z}} (-1)^{2k+2m} a_{1-2k-m} a_{1-m}$$

(by orthonormality)

$$= \frac{1}{2} \sum_{m \in \mathbb{Z}} a_{1-2k-m} a_{1-m}$$

$$= \frac{1}{2} \sum_{n \in \mathbb{Z}} a_{n-2k} a_n$$

(re-indexing the sum)

$$= \delta_{0,k}$$

(by Theorem 6.8),

as required.

For the third case, it suffices, again using Section 5.4, to show that

$$\langle \phi(x - k), \psi(x) \rangle = 0,$$

for all k. Now

$$\langle \phi(x-k), \psi(x) \rangle = \left\langle \phi(x-k), \sum_{n\in\mathbb{Z}} (-1)^n a_{1-n} \phi(2x-n) \right\rangle$$

$$= \sum_{n\in\mathbb{Z}} (-1)^n a_{1-n} \langle \phi(x-k), \phi(2x-n) \rangle$$

(by Theorem 4.19)

$$= \sum_{n\in\mathbb{Z}} (-1)^n a_{1-n} \int_{-\infty}^{\infty} \phi(x-k)\phi(2x-n)\,dx$$

$$= \sum_{n\in\mathbb{Z}} (-1)^n a_{1-n} \int_{-\infty}^{\infty} \phi(x)\phi(2x-(n-2k))\,dx$$

(changing variables in the integral)

$$= \frac{1}{2} \sum_{n\in\mathbb{Z}} (-1)^n a_{1-n} a_{n-2k}.$$

But examination of the last sum shows that its terms cancel in pairs (the numerical values of the scaling coefficients play no role in the argument; see Exercise 6.13), and hence sum to 0, as required. We conclude that the set $\{\phi(x-k)\} \cup \{\psi(x-k)\}$ is orthonormal.

The final step required is to show that $\{\phi(x-k)\} \cup \{\psi(x-k)\}$ is maximal as an orthonormal set V_1; this is left as Exercise 6.14.   □

6.3.5 Smoothness and Support

Two important properties of wavelets, noted earlier, are their degree of smoothness and the boundedness or otherwise of their supports. Our analysis so far has shown that a wavelet ψ can be constructed in a direct fashion from a scaling function ϕ, so it is not surprising that we can largely resolve the smoothness and support issues for ψ by looking at the corresponding properties of ϕ. The facts are summarised in the following theorem.

Theorem 6.11 *Suppose that the scaling function ϕ has bounded support. Then we have the following.*

(i) *Only a finite number of the scaling coefficients are non-zero.*
(ii) *The corresponding wavelet ψ, constructed according to Theorem 6.9, has bounded support.*
(iii) *If ϕ is in $C^k(\mathbb{R})$, then ψ is also in $C^k(\mathbb{R})$.*

Proof   Recall from equations (6.4) and (6.5) that the scaling equation is

$$\phi(x) = \sum_{k \in \mathbb{Z}} a_k \phi(2x - k),$$

and that the scaling coefficients a_k are given by

$$a_k = 2 \int_{-\infty}^{\infty} \phi(x) \phi(2x - k)\, dx \quad \text{for } k \in \mathbb{Z}.$$

If ϕ has bounded support, then only a finite number of the coefficients a_k can be non-zero, since the supports of $\phi(x)$ and $\phi(2x - k)$ can only intersect for a finite number of values of k. This proves (i).

The expression for the wavelet ψ given in Theorem 6.9 is

$$\psi(x) = \sum_{k \in \mathbb{Z}} (-1)^k a_{1-k} \phi(2x - k).$$

Since only a finite number of the a_k are non-zero, ψ is a linear combination of a finite number of translates of $\phi(2x)$, each of which has bounded support, and this gives ψ itself bounded support, proving (ii).

If in addition ϕ is in $C^k(\mathbb{R})$, then each of the finite number of translates of $\phi(2x)$ defining ψ is in $C^k(\mathbb{R})$, and it follows that ψ is also in $C^k(\mathbb{R})$, which proves (iii).   □

We will see below that very precise information can be obtained on the size and location of the support of ψ relative to that of ϕ by analysing more carefully which of the coefficients a_k are non-zero.

This section has shown in a simple and definitive fashion that if we have a multiresolution analysis then we have a wavelet. This is very significant information, but we must note that it does not help us at all with the problem of finding a multiresolution analysis in the first place. In particular, it does not help us to find a multiresolution analysis that will yield a wavelet with some preassigned property or properties; our ultimate aim, of course, is the quite specific one of finding multiresolution analyses that yield the family of Daubechies wavelets. It will not be until Chapter 7 that we approach this problem directly. We must first develop more of the general theory.

6.4 Wavelets with Bounded Support

The background for this section continues to be that of a multiresolution analysis $\{V_j : j \in \mathbb{Z}\}$ with scaling function $\phi \in V_0$, as defined in Section 6.2. This is the point however where, as foreshadowed in Section 6.1, we restrict attention to the case of wavelets with bounded support or, more precisely, scaling functions with bounded support.

For convenience below, we introduce the following (non-standard) notation: If $f \in L^2(\mathbb{R})$, we write isupp f to denote the smallest closed *interval* containing the support of f. (In general, of course, the support of a function need not be an interval.) Note carefully that if isupp $f = [a, b]$, then there must exist points x arbitrarily near to a on the right and points x arbitrarily near to b on the left such that $f(x) \neq 0$. (Exercise 6.15 asks for confirmation of this. See also Exercise 6.27 and the several interesting but rather complex exercises that precede it for more information on the support of a scaling function.)

Theorem 6.12 *If the scaling function ϕ has bounded support, then*

$$\text{isupp } \phi = [m, m + 2n - 1]$$

for some $m \in \mathbb{Z}$ and some $n \in \mathbb{N}$.

Proof Write isupp $\phi = [b, c]$, for some $b, c \in \mathbb{R}$. We will show first, by examining the scaling equation carefully, that b and c are integers. Here it is convenient to use the original form of the scaling equation given by equation (6.3), namely,

$$\phi\left(\frac{x}{2}\right) = \sum_{k \in \mathbb{Z}} a_k \phi(x - k).$$

Since isupp $\phi = [b, c]$, we clearly have

$$\text{isupp } \phi(x/2) = [2b, 2c]. \tag{6.8}$$

Using the scaling equation, we now derive an alternative expression for isupp $\phi(x/2)$. Since, by Theorem 6.11, only a finite number of the scaling coefficients a_k are non-zero, we can find $r, s \in \mathbb{Z}$ such that $a_r, a_s \neq 0$ and $a_k = 0$ for $k < r$ and $k > s$, and we therefore have

$$\phi\left(\frac{x}{2}\right) = \sum_{k=r}^{s} a_k \phi(x - k).$$

Observe that

$$\text{isupp } \phi(x - r) = [b + r, c + r],$$
$$\text{isupp } \phi(x - (r + 1)) = [b + (r + 1), c + (r + 1)],$$
$$\vdots$$
$$\text{isupp } \phi(x - (s - 1)) = [b + (s - 1), c + (s - 1)],$$
$$\text{isupp } \phi(x - s) = [b + s, c + s].$$

Clearly, isupp $\phi(x - r)$ is the only one of these intervals that intersects the interval $[b+r, b+r+1)$. But we observed earlier that there are points x arbitrarily near to $b + r$ on the right where $\phi(x - r)$ is non-zero, so since $a_r \neq 0$, it follows that $\phi(x/2)$ is also non-zero at these points. Using a similar argument on the left of $c + s$, we can therefore conclude that

$$\text{isupp } \phi(x/2) = [b + r, c + s]. \tag{6.9}$$

Comparison of the two expressions (6.8) and (6.9) for isupp $\phi(x/2)$ now immediately gives

$$b = r \quad \text{and} \quad c = s,$$

so that b and c, the endpoints of the interval isupp ϕ, are integers. Thus, if we set $t = s - r$, we have isupp $\phi = [r, r + t]$, where $r, t \in \mathbb{Z}$ and

$$a_r, a_{r+t} \neq 0 \quad \text{and} \quad a_k = 0 \text{ for } k < r \text{ and } k > r + t. \tag{6.10}$$

It remains to show that t is odd. Suppose that t is in fact even, so that $t = 2\ell$ for some ℓ. Now by Theorem 6.8, we have

$$\sum_{k \in \mathbb{Z}} a_k a_{k-2\ell} = 2\delta_{0,\ell},$$

and since, by (6.10), a_k can be non-zero only for the indices $r, \ldots, r + t$, only one term in this summation can be non-zero, namely, the term corresponding to the value $k = r + t$, and the equation therefore reduces to the statement that $a_{r+t} a_r = 0$. But this equation contradicts the fact that $a_r, a_{r+t} \neq 0$, and it follows that t is odd, as required. □

Since a translate of a scaling function by an integer is again a scaling function (this is immediate from the definition of a scaling function), we can assume essentially without loss of generality that $m = 0$ in the theorem. That is, we can assume in the case of bounded support that isupp $\phi = [0, 2N - 1]$ for some $N \in \mathbb{N}$, and we will always work from here on under this assumption, signalling the fact by the use of an 'N' in place of the 'n' used in the theorem. The Haar scaling function of course exemplifies the simplest instance of this, with $N = 1$.

On this assumption, the conditions (6.10) in the proof of the previous theorem yield the following precise statement about the form of the scaling equation and its associated wavelet. (We revert now to our preferred form for the scaling equation, $\phi(x) = \sum_{k \in \mathbb{Z}} a_k \phi(2x - k)$.)

Corollary 6.13 *When the scaling function ϕ satisfies* isupp $\phi = [0, 2N - 1]$ *for some $N \in \mathbb{N}$, the scaling coefficients satisfy*

$$a_0, a_{2N-1} \neq 0 \quad and \quad a_k = 0 \text{ for } k < 0 \text{ and } k > 2N - 1$$

and the scaling equation has the form

$$\phi(x) = \sum_{k=0}^{2N-1} a_k \phi(2x - k).$$

Recall from Theorem 6.9 that the scaling equation

$$\phi(x) = \sum_{k \in \mathbb{Z}} a_k \phi(2x - k)$$

gives rise to a wavelet ψ defined by

$$\psi(x) = \sum_{k \in \mathbb{Z}} (-1)^k a_{1-k} \phi(2x - k).$$

By replacing k by $1 - k$ in the last summation, we can rewrite this equation in the form

$$\psi(x) = \sum_{k \in \mathbb{Z}} (-1)^{k-1} a_k \phi(2x + k - 1).$$

On our assumption about the support of ϕ, the non-zero values among the scaling coefficients are, according to the corollary, at most

$$a_0, a_1, \ldots, a_{2N-1},$$

so we have the following result (except for the claim about isupp ψ, which is straightforward but left for Exercise 6.16).

Corollary 6.14 *When the scaling function ϕ satisfies* isupp $\phi = [0, 2N - 1]$ *for some $N \in \mathbb{N}$, the equation for the corresponding wavelet ψ can be written in the form*

$$\psi(x) = \sum_{k=0}^{2N-1} (-1)^{k-1} a_k \phi(2x + k - 1),$$

and isupp $\psi = [-N + 1, N]$.

➤ With ϕ as in the two corollaries, we may in later discussion write the scaling equation in either of the two forms

$$\phi(x) = \sum_{k \in \mathbb{Z}} a_k \phi(2x - k) \quad \text{and} \quad \phi(x) = \sum_{k=0}^{2N-1} a_k \phi(2x - k),$$

depending on convenience. In the proof of the next theorem, for example, where a sum and an integral need to be interchanged, it is important to know that the sum is finite, but it is not important to know precisely which terms are non-zero, and it is more notationally convenient to leave the sum in its formally infinite form.

Similarly, we are free to write the wavelet equation (6.7) in either of the two forms

$$\psi(x) = \sum_{k \in \mathbb{Z}} (-1)^k a_{1-k} \phi(2x - k) \quad \text{and} \quad \psi(x) = \sum_{k=2-2N}^{1} (-1)^k a_{1-k} \phi(2x - k)$$

or in either of the two re-indexed forms

$$\psi(x) = \sum_{k \in \mathbb{Z}} (-1)^{k-1} a_k \phi(2x + k - 1) \quad \text{and} \quad \psi(x) = \sum_{k=0}^{2N-1} (-1)^{k-1} a_k \phi(2x + k - 1).$$

Theorem 6.16 below lists several further conditions that must be satisfied by any scaling function of bounded support and the corresponding scaling coefficients and wavelet. We first note a simple algebraic result for use in the proof of the theorem and at one or two later points. (The proof is left for Exercise 6.17.)

Lemma 6.15 *The equations $x^2 + y^2 = 2$ and $x + y = 2$ have the unique common solution $x = y = 1$.*

Theorem 6.16 *If the scaling function ϕ and wavelet ψ have bounded support, then*

(i) $\int_{-\infty}^{\infty} \phi = \pm 1$,

(ii) $\int_{-\infty}^{\infty} \psi = 0$,

(iii) $\sum_{k \in \mathbb{Z}} a_k = 2$ *and*

(iv) $\sum_{k \in \mathbb{Z}} (-1)^k a_k = 0$.

Proof Part (i) of the theorem is the key to the other three parts; further, the proof of (i) is reasonably complex while the proofs of the other parts are comparatively simple. We will give proofs of parts (ii), (iii) and (iv) here, and for a proof of part (i) we refer to Exercise 6.18. We use the abbreviation $I = \int_{-\infty}^{\infty} \phi$, so that, assuming part (i), we have $I = \pm 1$.

We first prove part (iii). Consider the scaling equation

$$\phi(x) = \sum_{k \in \mathbb{Z}} a_k \phi(2x - k).$$

For each k, we have, using a change of variable,

$$\int_{-\infty}^{\infty} \phi(2x - k) \, dx = \frac{1}{2} \int_{-\infty}^{\infty} \phi(x) \, dx = \frac{1}{2} I, \tag{6.11}$$

so integrating through the scaling equation (and recalling that the sum involves only a finite number of non-zero terms), we have

$$I = \frac{1}{2} I \sum_{k \in \mathbb{Z}} a_k,$$

and the fact that $I \neq 0$ gives $\sum_{k \in \mathbb{Z}} a_k = 2$, which proves (iii).

We next prove part (iv). By part (iii), we have $\sum_{k\in\mathbb{Z}} a_k = 2$, and we wish to prove that $\sum_{k\in\mathbb{Z}}(-1)^k a_k = 0$. But we have

$$\sum_{k\in\mathbb{Z}} a_k + \sum_{k\in\mathbb{Z}}(-1)^k a_k = (\cdots + a_{-2} + a_{-1} + a_0 + a_1 + a_2 + \cdots)$$
$$+ (\cdots + a_{-2} - a_{-1} + a_0 - a_1 + a_2 - \cdots)$$
$$= 2 (\cdots + a_{-2} + a_0 + a_2 + \cdots)$$
$$= 2 \sum_{k\in\mathbb{Z}} a_{2k},$$

so it will suffice if we prove that $\sum_{k\in\mathbb{Z}} a_{2k} = 1$ (note by part (iii) that this equation is equivalent to the equation $\sum_{k\in\mathbb{Z}} a_{2k+1} = 1$).

Using the relation from Theorem 6.8 in the alternative form (6.6), we have

$$\sum_{k\in\mathbb{Z}} a_k a_{k+2\ell} = 2\delta_{0,\ell}.$$

Therefore,

$$\sum_{\ell\in\mathbb{Z}} \sum_{k\in\mathbb{Z}} a_k a_{k+2\ell} = 2,$$

and so we have

$$2 = \sum_{\ell\in\mathbb{Z}} \sum_{k\in\mathbb{Z}} a_k a_{k+2\ell}$$
$$= \sum_{\ell\in\mathbb{Z}} \left(\sum_{k\in\mathbb{Z}} a_{2k} a_{2k+2\ell} + \sum_{k\in\mathbb{Z}} a_{2k+1} a_{2k+1+2\ell} \right)$$

(splitting the summation over k into
separate sums over even and odd values)

$$= \sum_{k\in\mathbb{Z}} \left(a_{2k} \sum_{\ell\in\mathbb{Z}} a_{2k+2\ell} \right) + \sum_{k\in\mathbb{Z}} \left(a_{2k+1} \sum_{\ell\in\mathbb{Z}} a_{2k+1+2\ell} \right)$$
$$= \left(\sum_{k\in\mathbb{Z}} a_{2k} \right) \left(\sum_{\ell\in\mathbb{Z}} a_{2\ell} \right) + \left(\sum_{k\in\mathbb{Z}} a_{2k+1} \right) \left(\sum_{\ell\in\mathbb{Z}} a_{2\ell+1} \right)$$

(replacing ℓ by $\ell - k$)

$$= \left(\sum_{k\in\mathbb{Z}} a_{2k} \right)^2 + \left(\sum_{k\in\mathbb{Z}} a_{2k+1} \right)^2.$$

But from part (iii), we have $\sum_{k\in\mathbb{Z}} a_k = 2$, which we can rewrite as

$$\sum_{k\in\mathbb{Z}} a_{2k} + \sum_{k\in\mathbb{Z}} a_{2k+1} = 2.$$

Thus if we put $x = \sum_{k\in\mathbb{Z}} a_{2k}$ and $y = \sum_{k\in\mathbb{Z}} a_{2k+1}$, we have

$$x^2 + y^2 = 2 \quad \text{and} \quad x + y = 2,$$

and it follows by Lemma 6.15 that $x = y = 1$. Thus $\sum_{k \in \mathbb{Z}} a_{2k} = 1$, which proves (iv).

Finally, integrating through the wavelet equation

$$\psi(x) = \sum_{k \in \mathbb{Z}} (-1)^{k-1} a_k \phi(2x + k - 1),$$

using a simple change of variable and applying (6.11) gives

$$\int_{-\infty}^{\infty} \psi = \tfrac{1}{2} I \sum_{k \in \mathbb{Z}} (-1)^{k-1} a_k,$$

and part (iv) immediately gives (ii). □

By the equations (6.6) and Theorem 6.16, we have now established the following facts about the scaling coefficients in the case of bounded support:

$$\left. \begin{aligned} \sum_{k \in \mathbb{Z}} a_k a_{k+2\ell} &= 2\delta_{0,\ell} \quad \text{for all } \ell \in \mathbb{Z}, \\ \sum_{k \in \mathbb{Z}} a_k &= 2. \end{aligned} \right\} \tag{6.12}$$

For convenience, we will refer to these equations (one infinite family of quadratic equations and one individual linear equation) as the **fundamental relations** among the coefficients a_k.

From Theorem 6.16, we have the further relation

$$\sum_{k \in \mathbb{Z}} (-1)^k a_k = 0,$$

but since we derived this equation directly from the fundamental relations, it places no constraints on the a_k beyond those already imposed by those relations. A special case of the family of quadratic equations in (6.12) is also worth noting: putting $\ell = 0$, we have

$$\sum_{k \in \mathbb{Z}} a_k^2 = 2.$$

We inserted explicit summation bounds into the scaling equation and the wavelet equation earlier, to avoid mention of any coefficients a_k for which either $k < 0$ or $k > 2N - 1$, and we can do the same for the fundamental relations. Then, after we eliminate from the system the infinitely many relations of the form $0 = 0$, the remaining relations take the form of the $N + 1$ equations

$$\left. \begin{aligned} \sum_{k=0}^{2N-2\ell-1} a_k a_{k+2\ell} &= 2\delta_{0,\ell} \quad \text{for } \ell = 0, 1, \ldots, N - 1, \\ \sum_{k=0}^{2N-1} a_k &= 2, \end{aligned} \right\} \tag{6.13}$$

and we will refer to (6.12) and (6.13) indiscriminately as the fundamental relations. We will make much use of these equations, in one form or another. (The summation bounds in the equation (6.13) are optimal, in the sense that they include all the terms that *can* be non-zero but none that *must* be zero. The checking of this claim is left for Exercise 6.19.)

We noted earlier, in view of Theorem 6.12, that the interval [0,1] is the smallest interval that can be isupp ϕ for a scaling function ϕ. A natural question is whether the Haar scaling function is the only scaling function on [0,1]. The next result shows that the answer is essentially yes. (The proof here is a direct one, but the fact also follows from results in Chapter 7 below; see the start of Section 7.2, for example.)

Theorem 6.17   *If the scaling function ϕ satisfies* isupp $\phi = [0, 1]$, *then ϕ is either the scaling function of the Haar wavelet or its negative.*

Proof   By Corollary 6.13, the scaling equation for ϕ is

$$\phi(x) = a_0\phi(2x) + a_1\phi(2x - 1),$$

where $a_0, a_1 \neq 0$. Also, by Theorem 6.16, we have $a_0 + a_1 = 2$ and $a_0 - a_1 = 0$, and these equations give $a_0 = a_1 = 1$. Therefore, the scaling equation becomes $\phi(x) = \phi(2x) + \phi(2x - 1)$, and the wavelet equation, which defines the corresponding wavelet ψ, has the form $\psi(x) = \phi(2x) - \phi(2x - 1)$. (Observe that all this is exactly as found above in the specific case of the Haar wavelet, though the conclusion of the theorem is not quite immediate from this.)

Let $f \in L^2(\mathbb{R})$ be the scaling function of the Haar wavelet (so that f has value 1 on $[0, 1)$ and value 0 elsewhere). Now f has a wavelet series with respect to the wavelet ψ, which as usual takes the form

$$f = \sum_{j=-\infty}^{\infty} \sum_{k=-\infty}^{\infty} w_{j,k}\psi_{j,k},$$

where

$$w_{j,k} = \langle f, \psi_{j,k} \rangle = \int_{-\infty}^{\infty} f\psi_{j,k},$$

for all $j, k \in \mathbb{Z}$. It easy to see, because of the special form of f and the simple form of the equation above for ψ, that $w_{j,k} = 0$ for all $j \geq 0$.

Consider the ladder of spaces $\cdots \subset V_{-1} \subset V_0 \subset V_1 \subset \cdots$ in the multiresolution analysis of which ϕ is the scaling function. Recall that $\psi \in V_1$ and that property (MRA4) tells us that $\psi_{j,k} \in V_{j+1}$ for all j and k. But $w_{j,k} = 0$ for all $j \geq 0$, so the wavelet series for f can only contain a non-zero term corresponding to the function $\psi_{j,k}$ if $j < 0$, and all these functions lie in V_0. Therefore, f itself lies in V_0.

Now condition (MRA6) tells us that the set $\{\phi(x-k) : k \in \mathbb{Z}\}$ is an orthonormal basis for V_0, and since both f and ϕ are supported within $[0,1]$, the Fourier series for f with respect to this orthonormal basis is the trivial one-term series

$$f = \langle f, \phi \rangle \phi. \tag{6.14}$$

Taking norms of both sides gives

$$\|f\| = \left|\langle f, \phi \rangle\right| \|\phi\|,$$

and since $\|f\| = \|\phi\| = 1$, we have $\langle f, \phi \rangle = \pm 1$. Therefore, by (6.14), we finally have $\phi = \pm f$, as required. $\square$

We have one further general result for the case of bounded support, this time one that gives a relation on the values of the scaling function. Recall from Theorem 6.16 that $\int_{-\infty}^{\infty} \phi$ is either 1 or -1; we make the assumption in the next result that it is 1 specifically.

We need also a preliminary comment about Fourier series. Theorem 4.34 tells us that every Hilbert space has an orthonormal basis. A minor extension of that result is that any given orthonormal set in a Hilbert space can be *enlarged* to yield an orthonormal basis. Since the function on $[0,1]$ with constant value 1 has norm 1 in $L^2([0,1])$, there must therefore be an orthonormal basis for $L^2([0,1])$ that contains that constant function. In fact, it is easy to adapt the standard orthonormal basis of trigonometric functions for $L^2([-\pi,\pi])$ to give such a basis explicitly – see Exercise 4.24 – but for the following argument it turns out not to matter which specific functions we add to the constant function to give a complete set.

Theorem 6.18 *Let ϕ be the scaling function of a multiresolution analysis. Suppose that ϕ is continuous and has bounded support, and that $\int_{-\infty}^{\infty} \phi = 1$. Then $\sum_{n \in \mathbb{Z}} \phi(x+n) = 1$ for all $x \in \mathbb{R}$, and $\sum_{n \in \mathbb{Z}} \phi(n) = 1$ in particular.*

Proof Consider the function Φ on $\mathbb{R}$ defined by setting $\Phi(x) = \sum_{n \in \mathbb{Z}} \phi(x+n)$, and note that Φ is properly defined, since the summation involved is finite for each x. Now define the function Φ_0 on $[0,1]$ to be the restriction of Φ to $[0,1]$. Then $\Phi_0 \in L^2([0,1])$, since the restriction to $[0,1]$ of each function of the form $\phi(x+n)$ belongs $L^2([0,1])$ and Φ_0 is a finite sum of such restrictions. Choose any orthonormal basis for $L^2([0,1])$ that consists of the function f_0 on $[0,1]$ that has constant value 1 and some collection of other functions f_n for $n \geq 1$. We will determine the Fourier series of Φ_0 with respect to this basis.

For $n = 0, 1, 2, \ldots$, denote the Fourier coefficient of Φ_0 with respect to f_n by c_n. Then we have

$$c_0 = \sum_{n \in \mathbb{Z}} \int_0^1 \phi(x+n)\,dx = \sum_{n \in \mathbb{Z}} \int_n^{n+1} \phi(y)\,dy = \int_{-\infty}^{\infty} \phi(y)\,dy = 1.$$

Now Parseval's relation (Equation (4.10) in Theorem 4.39) tells us that

$$\sum_{n=0}^{\infty} c_n^2 = \|\Phi_0\|^2.$$

But by what we have shown, $c_0^2 = 1$, and it is also the case that $\|\Phi_0\|^2 = 1$ (see Exercise 6.20), so it follows that $c_n = 0$ for $n = 1, 2, \ldots$.

Hence the Fourier series of Φ_0 consists of one term only, namely, $c_0 f_0$, and we simply have $\Phi_0(x) = \sum_{n \in \mathbb{Z}} \phi(x+n) = 1$ on $[0, 1]$. Now the final equals sign here indicates equality in the L^2 sense, not necessarily pointwise equality, but if we use (for the first time) the assumption that ϕ, and therefore also Φ_0, is continuous, we do in fact get equality for each $x \in [0, 1]$ (see Exercise 4.16).

Finally, the function Φ on $\mathbb{R}$ is periodic of period 1 (see Exercise 6.20 again), and it follows that $\sum_{n \in \mathbb{Z}} \phi(x+n) = 1$ for all $x \in \mathbb{R}$. Taking $x = 0$ of course gives $\sum_{n \in \mathbb{Z}} \phi(n) = 1$.   □

Exercises

6.1   Suppose that the closed subspaces $\{V_j : j \in \mathbb{Z}\}$ of $L^2(\mathbb{R})$ and the function $\phi \in V_0$ form a multiresolution analysis. Prove the following straightforward generalisations of properties (MRA4), (MRA5) and (MRA6).

   (a) For all $j_1, j_2 \in \mathbb{Z}$, $f \in V_{j_1}$ if and only if $f_{j_2-j_1,0} \in V_{j_2}$.
   (b) For all $j, k \in \mathbb{Z}$, $f \in V_j$ and only if $f_{0,2^{-j}k} \in V_j$.
   (c) For all $j \in \mathbb{Z}$, $\{\phi_{j,k} : k \in \mathbb{Z}\}$ is an orthonormal basis for V_j.

   (Note that the subscript term $2^{-j}k$ in part (ii) is not always integer-valued, but that this does not affect the validity of the statement or its proof; compare Theorem 5.2 and, especially, the comment in Exercise 5.9.)

6.2   Show that condition (MRA6) implies condition (MRA5).

6.3   (a) For each $j \in \mathbb{Z}$, let V_j be the subspace of $L^2(\mathbb{R})$ consisting of the functions that are linear on each interval $[k/2^j, (k+1)/2^j)$. Show that conditions (MRA1)–(MRA5) hold, but that (MRA6) does not.
   (b) Show that if V_j consists of the functions that are linear on each interval $[k/2^j, (k+1)/2^j)$ *and continuous*, then conditions (MRA1)–(MRA5) hold. Show also that condition (MRA6) does not hold for

any $\phi \in V_0$ that has bounded support. (It can be shown that there exists $\phi \in V_0$ with unbounded support for which condition (MRA6) is satisfied, giving a multiresolution analysis, but the definition of ϕ is indirect.)

6.4 Show that the function whose value is x when $0 \le x < 1$ and is 0 elsewhere cannot be the scaling function for any multiresolution analysis. (Consider the form taken by elements of V_0, using (MRA6), and by elements of V_1, using (MRA4), and show that (MRA1) cannot hold. The result of the exercise actually follows immediately from Theorem 6.18, but it is interesting to construct a direct argument.)

6.5 (a) Suppose that the spaces $\{V_j\}$ form a multiresolution analysis with scaling function ϕ. Show using Exercise 4.29 that

$$V_j = \left\{ \sum_{k \in \mathbb{Z}} \alpha_k \phi_{j,k} : \sum_{k \in \mathbb{Z}} \alpha_k^2 \text{ converges} \right\}$$

for each $j \in \mathbb{Z}$.

(b) Conclude that if a function ϕ is known to be the scaling function of a multiresolution analysis, then the associated spaces V_j can be recovered.

6.6 For the multiresolution analysis corresponding to the Haar wavelet, use the previous exercise to write down an expression for V_0, and sketch the graph of a 'typical' element of V_0. Also, write down an expression for a specific function in V_0 that is non-zero at all points in $\mathbb{R}$.

6.7 Prove the assertion following Theorem 6.2 that the two conditions of that theorem are equivalent to conditions (MRA2) and (MRA3), respectively.

6.8 Verify the details of the proof of part (ii) of Theorem 6.3.

6.9 Prove Theorem 6.5 and fill in any details missing from the proof of Theorem 6.6.

6.10 Confirm the correctness of the scaling equation for the Haar multiresolution analysis (after Definition 6.7 in the main text) and of the corresponding wavelet equation (after Definition 6.10).

6.11 Following the definition of the scaling equation (Definition 6.7), we noted that the equation holds almost everywhere, but not necessarily everywhere. Recall that the scaling function ϕ of the Haar multiresolution analysis was defined in Subsection 5.2.2 to have the value 1 for $x \in [0, 1)$ and the value 0 elsewhere.

(a) Show that the variant definition in which ϕ is 1 for $x \in [0, 1]$ and 0 elsewhere is also a scaling function for the same collection of subspaces V_j. (This involves not much more than the observation that the two versions of ϕ are equal almost everywhere, and therefore represent the same element of $L^2(\mathbb{R})$.)

(b) Show that the Haar scaling function as defined in the main text satisfies the scaling equation everywhere.

(c) Show that variant form of the Haar scaling function given in part (a) satisfies the scaling equation almost everywhere but not everywhere.

6.12 Check the claim near the beginning of the proof of Theorem 6.9 that the formula used to define ψ in the theorem does produce a function in $L^2(\mathbb{R})$. (Hint: Use Exercise 4.26 and the fact that ϕ is in $L^2(\mathbb{R})$.)

6.13 Prove that the sum $\sum_{n\in\mathbb{Z}}(-1)^n a_{1-n} a_{n-2k}$ appearing in the proof of Theorem 6.9 equals 0, as claimed. (Check that each product term occurs twice in the sum, but has opposite signs on the two occurrences.)

6.14 Carry out the final step in the proof of Theorem 6.9, perhaps as follows. The collection $B_1 = \{\phi_{0,k} : k \in \mathbb{Z}\} \cup \{\psi_{0,k} : k \in \mathbb{Z}\}$ and the collection $B_2 = \{\phi_{1,k} : k \in \mathbb{Z}\}$ are both orthonormal subsets of V_1, and we know that B_2 is an orthonormal basis for V_1. Use the following steps to show that B_1 is also an orthonormal basis for V_1.

(a) Confirm, using Exercise 4.28, that B_1 is an orthonormal basis for a closed subspace, say V_1', of V_1.

(b) Show that to conclude that $V_1' = V_1$ it is sufficient to prove that $\phi_{1,k} \in V_1'$ for all $k \in \mathbb{Z}$.

(c) Show that for part (b) it is sufficient to prove that $\phi_{1,0}, \phi_{1,1} \in V_1'$.

(d) Show, using Exercise 4.37, that for part (c) it is sufficient to prove that $\|P\phi_{1,0}\| = \|\phi_{1,0}\|$ and $\|P\phi_{1,1}\| = \|\phi_{1,1}\|$, where P denotes orthogonal projection from V_1 into V_1'.

(e) Compute the norms and draw the desired conclusion.

6.15 Confirm that, as claimed before Theorem 6.12, if isupp $f = [a, b]$, then there must exist points x arbitrarily near to a on the right and points x arbitrarily near to b on the left such that $f(x) \neq 0$. Deduce that if f is also continuous, then $f(a) = f(b) = 0$.

6.16 Prove the statement that isupp $\psi = [-N + 1, N]$ in Corollary 6.14. (Proving that the first set is a subset of the second is easy; a little more thought is needed to prove equality.)

6.17 Prove Lemma 6.15. The lemma did not explicitly say that x and y are real numbers; prove that the result holds without change even if x and y are allowed to take complex values.

6.18 This exercise gives suggested steps towards a proof that the scaling function ϕ of a multiresolution analysis $\{V_j\}$ satisfies $\int_{-\infty}^{\infty} \phi = \pm 1$ under the condition that ϕ has bounded support, which is part (i) of

Theorem 6.16. For convenience, write $I = \int_{-\infty}^{\infty} \phi$, so that we aim to prove that $I = \pm 1$.

Suppose that the support of ϕ is contained within $[-M, M]$ for some integer M. We use the fact (part (i) of Theorem 6.2) that $P_j f \to f$ as $j \to \infty$ for all f in $L^2(\mathbb{R})$, from which it follows (Theorem 4.18) that $\|P_j f\| \to \|f\|$. By applying this to a specially chosen function f, we will be able to deduce that $I = \pm 1$.

For any set S, the **characteristic function** of S, denoted by χ_S, is the function defined to have value 1 at all points of S and 0 at all other points. The specific function we consider for the present problem is the function $f_M = \chi_{[-M,M]}$, the characteristic function of the interval $[-M, M]$.

(a) Confirm that $\| f_M \| = (2M)^{1/2}$.
(b) Using the fact for each fixed j the functions $\phi_{j,k}$ for $k \in \mathbb{Z}$ form an orthonormal basis for V_j, show that

$$\left\| P_j f_M \right\|^2 = 2^{-j} \sum_{k \in \mathbb{Z}} \left| \int_{-2^j M}^{2^j M} \phi(x - k)\, dx \right|^2 .$$

From this point on, assume that $j \geq 0$ (this is simply to ensure that the endpoints of the interval $[-2^j M, 2^j M]$ are whole numbers, which makes the rest of the analysis somewhat simpler). Note that the support of $\phi(x - k)$ is contained in $[-M + k, M + k]$. The aim is now to examine in detail how this interval intersects the intervals of integration $[-2^j M, 2^j M]$ as j varies.

(c) Confirm that exactly one of the three following assertions holds for the two intervals:

    (i) $[-M + k, M + k] \subseteq [-2^j M, 2^j M]$;
    (ii) the intervals are disjoint (except possibly for the endpoints); or
    (iii) the intervals overlap non-trivially (that is, at more than just the endpoints) but without either being contained in the other.

(d) For the cases (i), (ii) and (iii) from part (c), prove the following respective claims:

    (i) this case occurs for precisely $2(2^j - 1)M + 1$ values of k, and

$$\int_{-2^j M}^{2^j M} \phi(x - k)\, dx = I$$

    for each such k;

(ii) this case occurs for infinitely many values of k, and

$$\int_{-2^j M}^{2^j M} \phi(x - k)\, dx = 0$$

for each such k; and

(iii) this case occurs for precisely $4M - 2$ values of k, and there is a constant C such that

$$\left| \int_{-2^j M}^{2^j M} \phi(x - k)\, dx \right| \le C$$

for each such k (specifically, we can take C to be $\int_{-M}^{M} |\phi|$, which is finite by part (c) of Exercise 4.42).

(e) Deduce that $\left\| P_j f_M \right\| \to (2M)^{1/2} |I|$ as $j \to \infty$.

(f) Conclude that $|I| = 1$, and hence that $I = \pm 1$.

6.19 Derive the fundamental relations in the form (6.13) from those given in the form (6.12), on the assumption that $\operatorname{isupp} \phi = [0, 2N - 1]$. Confirm that the indexing in (6.13) is optimal, in the sense described in the main text.

6.20 (a) Give proofs of the two steps in the proof of Theorem 6.18 that were left as exercises.

(b) Show that the function Φ in the proof does not belong to $L^2(\mathbb{R})$.

6.21 (a) In the bounded support case, the wavelet ψ can be written as a finite sum of scaled, dilated, translated copies of the scaling function ϕ. Show that ϕ, though in $L^2(\mathbb{R})$ and therefore expressible as a wavelet series, can never be written as a finite sum of scaled, dilated, translated copies of ψ.

(b) Find the wavelet series for the scaling function of the Haar wavelet. (Hint: See Subsection 1.4.3.)

6.22 Let $f, g \in L^2(\mathbb{R})$ both have bounded support. Show that if $\{f_{0,k}\}$ and $\{g_{0,k}\}$ are both orthonormal bases for the same subspace of $L^2(\mathbb{R})$, then $f = g_{0,k}$ for some k. (Express f as a Fourier series $f = \sum_{k=-\infty}^{\infty} \alpha_k g_{0,k}$, and find from this a Fourier series expansion for $f_{0,\ell}$, for all $\ell \in \mathbb{Z}$. Now apply the assumption about bounded support, and consider suitably chosen inner products $\langle f, f_{0,\ell} \rangle$.)

Note that this result implies that in the bounded support case there is essentially only one scaling function belonging to a given multiresolution analysis. (There are examples showing that this statement may fail without the assumption of bounded support.)

6.23 Let $\{V_j\}$ be a multiresolution analysis with scaling function ϕ, and suppose that $\psi \in V_1$ is chosen as in Theorem 6.6 (and so in particular is a wavelet).

(a) Prove that if $f \in V_0$, then $f \perp \psi_{j,k}$ for all $j \geq 0$ and all $k \in \mathbb{Z}$.
(b) Prove that if $f \in V_1$, then $f \in V_0$ if and only if $f \perp \psi_{0,k}$ for all $k \in \mathbb{Z}$.
(c) Write down and prove appropriate generalisations of parts (a) and (b) to the case of a function f at any 'rung' on the ladder of spaces $\{V_j\}$.

6.24 Let ϕ be the scaling function of a multiresolution analysis $\{V_j\}$ and suppose that $\mathrm{isupp}\,\phi = [0, 2N - 1]$ for some $N \geq 1$. Consider $f \in V_0$.

(a) Suppose that $\mathrm{isupp}\,f = [a, b]$, where $a < b$. Show that $a, b \in \mathbb{Z}$ and $b - a \geq 2N - 1$. Show further that the only possible non-zero terms $\langle f, \phi_{0,k} \rangle$ for $k \in \mathbb{Z}$ are those for $k = a, a + 1, \ldots, b - 2N + 1$.
(b) Formulate and prove analogous statements for the cases when ϕ is supported in a semi-infinite interval $(-\infty, b]$ or $[a, \infty)$.

6.25 In this exercise and the next, we will use characteristic functions in the following way (see Exercise 6.18 for the definition and notation). If f is a function on $\mathbb{R}$ and S is a subset of $\mathbb{R}$, then $f\chi_S$ is the function that agrees with f at all points of S and has value 0 at all other points; thus, $f\chi_S$ 'zeroes out' f, except on S.

Let ϕ be the scaling function of a multiresolution analysis $\{V_j\}$ and suppose that $\mathrm{isupp}\,\phi = [0, 2N - 1]$ for some N. Consider $f \in V_0$. Show that if f is identically 0 on an open interval (a, b), where $b - a \geq 2N - 1$, then the functions $f\chi_{(-\infty,a]}$ and $f\chi_{[b,\infty)}$ belong to V_0. Generalise suitably to the case of $f \in V_j$, for arbitrary j.

6.26 Let ϕ be the scaling function of a multiresolution analysis $\{V_j\}$ and suppose that $\mathrm{isupp}\,\phi = [0, 2N - 1]$ for some N, so that the corresponding wavelet ψ satisfies $\mathrm{isupp}\,\psi = [-N + 1, N]$. Suppose that $f \in V_0$ and that f is identically 0 on some open interval (a, b). The aim of this exercise is to show, in several steps, that $f\chi_{(-\infty,a]}$ and $f\chi_{[b,\infty)}$ belong to V_0. Fix j such that $N/2^j \leq b - a$.

(a) Use Exercise 6.25 to show that $f\chi_{(-\infty,a]}$ and $f\chi_{[b,\infty)}$ belong to V_j.
(b) Use part (a) of Exercise 6.23 to show that $\langle f, \psi_{j-1,k} \rangle = 0$ for all k.
(c) By examining the series expansion of $f\chi_{(-\infty,a]}$ with respect to the orthonormal basis $\{\phi_{j,k} : k \in \mathbb{Z}\}$ in V_j and using the wavelet equation (6.7), show that $\langle f\chi_{(-\infty,a]}, \phi_{j,k} \rangle = 0$ whenever $2k \geq \ell_0$, where $\ell_0 = \lfloor 2^j a \rfloor$, the greatest integer less than or equal to $2^j a$.
(d) Show similarly that $\langle f\chi_{[b,-\infty)}, \phi_{j,k} \rangle = 0$ whenever $2k \leq \ell_1 - 2$, where $\ell_1 = \lceil 2^j b \rceil$, the smallest integer greater than or equal to $2^j b$.

(e) Deduce using the previous parts of the exercise that at least one of the terms $\langle f\chi_{[b,-\infty)}, \phi_{j,k}\rangle$ and $\langle f\chi_{[b,-\infty)}, \phi_{j,k}\rangle$ equals 0 for every $k \in \mathbb{Z}$.

(f) Conclude that $f\chi_{(-\infty,a]}$ and $f\chi_{[b,\infty)}$ belong to V_{j-1}.

(g) Repeat the argument, or, strictly, use induction, to conclude that $f\chi_{(-\infty,a]}$ and $f\chi_{[b,\infty)}$ belong to V_0.

6.27 Using the assumptions and the conclusion of Exercise 6.26, show that the support of ϕ contains no open interval on which ϕ is identically 0. Deduce that the support of ϕ, taken here to be the *closure* of the set of points at which ϕ is non-zero, is equal to isupp ϕ. (We have previously taken the support to be precisely the set of points at which ϕ is non-zero; using this convention, isupp ϕ is equal to the closure of the support of ϕ.)

6.28 Here we use again the assumptions and conclusion of Exercise 6.26. For any interval I, denote by $V_j(I)$ the vector space consisting of the restrictions to I of the functions in V_j.

(a) Show that the restrictions of the functions $\phi(x+k)$ to the interval $[0, 1]$, for $k = 0, 1, \ldots, 2N-2$, are linearly independent in $V_0([0, 1])$. (Hint: Suppose that there are constants $c_0, c_1, \ldots, c_{2N-2}$ such that

$$\sum_{k=0}^{2N-2} c_k \phi(x+k) = 0$$

for all $x \in [0, 1]$. Deduce using Exercise 6.26 that the functions

$$\chi_{[2-2N,0]} \sum_{k=0}^{2N-2} c_k \phi(x+k) \qquad \text{and} \qquad \chi_{[1,2N-1]} \sum_{k=0}^{2N-2} c_k \phi(x+k)$$

are in V_0, and then by part (a) of Exercise 6.24 that they are both identically 0. Finally, use the linear independence of the functions $\phi(x+k)$ in V_0.)

(b) Deduce that the restricted functions of part (a) form a basis for the vector space $V_0([0, 1])$, which therefore has dimension $2N-1$.

(c) Show similarly that if $0 \le a < b \le 1$, then the corresponding restrictions to $[a, b]$ form a basis for $V_0([a, b])$.

(d) Find a basis for the space $V_0([0, 2])$, and its dimension.

(e) Consider the collection of spaces $V_j([0, 1])$ for $j \in \mathbb{Z}$. Show that $V_j([0, 1]) \subseteq V_{j+1}([0, 1])$ for all j, and find a basis for $V_j([0, 1])$ and its dimension.

6.29 (a) Under the assumptions of Exercise 6.26, show that if $f \in V_0$ and isupp f has support length less than or equal to $4N-2$, then there is no open interval I within isupp f on which f is identically 0.

(b) What can be said when isupp f has length greater than $4N-2$?

6.30 This and the following two exercises are about wavelets in $\mathbb{R}^2$. (This means that we will be working in the space $L^2(\mathbb{R}^2)$. We have not previously discussed or mentioned L^2 spaces over domains other than intervals of the real line, but the basic ideas and results are in close analogy to those for $L^2(\mathbb{R})$ and no more will be needed here. Specifically, the elements of $L^2(\mathbb{R}^2)$ are the square-integrable real-valued functions on $\mathbb{R}^2$, that is, the functions $f : \mathbb{R}^2 \to \mathbb{R}$ such that $\int_{\mathbb{R}^2} f^2$ exists and is finite, and $L^2(\mathbb{R}^2)$ is a Hilbert space with the naturally defined inner product.)

Now let H denote the usual Haar wavelet on $\mathbb{R}$. Consider the functions H_0, H_1 and H_2 defined on $\mathbb{R}^2$ by

$$H_0(x, y) = H(x) \cdot H(y),$$
$$H_1(x, y) = H(x) \cdot \chi_{[0,1]}(y),$$
$$H_2(x, y) = \chi_{[0,1]}(x) \cdot H(y)$$

(where we again use the characteristic function notation introduced in Exercise 6.18). Show that the collection

$$\left\{ 2^j H_i(2^j x - k, 2^j x - \ell) : i = 0, 1, 2; \ j, k, l \in \mathbb{Z} \right\}$$

is an orthonormal set in $L^2(\mathbb{R}^2)$. (It seems reasonable to think of H_0, H_1 and H_2 as 'the Haar wavelets in $\mathbb{R}^2$'.)

6.31 Continuing the previous question, work out the structure of the corresponding multiresolution analysis in $L^2(\mathbb{R}^2)$, and use the fact that H is a wavelet in $L^2(\mathbb{R})$ to deduce that the collection of functions in the previous question is an orthonormal basis for $L^2(\mathbb{R}^2)$.

6.32 We can take the idea of the last two questions further. Write ϕ for the scaling function of the Haar wavelet and ψ for the Haar wavelet, and observe that the functions H_0, H_1 and H_2 can be rewritten as

$$\psi_0(x, y) = \psi(x) \cdot \psi(y), \quad \psi_1(x, y) = \psi(x) \cdot \phi(y), \quad \psi_2(x, y) = \phi(x) \cdot \psi(y).$$

Show that if ϕ and ψ are the scaling function and the wavelet for *any* multiresolution analysis in $L^2(\mathbb{R})$, then a multiresolution analysis for $L^2(\mathbb{R}^2)$ can be constructed following the procedure used above in the case of the Haar wavelet. (Further generalisations can be pursued: extend the idea to $\mathbb{R}^n$ for any n; allow independent multiresolution analyses, scaling functions and wavelets to be used in each coordinate.)

7

The Daubechies Wavelets

7.1 Introduction

Let us assess our progress towards the goal of developing the theory needed for the construction of wavelets of bounded support, and for the construction specifically of the Daubechies wavelets. In Chapter 6, we introduced the notion of a multiresolution analysis, including the associated scaling function ϕ, and we saw how these can be used to construct a wavelet ψ. On the assumption that ϕ has bounded support, which we made for the later results of the chapter, Corollary 6.13, Corollary 6.14 and equation (6.13) gave us the following detailed information.

- Without loss of generality we can assume that $\mathrm{isupp}(\phi) = [0, 2N - 1]$ for some $N \in \mathbb{N}$ and that ϕ satisfies the scaling equation

$$\phi(x) = \sum_{k=0}^{2N-1} a_k \phi(2x - k) \tag{7.1}$$

  for a suitable sequence $\{a_k : k \in \mathbb{Z}\}$ of scaling coefficients.
- The scaling coefficients a_k satisfy

$$a_0, a_{2N-1} \neq 0 \quad \text{and} \quad a_k = 0 \text{ for } k < 0 \text{ and } k > 2N - 1 \tag{7.2}$$

  and are constrained by the fundamental relations

$$\left.\begin{array}{c} \displaystyle\sum_{k=0}^{2N-2\ell-1} a_k a_{k+2\ell} = 2\delta_{0,\ell} \quad \text{for } \ell = 0, 1, \ldots, N - 1, \\[3mm] \displaystyle\sum_{k=0}^{2N-1} a_k = 2. \end{array}\right\} \tag{7.3}$$

- If ψ is the corresponding wavelet, then $\mathrm{isupp}\,\psi = [-N + 1, N]$ and ψ is defined by the wavelet equation

$$\psi(x) = \sum_{k=0}^{2n-1} (-1)^{k-1} a_k \phi(2x + k - 1). \tag{7.4}$$

This information forms the background to our discussion in this chapter, and we will refer to it repeatedly. In particular, the assumption of bounded support remains in force throughout.

For a given value of the parameter N, we might refer to the sequences $\{a_k : k \in \mathbb{Z}\}$ that satisfy the constraints (7.2) and the fundamental relations (7.3) as the sequences of *potential scaling coefficients* for that value of N. One main aim of this chapter is to determine when a sequence of potential scaling coefficients is a sequence of *actual* scaling coefficients – that is, when we can use it to construct a function ϕ that satisfies the corresponding scaling equation and is the scaling function of a multiresolution analysis, and therefore, by the results of Chapter 6, gives rise to a wavelet. A second important aim, bearing in mind that our ultimate goal is specifically to construct the Daubechies wavelets, is to find constraints additional to those of (7.3) whose imposition on a sequence of scaling coefficients will be sufficient to guarantee that the corresponding wavelet has some desired degree of smoothness.

The outline of the chapter is as follows. In Section 7.2, we briefly discuss some of the geometric properties of the collection of sequences of potential scaling coefficients; although these ideas stand somewhat apart from our main concerns in the chapter, they are interesting enough to justify a short digression.

Our aim in Sections 7.3 and 7.4 is to find the additional constraints mentioned just above that determine the Daubechies scaling coefficients for a given value of N. This involves two steps. In Section 7.3 we derive certain equations involving the scaling coefficients that reflect the smoothness that we desire in the Daubechies wavelets; we will see that this drastically narrows the collection of sequences of scaling coefficients within which we need to search from an infinite set to a small finite set. Following this, a further constraint introduced in Section 7.4 narrows the search-space again – leaving, in fact, precisely one sequence of scaling coefficients, which is the Daubechies sequence for the value of N under consideration. The route followed in these sections is equivalent to that taken by Daubechies herself.

In Section 7.5 introduces two important results, Theorems 7.4 and 7.5, from the general theory of wavelets. Although the statements of these results are easy enough to understand, they are both cases where the required arguments are out of reach of the methods adopted in this book (in the second case, in fact, we do not even state the result completely rigorously, because doing so would unavoidably involve using the language of general measure theory). The results tell us among other things that 'most' sequences of potential scaling coefficients, including the Daubechies sequence in particular, give rise to a function ϕ that satisfies the corresponding scaling equation and then to a multiresolution analysis of which ϕ is the scaling function. (Note that two logically separate issues are involved here: a function ϕ can satisfy the scaling equation and yet fail to be the scaling function of any multiresolution analysis. Part (c) of Exercise 7.3 gives an example of this phenomenon.)

The remaining sections of the chapter are devoted to a reasonably detailed discussion of the properties of the Daubechies wavelet and scaling function for the case $N = 2$ in Section 7.6 and a much briefer discussion for general N in Section 7.7.

7.2 The Potential Scaling Coefficients

Fix $N \in \mathbb{N}$. Consider all the sequences $\{a_k : k \in \mathbb{Z}\}$ that satisfy the constraints (7.2) and (7.3) – that is, all the sequences of potential scaling coefficients. If for each such sequence we disregard the terms for $k < 0$ and $k > 2N - 1$, which must be 0, we can 'package' the remaining terms into a $2N$-tuple $(a_0, \ldots, a_{2N-1})$, and regard this as a vector in $\mathbb{R}^{2N}$. If we denote the collection of all such $2N$-tuples by A_N, then $A_N \subseteq \mathbb{R}^{2N}$, and we can enquire how to describe this set geometrically as a subset of Euclidean space. A great deal of very interesting information is known about this. Let us briefly examine the first two values of N.

For $N = 1$, the fundamental relations reduce to the two equations $a_0^2 + a_1^2 = 2$ and $a_0 + a_1 = 2$, and Lemma 6.15 tells us that these have the unique solution $a_0 = a_1 = 1$, so the set A_1 consists of the single point $(1, 1) \in \mathbb{R}^2$. This is of course consistent with the fact (see Theorem 6.17) that the Haar scaling function is the unique scaling function ϕ for which $\text{isupp}\,\phi = [0, 1]$, up to a change of sign. (Note that the fact that the fundamental relations have a unique solution does not imply that only one function ϕ satisfies the corresponding scaling equation (7.1); for example, it is obvious on inspection that if ϕ satisfies the equation then so does $-\phi$. Theorem 7.5 below deals with the issues raised by this observation.)

For $N = 2$, the fundamental relations yield three equations for the four scaling coefficients a_0, a_1, a_2 and a_3, so we might reasonably anticipate that A_2 will be a 1-dimensional subset of $\mathbb{R}^4$. This is indeed the case. It turns out that

A_2 can be parametrised in the form

$$\left.\begin{aligned}
a_0 &= \tfrac{1}{\sqrt{2}}\cos\alpha + \tfrac{1}{2}, \\
a_1 &= \tfrac{1}{\sqrt{2}}\sin\alpha + \tfrac{1}{2}, \\
a_2 &= -\tfrac{1}{\sqrt{2}}\cos\alpha + \tfrac{1}{2}, \\
a_3 &= -\tfrac{1}{\sqrt{2}}\sin\alpha + \tfrac{1}{2}
\end{aligned}\right\} \tag{7.5}$$

for $0 \le \alpha < 2\pi$, and that A_2 is a circle lying in a (2-dimensional) plane in $\mathbb{R}^4$. (This and more is covered in Exercises 7.1 and 7.2.)

Parametrisations of A_N for $N \ge 3$ are also known, again in terms of trigonometric functions, but they become increasingly complicated as N increases.

Although, as noted in Section 7.1, we will shortly add extra constraints on the scaling coefficients (at least when $N > 1$), we cannot expect to be able to make *arbitrary* choices of the extra constraints, since that might give us an inconsistent system of equations. This is clear enough from our experience in solving systems of linear equations, but we can also see it geometrically. Observe that if we take $\ell = 0$ in the first of the fundamental relations, we get the equation

$$\sum_{k=0}^{2N-1} a_k^2 = 2;$$

this is the equation of a sphere (or, more correctly, a **hypersphere**) of radius $\sqrt{2}$ in $\mathbb{R}^{2N}$. All solutions to the system must lie on this sphere, so the set A_N is always a *bounded* region in $\mathbb{R}^{2N}$ (specifically, $|a_k| \le \sqrt{2}$ for all k). Hence extra constraints that forced any of the scaling coefficients to have large absolute values would lead to inconsistency.

Thus any extra constraints used will need to be determined in a *systematic* way; and, as Section 7.1 suggests, they will be found through an analysis of the relation between the scaling coefficients and the desired smoothness properties of the associated wavelet. This analysis is the main topic of the next section.

7.3 The Moments of a Wavelet

Definition 7.1 If f is a function on $\mathbb{R}$ and m is a non-negative integer, then the integral

$$\int_{-\infty}^{\infty} x^m f(x)\, dx,$$

if it exists, is called the **moment of** f **of order** m. For a non-negative integer M, we say that f has **vanishing moments to order** M if the moments of f of orders $m = 0, 1, \ldots, M$ are all 0.

We will see shortly how the imposition of vanishing moment conditions on a wavelet leads to specific constraints on the scaling coefficients that will ultimately allow us to see how to construct the Daubechies wavelets, but we will look first at some of the theoretical and practical considerations that help to explain why moment conditions are appropriate and useful ones to impose when smoothness of a wavelet is desired.

There are at least two aspects to this. The first is that in some applications it is convenient to work with a wavelet ψ that has a number of vanishing moments, because this will lead to many wavelet coefficients that are equal or close to 0. Suppose that ψ has vanishing moments to order M. Then if f were a polynomial of degree $n \leq M$, its wavelet coefficients would all be 0. In detail, if $f(x) = c_0 + c_1 x + \cdots + c_n x^n$, say, then

$$\langle f, \psi \rangle = c_0 \int_{-\infty}^{\infty} \psi(x)\,dx + c_1 \int_{-\infty}^{\infty} x\psi(x)\,dx + \cdots + c_n \int_{-\infty}^{\infty} x^n \psi(x)\,dx = 0,$$

and similarly for $\langle f, \psi_{j,k} \rangle$ for arbitrary $j, k \in \mathbb{Z}$. Now things cannot be quite as simple as this, because the only polynomial function *that is in* $L^2(\mathbb{R})$ is the function that is identically 0. But a function $f \in L^2(\mathbb{R})$ may certainly be *piecewise polynomial* – for example, given by a polynomial on a bounded interval and identically 0 everywhere else. In such a case, if the polynomials involved were of degree at most M then the inner product $\langle f, \psi_{j,k} \rangle$ would be 0 whenever the support of $\psi_{j,k}$ lay entirely within a region where f was given by a single polynomial expression, and non-zero values could only occur where the support straddled the boundary between two regions where f was given by two different polynomial expressions. Extending this idea a little, f might be given, piecewise again, as the sum of a polynomial and another more complicated function, and in this case the polynomial terms would make no contribution to the wavelet coefficients, except again at boundaries. All this implies that vanishing moments may in principle lead to a saving in the quantity of data that needs to be dealt with.

Somewhat more than this is true. There are results, though we will not state them here, that precisely quantify the way in which if a wavelet ψ has many vanishing moments then the wavelet coefficients $\langle f, \psi_{j,k} \rangle$ of a smooth function f tail off quickly as $j \to \infty$. Again, this is desirable in some applications, such as compression, since small wavelet coefficients are ones that, in principle, might be set to 0, and whose absence might not significantly affect reconstruction of f.

There is a second rather different reason for considering the imposition of vanishing moment conditions which relates to the smoothness of the wavelets themselves. The following result gives information on the relation between vanishing moments and the smoothness of a wavelet.

Theorem 7.2   *Suppose that $\psi \in L^2(\mathbb{R})$ has the property that the collection of scaled, dilated, translated functions $\{\psi_{j,k} : j, k \in \mathbb{Z}\}$ is an orthonormal set. If ψ has bounded support and is in $C^M(\mathbb{R})$ for some $M \geq 0$, then ψ has vanishing moments up to order M; that is,*

$$\int_{-\infty}^{\infty} x^m \psi(x)\, dx = 0 \quad for\ m = 0, 1, \ldots, M.$$

Note that the theorem does not require that ψ is the wavelet corresponding to a multiresolution analysis; in fact, it does not even require that the $\psi_{j,k}$ form an orthonormal basis for $L^2(\mathbb{R})$, but merely that they form an orthonormal set. However, the result certainly applies to the case of a wavelet of bounded support arising from a multiresolution analysis. (The theorem also holds in a more general form than ours, without the assumption of bounded support, but more complicated hypotheses then need to be assumed. We will in any case omit the proof.)

An immediate and highly significant implication of this result is, informally, that if we are looking for smooth wavelets then we will only be able to find them among functions that have vanishing moments. More precisely, to find wavelets that have continuous derivatives up to order M we need to look only among the functions whose moments vanish up to order M. As noted earlier, this is the approach followed by Daubechies; in fact, the Daubechies wavelets can be characterised as those which have the largest number of vanishing moments consistent with their support length.

7.3.1 Vanishing Moment Conditions

We will now investigate how vanishing moment conditions give rise to constraints on scaling coefficients. Consider a multiresolution analysis with a scaling function ϕ that has bounded support. With this condition, Theorem 6.16 shows that

$$\int_{-\infty}^{\infty} \psi = 0 \qquad \text{and} \qquad \sum_{k \in \mathbb{Z}} (-1)^k a_k = 0; \qquad (7.6)$$

in fact, the last few lines of the proof show that

$$\int_{-\infty}^{\infty} \psi = \tfrac{1}{2} I \sum_{k \in \mathbb{Z}} (-1)^{k-1} a_k,$$

where $I = \int_{-\infty}^{\infty} \phi$, and since part (i) of the same result tells us that $I \neq 0$, the integral and the sum in (7.6) are either both zero or both non-zero. But $\int_{-\infty}^{\infty} \psi$ is just the zeroth moment of ψ, and so, in the bounded support case, *the zeroth moment of ψ is always* 0. Hence if we impose vanishing moment conditions, we will only gain new constraints on the scaling coefficients for moments beyond the zeroth.

Suppose now that $\text{isupp}\,\phi = [0, 2N - 1]$ for some $N \in \mathbb{N}$, and that ϕ and the corresponding wavelet ψ satisfy the usual scaling equation and wavelet equation

$$\phi(x) = \sum_{k=0}^{2N-1} a_k \phi(2x - k) \qquad \text{and} \qquad \psi(x) = \sum_{k=0}^{2N-1} (-1)^{k-1} a_k \phi(2x + k - 1),$$

respectively. Then we know that the scaling coefficients $a_0, a_1, \ldots, a_{2N-1}$ satisfy the $N + 1$ fundamental relations. Since we need to solve for $2N$ unknowns and have only $N + 1$ equations, it is natural as a first step to seek for $N - 1$ suitable additional constraints to impose.

➤ Though this is indeed a natural first step, we noted in Section 7.1 that the imposition of one further constraint is necessary to determine $a_0, a_1, \ldots, a_{2N-1}$ uniquely, as we will need to do if we are to construct the Nth Daubechies wavelet (see Section 7.4 below).

A single quadratic equation can of course have two real solutions, depending on the choice of sign made in a certain square root, and a similar phenomenon occurs in a more complex way in the solution of a set of equations some of which are linear and some quadratic. We plan to add $N - 1$ new constraints to the $N + 1$ fundamental relations, giving $2N$ constraints on the $2N$ unknowns $a_0, a_1, \ldots, a_{2N-1}$, but since there are N quadratic equations among the fundamental relations, $2N$ equations will not be quite enough to determine a unique solution.

For the Nth Daubechies scaling function and wavelet, we obtain the extra $N - 1$ constraints that we want by setting the ℓth moment of ψ to 0 for $1 \le \ell \le N - 1$ (note that we omit the case $\ell = 0$, since this would give us the zeroth moment, which as we saw above is always 0). The following result tells us explicitly what constraints on the scaling coefficients these conditions lead to.

Theorem 7.3 *If* $\int_{-\infty}^{\infty} x^\ell \psi(x)\,dx = 0$ *for* $1 \le \ell \le N - 1$, *then*

$$\sum_{k=0}^{2N-1} (-1)^k k^\ell a_k = 0$$

for $1 \le \ell \le N - 1$.

The complete proof, which proceeds by induction, is not difficult, and is left as an exercise (see Exercise 7.5). We will outline the proof for $N = 2$, since this already shows most of the features of the general case.

For $N = 2$, just one additional constraint is imposed on the scaling coefficients, since $N = 2$ corresponds to setting the first moment, $\int_{-\infty}^{\infty} x\psi(x)\,dx$, to 0. The wavelet equation in this case is

$$\psi(x) = a_0\phi(2x - 1) - a_1\phi(2x) + a_2\phi(2x + 1) - a_3\phi(2x + 2),$$

and multiplying through this equation by x and integrating gives

$$0 = \int_{-\infty}^{\infty} x\psi(x)\,dx$$
$$= a_0\int_{-\infty}^{\infty} x\phi(2x - 1)\,dx - a_1\int_{-\infty}^{\infty} x\phi(2x)\,dx$$
$$+ a_2\int_{-\infty}^{\infty} x\phi(2x + 1)\,dx - a_3\int_{-\infty}^{\infty} x\phi(2x + 2)\,dx.$$

We now change variables in the four integrals in the last expression. Setting $y = 2x$ in the first integral gives

$$\int_{-\infty}^{\infty} x\phi(2x - 1)\,dx = \tfrac{1}{4}\int_{-\infty}^{\infty} y\phi(y - 1)\,dy,$$

and if we make the changes of variable $y = 2x + 1$, $y = 2x + 2$ and $y = 2x + 3$ in the other three integrals, respectively, we obtain

$$\int_{-\infty}^{\infty} (a_0 - a_1 + a_2 - a_3)y\phi(y - 1)\,dy + \int_{-\infty}^{\infty} (a_1 - 2a_2 + 3a_3)\phi(y - 1)\,dy = 0.$$

But the second equation in (7.6) shows that

$$a_0 - a_1 + a_2 - a_3 = \sum_{k\in\mathbb{Z}}(-1)^k a_k = 0,$$

and since $\int_{-\infty}^{\infty} \phi(y - 1)\,dy = \int_{-\infty}^{\infty} \phi(x)\,dx \neq 0$ by Theorem 6.16, we have

$$a_1 - 2a_2 + 3a_3 = 0,$$

which, after a change of sign, is the equation given by Theorem 7.3 when $N = 2$.

▶ Note that the changes of variable used above are not the most obvious ones. The more obvious choice of $y = 2x - 1$ in the first integral, and so on, will work perfectly well, but the substitutions chosen above lead most directly to the desired answer; the more obvious choices make a little extra manipulation necessary at the end of the argument.

Listing together the $N + 1$ fundamental relations and the $N - 1$ new relations we have just derived, we arrive at the following system of $2N$ equations for the $2N$ scaling coefficients $a_0, a_1, \ldots, a_{2N-1}$ of the Nth Daubechies scaling function:

$$\left.\begin{array}{c}
\displaystyle\sum_{k=0}^{2N-2\ell-1} a_k a_{k+2\ell} = 2\delta_{0,\ell} \quad \text{for } \ell = 0, 1, \ldots, N-1, \\[1em]
\displaystyle\sum_{k=0}^{2N-1} a_k = 2, \\[1em]
\displaystyle\sum_{k=0}^{2N-1} (-1)^k k^\ell a_k = 0 \quad \text{for } \ell = 1, 2, \ldots, N-1.
\end{array}\right\} \qquad (7.7)$$

We will refer to this as the **augmented system of fundamental relations**.

One small but worthwhile improvement to the system (7.7) is possible. Exercises 7.6 and 7.7 show that one of the quadratic equations in (7.7) can be replaced by a linear equation to give an *equivalent* system, that is, a system that has exactly the same set of solutions. Specifically, when we take $\ell = 0$ in the first subsystem of (7.7), we get the quadratic equation

$$\sum_{k=0}^{2N-1} a_k^2 = 2,$$

and this can be replaced by the linear equation

$$\sum_{k=0}^{2N-1} (-1)^k a_k = 0.$$

If we make this replacement in the system (7.7), we obtain a new system with $N-1$ quadratic equations and $N+1$ linear equations, which is an improvement (if only a modest one), since linear equations are easier to deal with than quadratic equations.

Note, however, that by the discussion at the start of the subsection the new linear equation is equivalent to the zeroth moment condition, which we specifically excluded from the moment conditions in the third subsystem of (7.7), by starting from the index value $\ell = 1$. We can therefore remove the old quadratic equation and incorporate the new linear equation into the augmented system by simply eliminating the case $\ell = 0$ from the first subsystem and including it in the third subsystem. Thus our improved and final system of $2N$ equations, which again we will refer to as the augmented system, is the following:

$$\left.\begin{array}{c} \displaystyle\sum_{k=0}^{2N-2\ell-1} a_k a_{k+2\ell} = 2\delta_{0,\ell} \quad \text{for } \ell = 1, 2, \ldots, N-1, \\[2ex] \displaystyle\sum_{k=0}^{2N-1} a_k = 2, \\[2ex] \displaystyle\sum_{k=0}^{2N-1} (-1)^k k^\ell a_k = 0 \quad \text{for } \ell = 0, 1, \ldots, N-1. \end{array}\right\} \tag{7.8}$$

(Note carefully that the new system differs from the old *only* in the ranges of values taken by the index ℓ.)

7.4 The Daubechies Scaling Coefficients

Taking $N = 1$ in the augmented system (7.8) yields the system

$$\left.\begin{array}{c} a_0 + a_1 = 2, \\ a_1 - a_2 = 0, \end{array}\right\}$$

which trivially has the unique solution $a_0 = a_1 = 1$; these are the Daubechies scaling coefficients for $N = 1$. (Compare this with the start of Section 7.2, where in effect we solved for the same coefficients using the augmented system in its initial form (7.7).)

For $N = 2$, the augmented system (7.8) is

$$\left.\begin{array}{c} a_0 a_2 + a_1 a_3 = 0, \\ a_0 + a_1 + a_2 + a_3 = 2, \\ a_0 - a_1 + a_2 - a_3 = 0, \\ a_1 - 2a_2 + 3a_3 = 0. \end{array}\right\} \tag{7.9}$$

Solving these equations (see Exercise 7.8), we find

$$\left.\begin{array}{cc} a_0 = \frac{1}{4}(1 \pm \sqrt{3}), & a_1 = \frac{1}{4}(3 \pm \sqrt{3}), \\[1ex] a_2 = \frac{1}{4}(3 \mp \sqrt{3}), & a_3 = \frac{1}{4}(1 \mp \sqrt{3}). \end{array}\right\} \tag{7.10}$$

Note that because one of our equations is quadratic, we do not obtain a unique solution, but rather two solutions that differ in the sign of the radical term. Note also, however, that the two solutions are mirror-images: each sequence is the reverse of the other. It follows (see Exercise 7.9) that scaling functions and wavelets constructed using these sequences are also mirror-images, so we lose nothing essential by restricting ourselves to just one of the two possibilities. We will choose the $+$ sign in the expression for a_0 (we discuss the basis for this specific choice below), which of course fixes the other signs. Thus, we have

$$a_0 = \tfrac{1}{4}(1 + \sqrt{3}), \quad a_1 = \tfrac{1}{4}(3 + \sqrt{3}),$$
$$a_2 = \tfrac{1}{4}(3 - \sqrt{3}), \quad a_3 = \tfrac{1}{4}(1 - \sqrt{3}),$$
$$\left. \right\} \tag{7.11}$$

or, to 10 places,

$$a_0 = 0.6830127019, \quad a_1 = 1.1830127019,$$
$$a_2 = 0.3169872981, \quad a_3 = -0.1830127019.$$

These are the Daubechies scaling coefficients for $N = 2$.

For $N = 3$, the system (7.8) is

$$
\left.
\begin{aligned}
a_0 a_2 + a_1 a_3 + a_2 a_4 + a_3 a_5 &= 0, \\
a_0 a_4 + a_1 a_5 &= 0, \\
a_0 + a_1 + a_2 + a_3 + a_4 + a_5 &= 2 \\
a_0 - a_1 + a_2 - a_3 + a_4 - a_5 &= 0, \\
a_1 - 2a_2 + 3a_3 - 4a_4 + 5a_5 &= 0 \\
a_1 - 4a_2 + 9a_3 - 16a_4 + 25a_5 &= 0.
\end{aligned}
\right\} \tag{7.12}
$$

In this case, there are four solutions, but two are real and two are complex (see Exercise 7.10). Since we are interested only in real scaling coefficients, we can disregard the complex solutions, and then the two remaining real solutions are related like those for $N = 2$: one sequence of values is the mirror-image of the other. One of these two sequences, chosen as discussed below, consists of the Daubechies scaling coefficients for $N = 3$.

For $N = 4$, there are 8 solutions, 4 real and 4 complex, and for $N = 5$, there are 16 solutions, 4 real and 12 complex, and in both cases the 4 real solutions form two mirror-image pairs. For general N, the number of real solutions to the system (7.8) is $2^{\lfloor N/2 \rfloor}$, where $\lfloor x \rfloor$ (read as the 'floor of x'), for a real number x, is the largest integer less than or equal to x.

Exact expressions are known for the real solutions for $N = 3$, 4 and 5; they are fairly complicated for $N = 3$ and very complicated for $N = 4$ and 5. However, computing the solutions numerically to any desired accuracy using a suitable computer system is straightforward, and numerical values are all that is required for applications. (Exercise 7.10 outlines a method for computing the exact expressions for $N = 3$; the expressions themselves are given in Subsection 8.5.3, though they are computed there by an entirely different method.)

For $N \geq 2$, the real solutions always occur in mirror-image pairs (note that for $N = 1$, the unique solution is its own mirror-image), and for $N \geq 4$, there is more than one mirror-image pair, so the question arises of how to pick out the Daubechies scaling coefficients from the set of solutions.

The specific choice that is made to single out the Daubechies scaling coefficients is determined as follows. Suppose that the sequence

$$a_0(\ell), a_1(\ell), \ldots, a_{2N-1}(\ell)$$

forms a solution to the system of equations (7.8) for each value of ℓ in a certain range. Then the particular sequence $a_0(\ell_0), a_1(\ell_0), \ldots, a_{2N-1}(\ell_0)$ is the **extremal phase** or **minimum phase solution** if

$$\sum_{k=0}^{n} a_k(\ell)^2 \le \sum_{k=0}^{n} a_k(\ell_0)^2 \quad \text{for } n = 0, 1, \ldots, 2N - 1$$

for each value of ℓ. Roughly speaking, the extremal phase solution is the one whose largest values occur as early as possible in the sequence; it is sometimes said to be 'front-loaded'. (The *extremal phase* and *minimum phase* terminology comes from the theory of signal processing.)

If for each fixed N we make the extremal phase choice, then by definition we obtain the Daubechies scaling coefficients $a_0, a_1, \ldots, a_{2N-1}$, which are shown numerically for $N = 2, 3, \ldots, 7$, rounded to 10 decimal places, in Table 7.1. Confirmation that the extremal phase solution does indeed yield the values in the table for $N = 2$ and $N = 3$ is left for Exercise 7.11.

Table 7.1. *The Daubechies scaling coefficients $a_0, a_1, \ldots, a_{2N-1}$ for $N = 2, 3, \ldots, 7$*

N	k	a_k	N	k	a_k	N	k	a_k
2	0	0.6830127019	5	0	0.2264189826		8	−0.0446637483
	1	1.1830127019		1	0.8539435427		9	0.0007832516
	2	0.3169872981		2	1.0243269443		10	0.0067560624
	3	−0.1830127019		3	0.1957669613		11	−0.0015235338
				4	−0.3426567154			
3	0	0.4704672078		5	−0.0456011319	7	0	0.1100994307
	1	1.1411169158		6	0.1097026586		1	0.5607912836
	2	0.6503650005		7	−0.0088268001		2	1.0311484916
	3	−0.1909344156		8	−0.0177918701		3	0.6643724822
	4	−0.1208322083		9	0.0047174279		4	−0.2035138225
	5	0.0498174997					5	−0.3168350113
			6	0	0.1577424320		6	0.1008464650
4	0	0.3258034281		1	0.6995038141		7	0.1140034452
	1	1.0109457151		2	1.0622637599		8	−0.0537824526
	2	0.8922001382		3	0.4458313229		9	−0.0234399416
	3	−0.0395750262.		4	−0.3199865989		10	0.0177497924
	4	−0.2645071674		5	−0.1835180641		11	0.0006075150
	5	0.0436163005		6	0.1378880930		12	−0.0025479047
	6	0.0465036011		7	0.0389232097		13	0.0005002269
	7	−0.0149869893						

7.5 The Daubechies Scaling Functions and Wavelets

In this section, we present results that allow us to take the final step in the construction of the Daubechies scaling functions and wavelets, though most of the results are stated in forms that apply more generally.

In Subsection 7.5.1 we examine results which tell us that almost all choices of potential scaling coefficients yield a multiresolution analysis that has a scaling equation with the given potential scaling coefficients as its actual scaling coefficients; these results apply in particular in the case of the Daubechies scaling coefficients, thus guaranteeing existence of the Daubechies scaling functions and, consequently, the Daubechies wavelets as well. In Subsection 7.5.2, we briefly discuss some of the smoothness properties of the Daubechies scaling functions and wavelets. In Subsection 7.5.3, we examine results which at a theoretical level show how the Daubechies scaling functions can be constructed as the limit functions of suitable procedures of successive approximation, and at a practical level allow us to approximate values of the functions at specific points with as much accuracy as we desire, and in particular to plot their graphs. As noted in Section 7.1, most of the results of this section are stated without proof.

7.5.1 Existence

Exercise 6.5 asked for a proof of the fact that if ϕ is the scaling function for a multiresolution analysis $\{V_j\}$, then

$$V_j = \left\{ \sum_{k \in \mathbb{Z}} \alpha_k \phi_{j,k} : \sum_{k \in \mathbb{Z}} \alpha_k^2 \text{ converges} \right\} \tag{7.13}$$

for each $j \in \mathbb{Z}$. In other words, if ϕ is known to be the scaling function of a multiresolution analysis, then the spaces V_j can be reconstructed from ϕ alone.

This raises the following question: If a function $\phi \in L^2(\mathbb{R})$ is given, under what conditions is ϕ the scaling function of a multiresolution analysis? That is, under what conditions on ϕ do the equations (7.13) define closed subspaces V_j that together with ϕ form a multiresolution analysis?

Since condition (MRA6) says that the translates $\{\phi_{0,k} : k \in \mathbb{Z}\}$ form an orthonormal basis for V_0, it is to be expected that one of the conditions that will have to be assumed for ϕ is that $\{\phi_{0,k}\}$ is an orthonormal set in $L^2(\mathbb{R})$, since this is a property of ϕ in isolation, and the problem is therefore to determine what further properties of ϕ are needed. An answer is given by the following theorem.

Theorem 7.4   *Let $\phi \in L^2(\mathbb{R})$ and suppose that*

(i) *ϕ has bounded support,*
(ii) *$\{\phi_{0,k} : k \in \mathbb{Z}\}$ is an orthonormal set,*
(iii) *ϕ satisfies an equation of the form $\phi(x) = \sum_{k \in \mathbb{Z}} a_k \phi(2x - k)$ for some constants $\{a_k : k \in \mathbb{Z}\}$, and*
(iv) *$\int_{-\infty}^{\infty} \phi = 1$.*

Then the spaces $\{V_j : j \in \mathbb{Z}\}$ defined by (7.13) form a multiresolution analysis of which ϕ is the scaling function.

A few remarks about this result are worth noting.

- It is easy to see (and has been noted earlier) that if ϕ is the scaling function of a multiresolution analysis with spaces $\{V_j\}$, then $-\phi$ is also a scaling function for the same sequence of spaces. It follows that the condition that the integral of ϕ over $\mathbb{R}$ is 1 could be replaced by the condition that the integral is ± 1. (If the value were -1, we could simply apply the theorem as stated to the function $-\phi$.)

- With the change of the value of the integral to ± 1, the conditions of the theorem are necessary as well as sufficient for ϕ to be a scaling function (see Exercise 7.13).

- Although it is far from explicit in the statement of the theorem, the conclusion, that the given function ϕ is the scaling equation of a multiresolution analysis, along with the assumption that ϕ has bounded support, allows the derivation of all of the very detailed information summarised in Section 7.1.

The next theorem tells us that sequences of potential scaling coefficients almost always give rise to functions ϕ with the properties listed in the previous theorem.

Theorem 7.5   *Let $n \in \mathbb{N}$ and suppose that $a_0, \ldots, a_{2N-1}$ satisfy the fundamental relations, or, equivalently, that $(a_0, \ldots, a_{2N-1}) \in A_N$, as defined in Section 7.2. Then there exists a unique function $\phi \in L^2(\mathbb{R})$ that satisfies the scaling equation $\phi(x) = \sum_{k=0}^{2N-1} a_k \phi(2x-k)$ and has the property that $\int_{-\infty}^{\infty} \phi = 1$. This function ϕ also has the property that $\mathrm{isupp}(\phi) = [0, 2N - 1]$.*

Moreover, for almost all choices of $(a_0, \ldots, a_{2N-1}) \in A_N$, and in particular for the choice representing the Daubechies scaling coefficients, the function ϕ has the additional property that $\{\phi_{0,k} : k \in \mathbb{Z}\}$ is an orthonormal set.

➤ The phrase 'almost all' here requires comment. Although its sense is essentially that introduced in Section 3.3 and used in many places following, it is used here in a more general sense.

The definition of measure 0 in Section 3.3 captures the idea of a set being 'negligible' *as a subset of* $\mathbb{R}$. In the theorem above, its meaning is that the set of $2N$-tuples in A_N that do not yield a multiresolution analysis is 'negligible' when considered *as a subset of* A_N (which is in turn a subset of $\mathbb{R}^{2N}$). A more precise explanation requires development of the theory of Lebesgue measure, and is beyond the scope of this book.

Concretely, in the case $N = 1$, the 'negligible' subset of A_1 is simply the empty set, because A_1 consists of a single point that does yield a function ϕ with the property that $\{\phi_{0,k} : k \in \mathbb{Z}\}$ is an orthonormal set; this ϕ is of course the Haar scaling function. In the case $N = 2$ where, according to Section 7.2 and Exercise 7.2, A_2 is a circle in $\mathbb{R}^4$, there is exactly one exceptional point (which is identified in Exercise 7.3); the singleton set with that point as element is the claimed 'negligible' subset of A_2.

What is of most interest here is the use of the two theorems above in combination, as follows, since this tells us directly about the Daubechies case.

Theorem 7.6 *Let $N \in \mathbb{N}$. Then almost all choices of $a_0, \ldots, a_{2N-1}$ satisfying the fundamental relations yield a multiresolution analysis, including the scaling function, and its associated wavelet. This is the case in particular when $a_0, \ldots, a_{2N-1}$ are the Daubechies scaling coefficients, and in that case the corresponding scaling function and wavelet are, by definition, the Nth Daubechies scaling function and wavelet, respectively.*

We adopt the following notational convention:

*The symbols $_N\phi$ and $_N\psi$ denote respectively
the Nth Daubechies scaling function and wavelet.*

Although notation for these functions is far from standardised, our notation, with its rather unusual placement of the subscripts before the main symbols, was introduced by Daubechies herself.

7.5.2 Smoothness

Theorem 7.2 shows that wavelets with continuous derivatives up to order M must have vanishing moments up to order M, and this suggested the program followed above for constructing the Daubechies wavelets. Unfortunately, however, the converse of the theorem does not hold: it is not the case that vanishing moments to order M entail continuous derivatives to order M. We nevertheless wish to understand something of the smoothness properties of the Daubechies wavelets, both individually and asymptotically, and we summarise some of what is known in the following result. (More detailed information, including a more precise version of the asymptotic statement, is given in Section 7.7 below.)

Theorem 7.7 (i) $_1\phi$ and $_1\psi$, the Haar scaling function and wavelet, are not continuous, and hence do not lie in $C^m(\mathbb{R})$ for any $m \geq 0$.

(ii) $_2\phi$ and $_2\psi$ are continuous but not differentiable, that is, are in $C^0(\mathbb{R})$ but not $C^1(\mathbb{R})$.

(iii) $_3\phi$ and $_3\psi$ are continuously differentiable but are not twice differentiable, that is, are in $C^1(\mathbb{R})$ but not $C^2(\mathbb{R})$.

(iv) When N is large, $_N\phi$ and $_N\psi$ belong to $C^m(\mathbb{R})$ for an integer m which is approximately $N/5$.

7.5.3 Calculations and Graphs

Now that we know that it is possible in principle to pass from properly chosen scaling coefficients to a scaling function ϕ, we will examine two closely related ways of producing sequences of functions that better and better approximate ϕ. Both are derived in a straightforward way from the scaling equation that we want ϕ to satisfy, and their main difference lies in the choice of the initial approximation. Both will allow us in particular to compute better and better approximations to the Daubechies scaling functions.

Our first approach to finding ϕ as the limit, in a suitable sense, of a sequence of successive approximations, is as follows. For a fixed N, corresponding to the sequence of $2N$ scaling coefficients $a_0, a_1, \ldots, a_{2N-1}$, we are seeking $\phi \in L^2(\mathbb{R})$ that satisfies the equation

$$\phi(x) = \sum_{k=0}^{2N-1} a_k\phi(2x - k).$$

We start by picking a suitable simple function $\phi^{(0)}$. Then we define $\phi^{(1)}$ by setting

$$\phi^{(1)}(x) = \sum_{k=0}^{2N-1} a_k\phi^{(0)}(2x - k).$$

Note that we have $\phi^{(0)}$ on the right-hand side here and $\phi^{(1)}$ on the left, so there is no problem in making the definition (whatever our choice of $\phi^{(0)}$). We can then repeat the process, defining $\phi^{(2)}$ by setting

$$\phi^{(2)}(x) = \sum_{k=0}^{2N-1} a_k\phi^{(1)}(2x - k),$$

and then continue in a similar way, so that we recursively define an infinite sequence of functions $\{\phi^{(n)} : n \in \mathbb{N}\}$. In principle, a sequence defined in this kind of way may not converge, or if it does converge may not converge to anything useful, but it turns out that under fairly mild conditions, and in particular with a well-chosen starting function $\phi^{(0)}$, this sequence does in fact converge to the scaling function we are seeking. We formalise this information as follows.

Theorem 7.8 *Let $\phi^{(0)}$ be the Haar scaling function (thus $\phi^{(0)}(x) = 1$ for $0 \leq x < 1$ and $\phi^{(0)}(x) = 0$ elsewhere). Define*

$$\phi^{(n)}(x) = \sum_{k=0}^{2N-1} a_k \phi^{(n-1)}(2x - k)$$

for all $n \in \mathbb{N}$. Then under suitable conditions on $a_0, a_1, \ldots, a_{2N-1}$, the sequence $\{\phi^{(n)}\}$ converges both pointwise and in $L^2(\mathbb{R})$ as $n \to \infty$ to a function ϕ that satisfies all of the hypotheses of Theorem 7.4, and is therefore the scaling function of a multiresolution analysis, as in that theorem. The relevant conditions are satisfied in particular in the Daubechies case, and ϕ is then the Nth Daubechies scaling function $_N\phi$.

▶ Broadly similar methods of successive approximation are used in many fields of mathematics, both at the theoretical level and for numerical approximation of the solutions to specific problems. They may be used, for example, both to prove the existence of solutions to certain families of differential equations and to approximate the solutions of specific differential equations. In most cases, the same general issues arise: what the initial input to the approximation should be; whether and in what sense the process converges; and whether the limit, given convergence, is actually a solution to the original problem.

Of course, we can graph the successive approximations produced by this method in any concrete case. If we compute approximations to the second Daubechies scaling function $_2\phi$ by applying Theorem 7.8, then the graphs of $_2\phi^{(0)}, \ldots, _2\phi^{(5)}$, the first six functions in the sequence, are as in Figure 7.1.

Our second approach is to use the scaling equation to approximate ϕ by progressively calculating its values at the dyadic numbers. We have implicitly used the dyadic numbers at many points in discussion of the scaling, dilation and translation of wavelets, though without introducing them formally. The formal definition is that a **dyadic number** is a number of the form $k/2^j$, for any $j, k \in \mathbb{Z}$. (Exercise 7.19 explores some simple properties of the dyadic numbers.) A **dyadic interval** is an interval with left-hand endpoint $k/2^j$ and right-hand endpoint $(k + 1)/2^j$ for any $j, k \in \mathbb{Z}$ (irrespective of whether the endpoints belong to the interval).

Although dyadic numbers and dyadic intervals were not mentioned explicitly in the context of Theorem 7.8, it is obvious from the graphs in Figure 7.1, and straightforward to confirm analytically, that for each $n \geq 0$ the approximation $\phi^{(n)}$ produced by the theorem is constant on the half-open dyadic interval $[k/2^j, (k + 1)/2^j)$ for $j = n + 1$.

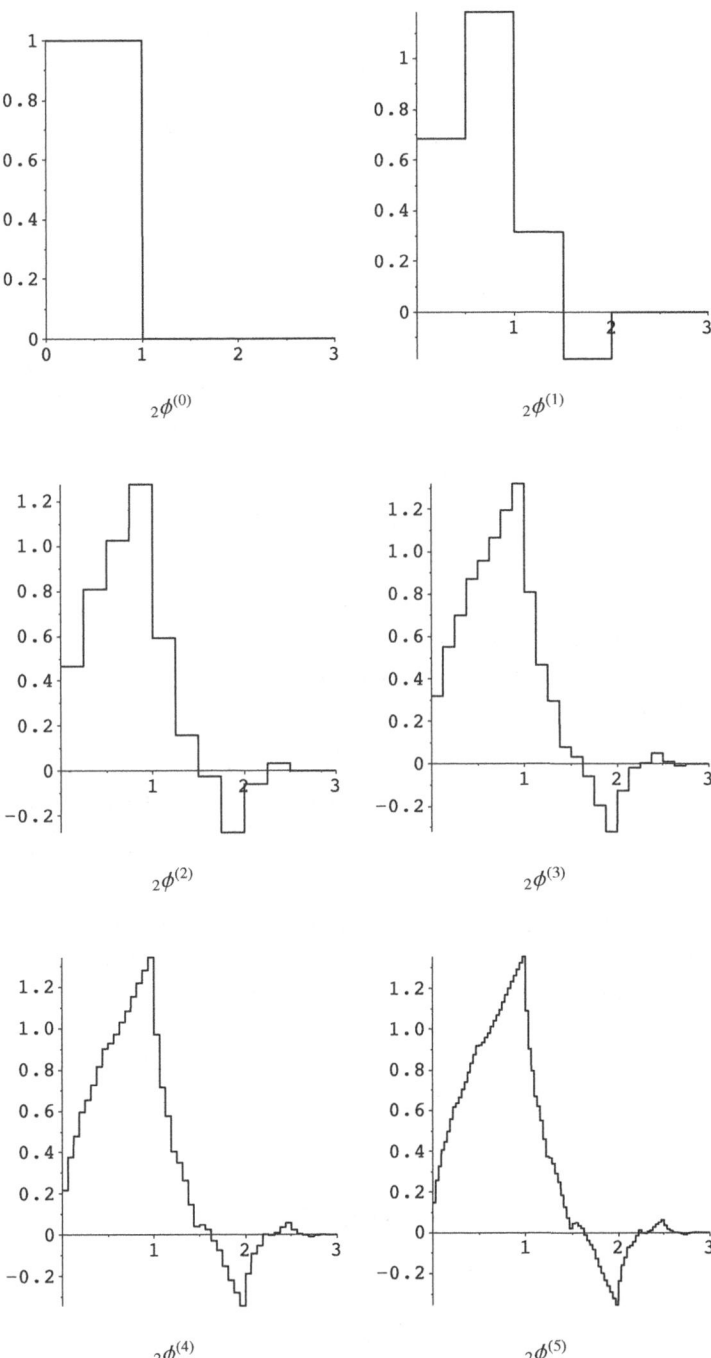

Figure 7.1 The graphs of the first six approximations to $_2\phi$ using Theorem 7.8

To illustrate the second approximation process, consider the case of the second Daubechies scaling function $_2\phi$. We will show how to find the value of $\phi = {}_2\phi$ at the dyadic numbers $k/2^j$ lying in $[0, 3]$, first at those with $j = 0$, namely the integers $0, 1, 2, 3$, then at those with $j = 1$, namely, the half-integers

$$0, \quad \tfrac{1}{2}, \quad 1, \quad \tfrac{3}{2}, \quad 2, \quad \tfrac{5}{2}, \quad 3,$$

then at those with $j = 2$, namely, the quarter-integers

$$0, \quad \tfrac{1}{4}, \quad \tfrac{1}{2}, \quad \tfrac{3}{4}, \quad 1, \quad \tfrac{5}{4}, \quad \tfrac{3}{2}, \quad \tfrac{7}{4}, \quad 2, \quad \tfrac{9}{4}, \quad \tfrac{5}{2}, \quad \tfrac{11}{4}, \quad 3,$$

and so on.

To find the values of ϕ at the integers in $[0, 3]$, note first that $\phi(0) = \phi(3) = 0$ by continuity (see Theorem 7.7), since ϕ is identically 0 outside of $[0, 3]$, so that we only need to compute $\phi(1)$ and $\phi(2)$. Substitution of $x = 1, 2$ into the scaling equation $\phi(x) = \sum_{k=0}^{3} a_k \phi(2x - k)$ yields the equations

$$\phi(1) = a_0 \phi(2) + a_1 \phi(1),$$
$$\phi(2) = a_2 \phi(2) + a_3 \phi(1).$$

It is easy to check that these two equations are dependent, and so another constraint is needed for us to solve for $\phi(1)$ and $\phi(2)$. This comes from Theorem 6.18, which gives us the relation $\phi(1) + \phi(2) = 1$, and we find that $\phi(1) = 2a_0$ and $\phi(2) = 2a_3$, so that we have

$$\phi(0) = 0, \quad \phi(1) = 2a_0, \quad \phi(2) = 2a_3, \quad \phi(3) = 0.$$

➤ For the case of a general scaling function ϕ, the equations for the value of ϕ at the integers is best set up as a matrix equation, and the solution process formulated in terms of a discussion of the eigenvalues and eigenvectors of the coefficient matrix (see Exercises 7.15 and 7.17).

Since the scaling equation expresses $\phi(x)$ in terms of values of $\phi(2x - k)$ for various values of k, we can now immediately compute expressions for the value of ϕ at all the multiples of $\tfrac{1}{2}$, and we find that

$$\phi(0) = 0,$$

$$\phi(\tfrac{1}{2}) = 2a_0^2,$$

$$\phi(1) = 2a_0,$$

$$\phi(\tfrac{3}{2}) = 2a_0 a_2 + 2a_1 a_3,$$

$$\phi(2) = 2a_3,$$

$$\phi(\tfrac{5}{2}) = 2a_3^2,$$

$$\phi(3) = 0.$$

We can clearly continue this process recursively, computing expressions for the values of ϕ at all dyadic points $k/2^j$ in $[0, 3]$, where j runs from 0 to any given bound n. For each fixed n, we can then use the known numerical values of a_0, a_1, a_2, a_3 to plot point–value pairs and join them by straight-line segments, giving a sequence of continuous, piecewise-linear approximations to ϕ. Given the fact that $\phi = {_2}\phi$ is continuous (see Theorem 7.7), it is not hard to show that these functions converge uniformly to ϕ on $[0, 3]$ (see Exercise 7.18).

➤ Note that it was convenient here to compute the values of ϕ symbolically, as functions of the scaling coefficients, rather than numerically; the scaling equation provides all the information needed to manipulate the expressions symbolically, and the system (7.11) gives us the numerical values of the coefficients once they are needed.

The symbolic form also highlights a formal symmetry in the values of ϕ at the dyadic points: systematic interchange of a_0 with a_3 and of a_1 with a_2 converts the expression for $\phi(x)$ into the expression for $\phi(3 - x)$ for every dyadic x (see Exercises 7.21 and 7.22). (As its graph clearly shows, the function ${_2}\phi$ has no *geometric* symmetry; see also the further comments on symmetry in Subsection 7.7.1.)

The same construction can be applied to the Daubechies scaling function ${_N}\phi$ for any $N \geq 2$ (and indeed to any scaling function that is continuous and has bounded support), and we summarise the procedure as follows, leaving the proof of its correctness for Exercise 7.18.

Theorem 7.9 *Suppose that ϕ is continuous, has support in $[0, 2N - 1]$ and is the scaling function of a multiresolution analysis. Calculate the values of ϕ at the integer points of $[0, 2N - 1]$, then, by repeated application of the scaling function, at the dyadic points $k/2^j$ for $j = 1, 2, \ldots, n$. For each $n \geq 0$, let $\phi^{(n)}$ be the continuous, piecewise-linear function obtained by interpolating between the values at those points. Then $\phi^{(n)}$ converges uniformly to ϕ as $n \to \infty$.*

In the case of ${_2}\phi$, Figure 7.2 shows graphs of the first few members of the sequence of approximations, denoted as in the previous figure by ${_2}\phi^{(0)}, \ldots, {_2}\phi^{(5)}$. (Note that in the last graph shown, for $n = 5$, there are $2^5 = 32$ subdivisions of each unit interval.)

➤ A difference between the sequence of approximations produced by Theorems 7.8 and 7.9, clearly visible in the graphs for $\phi = {_2}\phi$, is that although they both guarantee pointwise convergence to the scaling function, the first does not guarantee that the value of the approximate scaling function ever precisely equals the value of the scaling function, while the second proceeds directly by computing exact values of the scaling function at more and more dyadic points at each step.

However, it is fair to say that the two procedures are not as different as may appear at first sight, and that the main difference is in the starting function $\phi^{(0)}$ employed. It is easy to see by examining the statements of the two results that the choice of starting function causes the first theorem to produce piecewise-constant functions and the second to produce piecewise-linear functions. If Theorem 7.8 were given the initial function that had constant value $\phi(k)$ on $[k, k + 1)$ for each k, then its approximations would simply be piecewise-constant versions of those produced by Theorem 7.9.

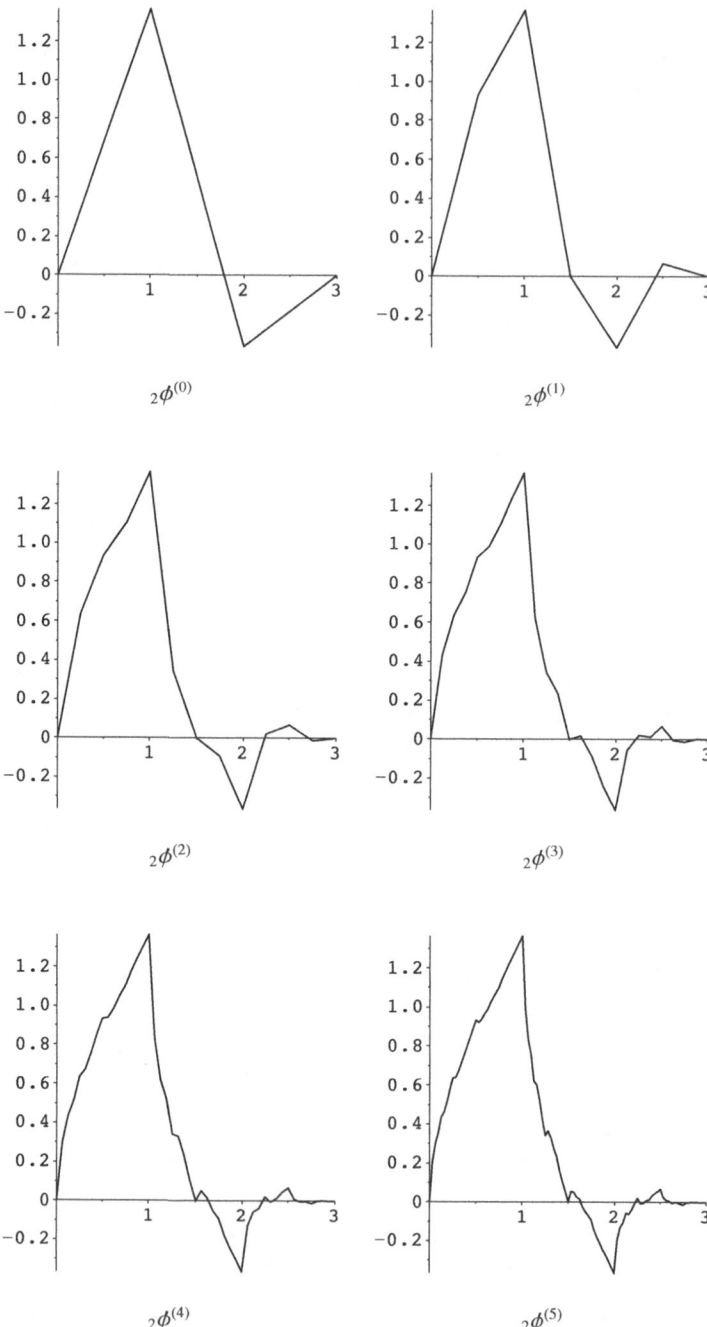

Figure 7.2 The graphs of the first six approximations to $_2\phi$ using Theorem 7.9

We used the algorithms implied by Theorems 7.8 and 7.9 for computing (approximate) values and drawing (approximate) graphs of scaling functions. These algorithms, however, have some disadvantages, and another algorithm which works in a more subtle way has been used to plot most of the later graphs of scaling functions in this chapter. We will not attempt to specify this algorithm here, but we will briefly discuss its distinctive difference from the algorithms used above.

The algorithms implied by Theorems 7.8 and 7.9 both use the scaling equation in an obvious and direct way to compute new approximations to values of ϕ from old ones. But the scaling equation expresses $\phi(x)$ in terms of the values

$$\phi(2x), \phi(2x - 1), \ldots, \phi(2x - 2N + 1),$$

and these values are at points that are widely spread through the support of ϕ: each one is 1 unit distant from its neighbours. The more subtle approach of the third algorithm allows each new approximate value $\phi(x)$ to be computed just from approximate values of ϕ at points that are close to x. This has the great advantage of allowing efficient 'zooming in' on the graph of ϕ. If we wanted a view of ϕ around some specific point x_0 at high magnification, the implied algorithms of Theorems 7.8 and 7.9 would both require the whole graph of ϕ to be found at high magnification, while for the third algorithm, high magnification would need to be used only in a small interval around x_0. At very high magnification, such as in some of the diagrams later in the chapter, this can represent a saving of many orders of magnitude in computing time and storage.

Once we have accurate approximations to ϕ, obtained by any of the methods discussed above, we can easily construct similarly accurate approximations to the corresponding wavelet ψ, because Corollary 6.14 tells us that ψ is given in terms of ϕ by the wavelet equation

$$\psi(x) = \sum_{k=0}^{2N-1} (-1)^{k-1} a_k \phi(2x + k - 1).$$

In Figures 7.3 and 7.4, we show graphs of the first few of the Daubechies scaling functions $_N\phi$ and wavelets $_N\psi$. The following points should be noted.

- We omit $_1\phi$ and $_1\psi$, since these are the familiar Haar scaling function and wavelet.
- Each function is plotted precisely over its support, that is, over $[0, 2N - 1]$ for $_N\phi$ and over $[-N + 1, N]$ for $_N\psi$.
- The scales on the axes differ, and become increasingly disproportionate as the support width increases with N.

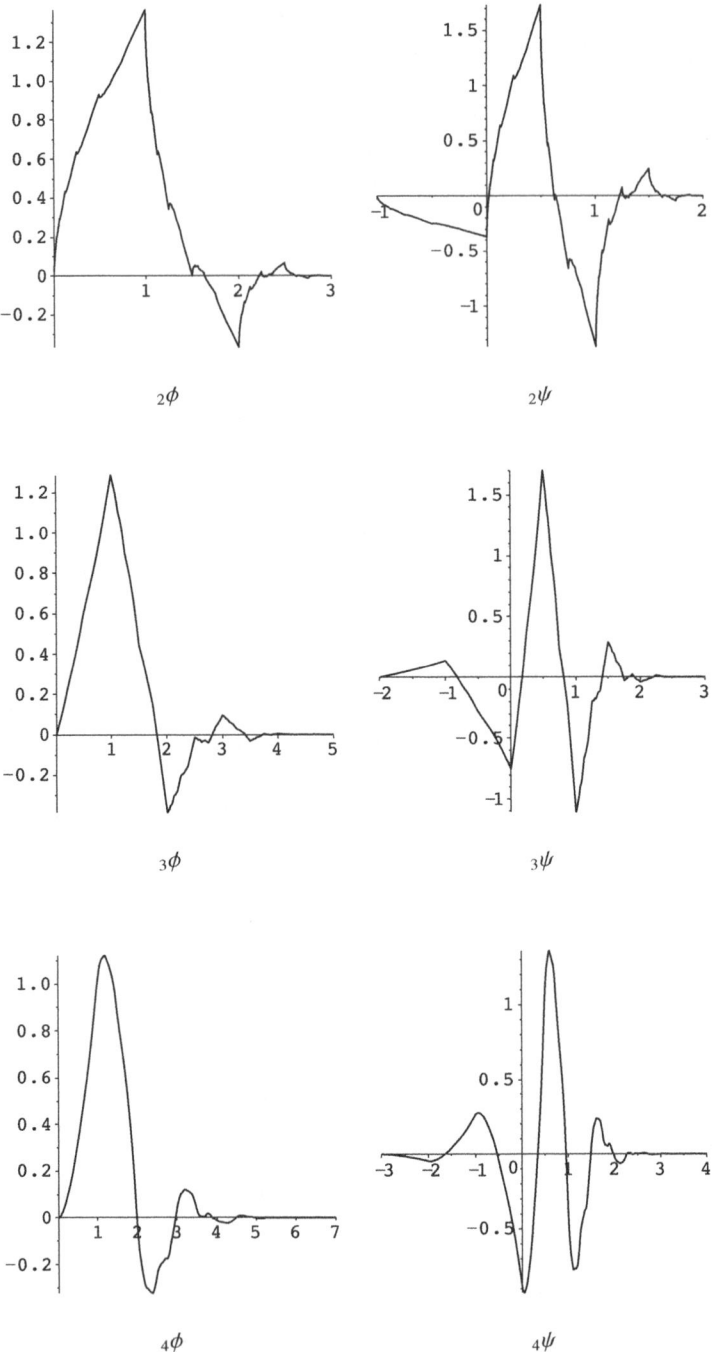

Figure 7.3 The Daubechies scaling functions $_N\phi$ and wavelets $_N\psi$ for $N = 2, 3, 4$

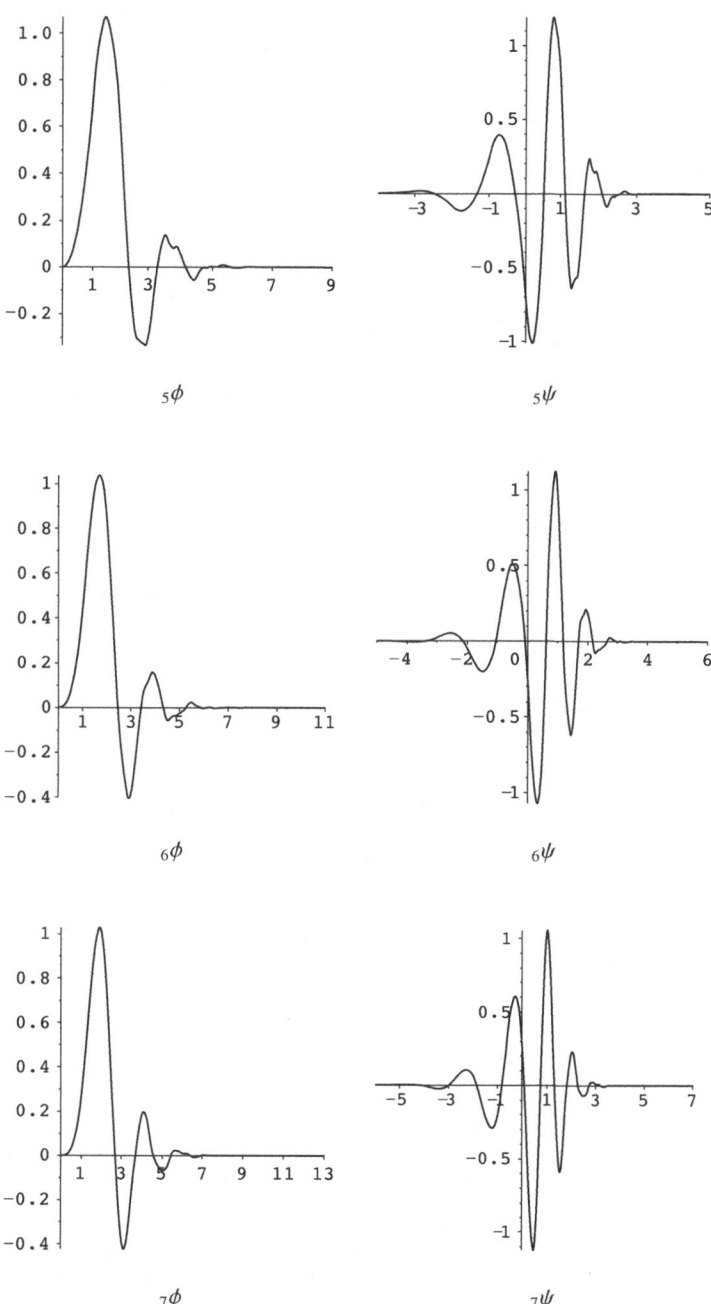

Figure 7.4 The Daubechies scaling functions $_N\phi$ and wavelets $_N\psi$ for $N = 5, 6, 7$

- An increase in smoothness seems apparent as N increases (though, as the following section hints, these functions are very complex analytically, and the information we can reliably read merely from their graphs is therefore limited; see Theorem 7.7 above and the further comments in Section 7.7 below).

7.6 Some Properties of $_2\phi$

A number of interesting observations remain to be made about the Daubechies scaling functions $_N\phi$ and wavelets $_N\psi$ in general, but before coming to these we will examine certain of the algebraic and analytic properties of the second Daubechies scaling function $_2\phi$ in some detail. Specifically, we will see that $_2\phi$ exhibits *self-similar* or *fractal* behaviour, we will investigate aspects of the smoothness of $_2\phi$ and we will introduce a few results that explain analytically some of the behaviour of $_2\phi$ that seems apparent from its graph. For the case of general N, the analogous issues are complex and require reasonably sophisticated analysis, but for $N = 2$ parts of the analysis can be done in an elementary way. For convenience, we will write 'ϕ' in place of '$_2\phi$' throughout this section.

First, we note that, given the continuity of ϕ, Theorem 6.18 tells us that for all $x \in \mathbb{R}$,

$$\sum_{n\in\mathbb{Z}} \phi(x + n) = 1.$$

This is often called the property of **partition of unity**.

Not only can the function with constant value 1 be expressed as a linear combination of translates of ϕ, but so can the function whose value at x is x, that is, the identity function, and this is often called the property of **partition of the identity**.

Lemma 7.10   *For all $x \in \mathbb{R}$,*

$$\sum_{n\in\mathbb{Z}} (2a_2 - n)\phi(x + n) = x.$$

We leave the proof of the lemma as an exercise (see Exercise 7.23).

➤ It is worthwhile noting that the two partition sums are not sums in $L^2(\mathbb{R})$. They *cannot* be L^2 sums, since the functions 1 and x are not members of $L^2(\mathbb{R})$. Rather, the sums express the simplest kind of equality, *pointwise* equality, between the two sides of the equations. (The fact of pointwise equality depends on the continuity of ϕ, as is shown by the statement and proof of Theorem 6.18 and by the argument suggested in Exercise 7.23 for Lemma 7.10.)

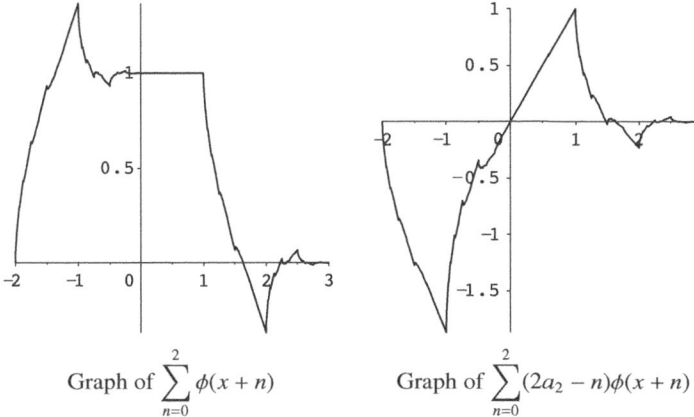

Graph of $\displaystyle\sum_{n=0}^{2} \phi(x+n)$ Graph of $\displaystyle\sum_{n=0}^{2} (2a_2 - n)\phi(x+n)$

Figure 7.5 Illustration of partition of unity and partition of the identity for $\phi = {}_2\phi$

Note that because isupp $\phi = [0, 3]$, only three terms contribute to these sums for any fixed value of x. In particular, the addition of exactly three consecutive terms from either sum permits a striking graphical illustration of the partition phenomena, as shown in Figure 7.5.

We now introduce several results that together allow us to explain what we meant when we said earlier that ϕ exhibits self-similar or fractal behaviour. All the proofs are straightforward, and are left for the exercises.

Lemma 7.11   *For all x such that $0 \le x \le 1$,*

$$2\phi(x) + \phi(x+1) = x + 2a_0,$$
$$\phi(x+1) + 2\phi(x+2) = -x + 2a_2$$

and

$$\phi(x) - \phi(x+2) = x - 2a_3.$$

Observe in particular how the first and third of these equations allow us to express the values of ϕ in the integer-length intervals $[1, 2]$ and $[2, 3]$ in terms of its values in $[0, 1]$:

$$\phi(x+1) = -2\phi(x) + x + 2a_0,$$
$$\phi(x+2) = \phi(x) - x + 2a_3.$$

The next result will allow us to express the values of ϕ in the intervals of *half-integer* length within $[0, 3]$ in terms of its values in $[0, 1]$.

Lemma 7.12 *For all x such that $0 \le x \le 1$,*

$$\phi\left(\frac{x}{2}\right) = a_0\phi(x),$$

$$\phi\left(\frac{x+1}{2}\right) = a_3\phi(x) + a_0x + \frac{2 + \sqrt{3}}{4},$$

$$\phi\left(\frac{x+2}{2}\right) = a_0\phi(x+1) + a_3x + \frac{\sqrt{3}}{4},$$

$$\phi\left(\frac{x+3}{2}\right) = a_3\phi(x+1) - a_0x + \frac{1}{4},$$

$$\phi\left(\frac{x+4}{2}\right) = a_0\phi(x+2) - a_3x + \frac{3 - 2\sqrt{3}}{4}$$

and

$$\phi\left(\frac{x+5}{2}\right) = a_3\phi(x+2).$$

Now, by combining this lemma with the two equations that just precede it, we can express the values of ϕ in, say, the interval $[1, 1.5]$ in terms of its values in $[0, 1]$, as follows:

$$\phi\left(\frac{x+2}{2}\right) = a_0\phi(x+1) + a_3x + \frac{\sqrt{3}}{4}$$

$$= a_0(x + 2a_0 - 2\phi(x)) + a_3x + \frac{\sqrt{3}}{4}$$

for all $x \in [0, 1]$, and a regrouping of terms in the last expression leads easily to an expression of the form $\alpha\phi(x) + \beta x + \gamma$. Pursuing this idea inductively leads to a proof of the following general statement about the values of ϕ in arbitrary dyadic intervals within $[0, 3]$ (see Exercises 7.27 and 7.28).

Theorem 7.13 *For each $j = 0, 1, 2, \ldots$ and each $k = 0, 1, \ldots, 3 \cdot 2^j - 1$, there exist α, β and γ (dependent on j and k), with $\alpha \neq 0$, such that*

$$\phi\left(\frac{x+k}{2^j}\right) = \alpha\phi(x) + \beta x + \gamma \quad \text{for } 0 \le x \le 1.$$

We commented above that ϕ exhibits self-similar or fractal behaviour, and the theorem explains partly what this means: on each dyadic interval

$$[k/2^j, (k + 1)/2^j]$$

lying in isupp $\phi = [0, 3]$, ϕ 'looks the same as on $[0, 1]$', the only difference being that it is *scaled* by some non-zero factor α and *sheared* by some linear term $\beta x + \gamma$. Note carefully that this self-similarity occurs at every dyadic scale.

For example, the most prominent feature of the graph of ϕ in Figure 7.3 is the large spike at $x = 1$, but the theorem implies that this spike actually appears, in some scaled and sheared form, at *every dyadic point* in $(0,3)$. (Note that we need to exclude the dyadic points 0 and 3 from the discussion here, and at similar points below, since $\phi(x)$ is identically 0 when $x \leq 0$ and when $x \geq 3$.) At the scale of the graph, this is clearly visible at all the integer and half-integer points, at most of the quarter-integer points and at a few of the eighth-integer points (though at some of these values, such as 1.5 and 2, the spike points downwards, corresponding to cases where $\alpha < 0$ in Theorem 7.13). The two plots in Figure 7.6 zoom in on the graph of ϕ around 0.5 and 2.5, with horizontal scales chosen carefully so as to bring out the similarity in behaviour of the function near each of those points, as well as near 1.

Another way of viewing the self-similarity property is given by rewriting the equation from the theorem in the form

$$\phi(x) = (1/\alpha)\phi\left(\frac{x+k}{2^j}\right) - (\beta/\alpha)x - (\gamma/\alpha);$$

this equation shows that ϕ on $[0, 1]$ (and hence on $[0, 3]$) can be reconstructed from its values on any one fixed dyadic interval within $[0, 3]$, however small.

The theorem implies, moreover, that any analytic condition that ϕ satisfies at a particular dyadic point in $[0, 3)$ will hold, in suitably modified form, at every other dyadic point in $[0, 3)$. The most interesting such properties are those related to smoothness.

In the graph of Figure 7.3, the spike at 1 is conspicuously lop-sided, and this is a graphical reflection of the mathematical fact that ϕ is differentiable from the left but not from the right at 1. More precisely, the limit that would define the right derivative at 1 if it existed and had finite value actually diverges

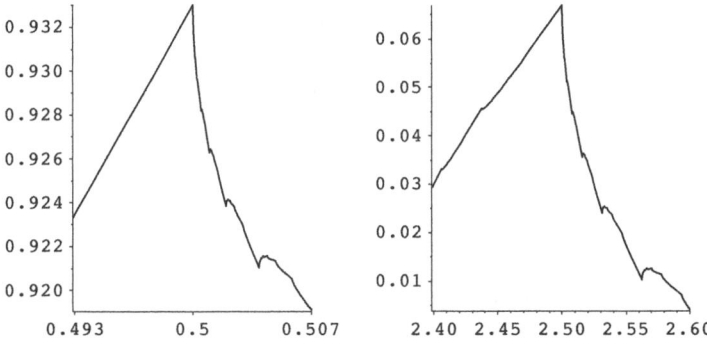

Figure 7.6 Magnified views of the graph of $_2\phi(x)$ around $x = 0.5$ and $x = 2.5$

to ∞, so that the curve has 'infinite slope' immediately to the right of 1. By the theorem, this behaviour must be reflected at every dyadic point: ϕ *is differentiable from the left but not from the right at every dyadic point in* $[0, 3)$, and moreover has 'infinite slope' on the right of each such point, though the slope can be ∞ or $-\infty$ depending on the sign of the coefficient α from Theorem 7.13. We will confirm this analytically below and in the exercises, but meanwhile we can gain confirmation of the fact graphically. Specifically, since ϕ has finite slope on the left and infinite slope on the right at every dyadic point, increasing levels of magnification on the x-axis at a dyadic point should cause the graph of ϕ to appear increasingly flat to the left of the point, but always to appear infinitely steep to the right of the point. Figure 7.7 confirms this expectation by zooming in at high magnification around 0.5 and 2.5.

We will now confirm the claim that ϕ is not differentiable from the right at dyadic points in $[0, 3)$. By Theorem 7.13 it suffices to deal with any one dyadic point, and for convenience we prove the claim at $x = 0$. (The stronger claims – that the derivative from the right is infinite and the derivative from the left is finite at every dyadic point – are left as exercises; see Exercises 7.29 and 7.31.)

By the first equation of Lemma 7.12 (or directly from the scaling equation), we have $\phi(x) = a_0 \phi(2x)$ for $0 \leq x \leq \frac{1}{2}$. Therefore, $\phi(\frac{1}{2}) = a_0 \phi(1)$, $\phi(\frac{1}{4}) = a_0 \phi(\frac{1}{2})$, $\phi(\frac{1}{8}) = a_0 \phi(\frac{1}{4})$, . . ., and it follows that

$$\phi(\tfrac{1}{2}) = a_0 \phi(1), \quad \phi(\tfrac{1}{4}) = a_0^2 \phi(1), \quad \phi(\tfrac{1}{8}) = a_0^3 \phi(1), \ldots,$$

or, in general,

$$\phi\left(\frac{1}{2^j}\right) = a_0^j \phi(1) \quad \text{for } j \geq 0.$$

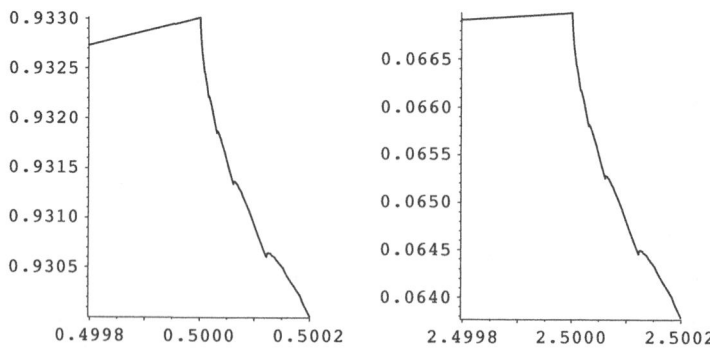

Figure 7.7 Highly magnified views of the graph of $_2\phi(x)$ around $x = 0.5$ and $x = 2.5$

Now differentiability of ϕ at 0 from the right would require existence of the limit

$$\lim_{h\to0^+}\frac{\phi(h)-\phi(0)}{h-0}=\lim_{h\to0^+}\frac{\phi(h)}{h}.$$

But when $h=1/2^j$, we have

$$\frac{\phi(h)}{h}=2^j\phi\left(\frac{1}{2^j}\right)=2^ja_0^j\phi(1)=(2a_0)^{j+1}\to\infty$$

as $j\to\infty$, since $\phi(1)=2a_0>1$, so the required limit does not exist. Hence ϕ is not differentiable from the right at 0 or, by Theorem 7.13, at any other dyadic point in $[0,3)$.

7.7 Some Properties of $_N\phi$ and $_N\psi$

In this section, we will survey a number of interesting and important properties of the Daubechies scaling functions and wavelets in general, as well as a few properties of arbitrary scaling functions and wavelets of bounded support.

Our measure of the smoothness of a function f to this point has been the largest integer $k\geq0$ for which $f\in C^k(\mathbb{R})$. It is in fact possible to define a notion of 'fractional differentiability', and corresponding classes $C^\alpha(\mathbb{R})$ for all real numbers $\alpha\geq0$, in such a way that when $\alpha=k$ is an integer, $C^\alpha(\mathbb{R})=C^k(\mathbb{R})$ is the collection of functions continuously differentiable to order k as already defined.

➤ The notion of fractional differentiability we are referring to specifically is the notion of **Hölder continuity**; the functions in $C^\alpha(\mathbb{R})$ are said to be **Hölder continuous with exponent** α. (We should also remark that there are actually several inequivalent ways of defining fractional differentiability.) We will not give the definition of Hölder continuity here, although it is quite straightforward. It is useful to be able to refer to the notion here, even if only in a non-technical way, since it allows us interesting insight into the behaviour of $_N\phi$ and $_N\psi$.

We start by drawing together a number of facts about the smoothness properties of the Daubechies scaling functions and wavelets. These include, but refine, the information given in Theorem 7.7.

- A precise statement about the asymptotic rate of growth in the smoothness of $_N\phi$ and $_N\psi$ as N grows is as follows. When N is large, $_N\phi,_N\psi\in C^{\mu N}(\mathbb{R})$, where $\mu=1-\ln3/(2\ln2)\approx0.2075$. Thus, for large N, roughly one extra continuous derivative is gained for each increase of 5 in the value of N.

This asymptotic information tells us nothing about how smooth $_N\phi$ and $_N\psi$ are for specific, small values of N, and for this case-by-case arguments are required. Some details are as follows.

- When $N = 1$, we have the Haar scaling function and wavelet $_1\phi$ and $_1\psi$, which are not in $C^0(\mathbb{R})$.
- For $N = 2$, the functions $_2\phi$ and $_2\psi$ are continuous but not differentiable, and hence are in $C^0(\mathbb{R})$ but not $C^1(\mathbb{R})$.
- The exponent α of Hölder continuity for $_2\phi$ and $_2\psi$ is about 0.5500, indicating that although the functions are continuous they are 'not very close' to being continuously differentiable.
- The functions $_2\phi$ and $_2\psi$ are not differentiable (in the ordinary two-sided sense) at any dyadic point in $[0, 3)$, because the right derivative does not exist at those points. Despite this, and despite the low Hölder exponent, the functions are known to be differentiable at many non-dyadic points, and in fact are differentiable at almost all points in $[0, 3]$.
- For $N = 3$, the functions $_3\phi$ and $_3\psi$ are continuously differentiable, but are not twice differentiable at any point in their support, and hence are in $C^1(\mathbb{R})$ but not $C^2(\mathbb{R})$. However, they are 'only just' in $C^1(\mathbb{R})$: their Hölder exponent α is about 1.0878. This is illustrated informally by the plot of $_3\phi$ in Figure 7.3, which at the scale used appears to have a sharp spike at $x = 1$. If this spike were like the spike of $_2\phi$ at $x = 1$, then the function $_3\phi$ would not be differentiable at $x = 1$. However, under suitable magnification, we find that the maximum of $_3\phi$ occurs a little to the right of 1, and that there is actually no sharp spike at 1 after all, as shown in Figure 7.8. (Similar comments apply to $_3\psi$.)

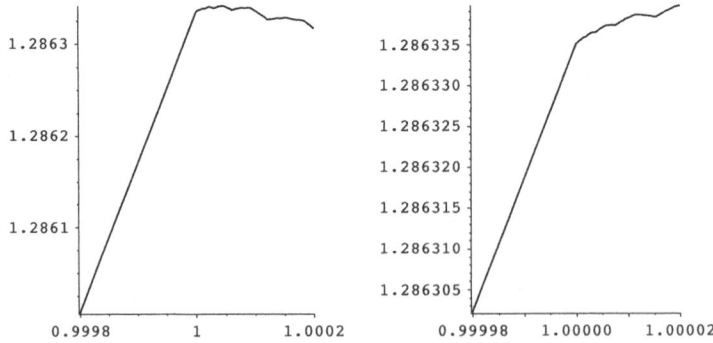

Figure 7.8 Highly magnified views of the graph of $_3\phi(x)$ around $x = 1$

We also note a few facts about the relation between the smoothness and the support length of a scaling function in the general case.

- Given a multiresolution analysis with a scaling function ϕ that has support $[0, 2N - 1]$, it is known that if $\phi \in C^k(\mathbb{R})$, then $2N - 1 \geq k + 2$. Note, however, that this gives only fairly weak information in the Daubechies case. For example, it implies that $_2\phi$ is not in $C^2(\mathbb{R})$, while we have already noted that $_2\phi$ is not in $C^1(\mathbb{R})$, and it implies that $_3\phi$ is not in $C^3(\mathbb{R})$, while we have already noted that $_3\phi$ is not in $C^2(\mathbb{R})$. Likewise, the asymptotic information it gives is much weaker than the asymptotic information already noted for the Daubechies case.

- However, the above result does have the interesting and significant corollary that a scaling function of bounded support cannot be in $C^\infty(\mathbb{R})$.

7.7.1 Other Wavelet Families

The Daubechies scaling function and wavelet for a given support length of $2N - 1$ are obtained by imposing the $N + 1$ fundamental relations (6.13) on the scaling coefficients, then the further $N - 1$ relations derived by setting the wavelet moments up to order N to 0 (see Theorem 7.3), solving either of the resultant equivalent systems (7.7) and (7.8) of $2N$ equations for the coefficients, and finally picking the specific sequence of coefficient values given by the extremal phase condition (see the end of Section 7.4 and Exercise 7.11). Once we have this sequence of coefficients, the scaling function and then the wavelet are uniquely determined (see Theorems 7.5 and 7.6).

Given N, the fundamental relations are non-negotiable: the scaling coefficients for all scaling functions and wavelets with support length $2N - 1$ must satisfy them (this was a major conclusion of Chapter 6). However, choices have to be made to pick out specific sequences of scaling coefficients from the collection A_N (see the opening paragraphs of Section 7.2) of all sequences of solutions to the fundamental relations, which is an infinite set except when $N = 1$. Moreover, according to Theorem 7.5, almost all choices of extra constraints will yield scaling functions and hence wavelets.

The original choice made by Daubechies was as just summarised: impose vanishing moment conditions and then make the extremal phase choice. However, Daubechies herself, and others since, have also developed other families of wavelets by making different choices, driven both by the requirements of applications and by theoretical interest. Some of these are as follows.

- The extremal phase choice results in an asymmetric, 'front-loaded' sequence of scaling coefficients (see the end of Section 7.4 and Exercise 7.11). This in turn results in an asymmetric scaling function, where the large values of the function are all at the left-hand end of its support and the values at the right-hand end are very close to 0. This effect becomes increasingly pronounced as N increases, and is clearly visible already in the scaling coefficients for the case $N = 7$ shown in Table 7.1 and in the graph of $_7\phi$ shown in Figure 7.4. The asymmetry diminishes a little in the corresponding wavelets (see Exercise 7.33), but is still noticeable.

  For some applications, it is useful to have wavelets that are as symmetric as possible. Daubechies has shown that a wavelet can be neither completely symmetric nor completely antisymmetric, except in the case of the Haar wavelet. (Formally, a function f is symmetric if some translate of f is an even function and is antisymmetric if some translate is an odd function.) However, Daubechies has also constructed the *least asymmetric wavelet* that has support length $2N - 1$ and N vanishing moments (where the phrase 'least asymmetric' is used in a well-defined technical sense), and has shown that somewhat better symmetry still can be achieved if a larger support length is permitted.

- With support length fixed as before at $2N - 1$, better smoothness may be sought than is given by the Daubechies family of wavelets. Theorem 7.2 shows that smoothness forces moments to vanish: if $\psi \in C^M(\mathbb{R})$, then all the moments of ψ up to order M are 0. On the other hand, as noted above, the Nth Daubechies wavelet $_N\psi$ is asymptotically only in $C^{\mu N}(\mathbb{R})$, where $\mu \approx 0.2075$, which suggests that it might be reasonable to impose fewer vanishing moment conditions and to try to use the resulting freedom to impose other conditions that directly increase smoothness. Daubechies has developed an approach for doing this.

  Other authors have made attempts to maximise smoothness directly, although there appears to be little general theory showing how much additional smoothness can be obtained.

- Finally, we note briefly another idea developed by Daubechies: that of exchanging some of the vanishing moment conditions on the wavelet for vanishing moment conditions on the scaling function.

➤ The last idea mentioned was suggested by Ronald Coifman for specific applications, and Daubechies whimsically named the resulting wavelets 'coiflets'. Since then, other families of wavelets have sometimes been referred to by similar nicknames – the Daubechies wavelets themselves as 'daublets' and the least asymmetric wavelets as 'symmlets', for example.

Exercises

7.1 When $N = 2$, the fundamental relations (6.13) form the system

$$a_0^2 + a_1^2 + a_2^2 + a_3^2 = 2,$$
$$a_0a_2 + a_1a_3 = 0,$$
$$a_0 + a_1 + a_2 + a_3 = 2.$$

Derive the trigonometric parametrisation (7.5) of the set $A_2 \subseteq \mathbb{R}^4$ defined by the relations. A possible approach is as follows.

(a) Write down the form the equations take after the substitution of $b_k = 2a_k - 1$ for $k = 0, 1, 2, 3$.

(b) Show that $b_0^2 + b_1^2 = b_2^2 + b_3^2 = 2$ and that $b_0 + b_2 = b_1 + b_3 = 0$.

(c) Hence write down a trigonometric parametrisation for b_0, b_1, b_2, b_3 and then convert this to a parametrisation of a_0, a_1, a_2, a_3. (Some minor trigonometric manipulation may be needed to obtain expressions in exactly the form (7.5).)

(d) Use suitable trigonometric manipulations to derive the alternative parametrisation

$$a_0 = (1 - \cos\alpha + \sin\alpha)/2,$$
$$a_1 = (1 + \cos\alpha + \sin\alpha)/2,$$
$$a_2 = (1 + \cos\alpha - \sin\alpha)/2,$$
$$a_3 = (1 - \cos\alpha - \sin\alpha)/2,$$

for $0 \le \alpha < 2\pi$.

7.2 Show that A_2 is a circle that lies in a plane in $\mathbb{R}^4$, as claimed in Section 7.2. (Change from the given coordinates (a_0, a_1, a_2, a_3) to the coordinates (b_0, b_1, b_2, b_3) of the previous question. This transformation represents nothing more than a change of scale and a translation in $\mathbb{R}^4$, so if we can show that B_2, the image of A_2 under the transformation, is a circle lying in a plane, the same will be true of A_2. Given the work done in the previous question, little extra should be needed to find a 2-dimensional subspace of $\mathbb{R}^4$ in which B_2 lies and forms a circle.)

7.3 (a) In terms of the parametrisation (7.5) of the set A_2, find the values of α that yield the following quadruples (a_0, a_1, a_2, a_3):

$$(1, 1, 0, 0), \quad (0, 1, 1, 0), \quad (0, 0, 1, 1), \quad (1, 0, 0, 1).$$

(b) Confirm that the first three of these sequences of scaling coefficients all yield the Haar scaling function or a translate of it. (Corollary 6.13, specialised to the case $N = 2$, shows that if we are seeking a scaling function ϕ with isupp $\phi = [0, 3]$, or a translate of it, then the condition $a_0, a_3 \neq 0$ holds, and this would rule out the three sequences in question from further consideration. However, if we require only that isupp ϕ be *at most* 3 units in length (and therefore either exactly 1 unit or exactly 3 units in length, by the result just cited), then a_0 and a_3 are not forced to be non-zero, and we can argue as above. We therefore see that the $N = 1$ case can be viewed as a degenerate instance of the $N = 2$ case. A similar statement in fact holds for general N.)

(c) Show that the fourth quadruple above, $(1, 0, 0, 1)$, yields the function $\phi = \frac{1}{3}\chi_{[0,3]}$ (see Exercise 6.18 for the notation) as the only solution to the corresponding scaling equation (note that uniqueness is given by Theorem 7.5). Show that the corresponding collection $\{\phi_{0,k} : k \in \mathbb{Z}\}$ is not an orthonormal set, so that ϕ is not in fact a scaling function. (In the comments following Theorem 7.5, we noted that for $N = 2$ there is exactly one quadruple of solutions to the fundamental relations that fails to yield a scaling function, and this exception is the quadruple $(1, 0, 0, 1)$.)

(d) Find the value of α in the system (7.5) that yields the Daubechies scaling coefficients.

7.4   Check in detail the derivation of the vanishing moment constraints for the third Daubechies scaling function.

7.5   Prove Theorem 7.3. A suitable strategy is to prove by induction on L that if $\int_{-\infty}^{\infty} x^\ell \psi(x)\, dx = 0$ for $0 \le \ell \le L$ then $\sum_{k=0}^{2N-1}(-1)^k k^\ell a_k = 0$ for $0 \le \ell \le L$. The base case, for $L = 0$, reduces to the known equation $\sum_{k=0}^{2N-1}(-1)^k a_k = 0$ (see Theorem 6.16), while the case $N = 2$ outlined in the main text and the case $N = 3$ of the previous exercise indicate the pattern for the inductive step.

7.6   The aim of this exercise and the next is to show that one of the quadratic equations in the system of fundamental relations can be replaced by a linear equation. (See the discussion around the systems (7.7) and (7.8) in Section 7.3. Note that although this discussion was in the context of the augmented system consisting of the fundamental relations together with the relations derived from vanishing moment conditions, the moment conditions play no role in the argument.)

For a scaling function ϕ with isupp$(\phi) = [0, 2N - 1]$, the only non-zero scaling coefficients are $a_0, \ldots, a_{2N-1}$, and we can write the fundamental relations (6.13) in the form

$$\sum_{k=0}^{2N-1} a_k a_{k+2\ell} = 2\delta_{0,\ell} \quad \text{for} \quad \ell = 0, 1, \ldots, N-1, \tag{7.14}$$

$$\sum_{k=0}^{2N-1} a_k = 2, \tag{7.15}$$

where, for the purposes of this exercise, we have uniformly changed the upper index of the first summation from $2N - 2\ell - 1$ to $2N - 1$. (It is convenient here to have all of the summations in the fundamental relations over the same range of k.)

(a) Confirm that this change of summation range is correct, in that all of the extra terms included are 0.

For $\ell = 0$, (7.14) gives the quadratic equation

$$\sum_{k=0}^{2N-1} a_k^2 = 2. \tag{7.16}$$

Our aim is to show that (7.16) can be replaced by the linear equation

$$\sum_{k=0}^{2N-1} (-1)^k a_k = 0, \tag{7.17}$$

giving an *equivalent* system, that is, a system that has exactly the same set of solutions.

(b) Show, by examining the proof of Theorem 6.16, that Equation (7.17) follows from the fundamental relations (7.14) and (7.15).

For the equivalence of the systems, we now need to show that if we assume that (7.14) holds for $\ell = 1, \ldots, N - 1$ and that (7.15) and (7.17) hold, then Equation (7.16) follows. Before the general case in the following exercise, it is probably worth working through one or two specific cases.

(c) Consider the case $N = 2$, where, either from (6.13) or from (7.14) and (7.15), the fundamental relations are

$$a_0^2 + a_1^2 + a_2^2 + a_3^2 = 2, \tag{7.18}$$

$$a_0 a_2 + a_1 a_3 = 0, \tag{7.19}$$

$$a_0 + a_1 + a_2 + a_3 = 2. \tag{7.20}$$

Assume that (7.19) and (7.20) hold, together with

$$a_0 - a_1 + a_2 - a_3 = 0. \tag{7.21}$$

Multiply through (7.20) by each of a_0, a_1, a_2, a_3 separately and add the four equations that result. Then multiply through (7.21) in a similar way and add. Finally, combine the two equations obtained, and use (7.19) to show that (7.18) holds.

(d) Work through a similar argument for the case $N = 3$.

7.7 Deal with the general case of the argument in the previous exercise as follows.

(a) Show that

$$\left(\sum_{k=0}^{2N-1} a_k^2 \right) = \left(\sum_{k=0}^{2N-1} a_k \right)^2 - 2 \left(\sum_{\substack{0 \le i < j \le 2N-1 \\ j-i \text{ even}}} a_i a_j \right) - 2 \left(\sum_{\substack{0 \le i < j \le 2N-1 \\ j-i \text{ odd}}} a_i a_j \right).$$

(b) Show that

$$\left(\sum_{\substack{0 \le i < j \le 2N-1 \\ j-i \text{ odd}}} a_i a_j \right) = \left(\sum_{\substack{k=0 \\ k \text{ even}}}^{2N-1} a_k \right) \left(\sum_{\substack{k=0 \\ k \text{ odd}}}^{2N-1} a_k \right).$$

(To deal with the two summations on the right-hand side, it may be useful to refer to the proof of Theorem 6.16.)

(c) Now assume that (7.14) holds for $\ell = 1, \ldots, N - 1$ and that (7.15) and (7.17) hold, and prove that Equation (7.16) holds.

7.8 Find both solutions to the system (7.9) for a_0, a_1, a_2, a_3, the non-zero scaling coefficients for the second Daubechies scaling function $_2\phi$. (Extract as much information as possible from the linear equations before making use of the quadratic equation.)

7.9 Investigate what happens if we make the other possible choice for the $\pm$ signs in the equations (7.11) for the scaling coefficients of $_2\phi$. In particular, justify the assertion in the main text that we lose nothing essential by restricting ourselves to just one of the two choices.

7.10 Find explicit expressions for the solutions of the system (7.12) for the Daubechies scaling coefficients for $N = 3$. Confirm in particular that the .equations have two real and two complex solutions. (Solve the four linear equations of the system to find expressions for, say, a_0, a_1, a_2, a_3 in terms of a_4, a_5. Then substitute these into the simpler of the two quadratic equations and solve for, say, a_4 in terms of a_5. Next, substitute the expression for a_4 into the remaining quadratic equation and solve for a_5. Finally, substitute back to find explicit expressions for all the coefficients (compare with the results of Subsection 8.5.2 and Exercise 8.40 below). Note that this solution process is quite complicated if done entirely by hand, and the use of a computer algebra system is recommended – though

of course such a system should be able to carry out the whole of the
solution process without the need to break it into steps at all.)

7.11 (a) For $N = 2$, the two solutions to the system (7.9) are given by
the equations (7.10), and the corresponding Daubechies scaling
coefficients are given by the equations (7.11) as well as numerically
in Table 7.1 (see also Exercise 7.8). Verify that the Daubechies
solution is the extremal phase solution.

(b) For the case $N = 3$, Exercise 7.10 was devoted to finding explicit
expressions for the two real-valued solutions to the system (7.12).
Find which of these two solutions is the extremal phase solution,
and verify that this corresponds with the solution given numerically
in Table 7.1.

7.12 Using the hypotheses of Theorem 7.4, prove that the sets V_j of the
theorem are closed subspaces of $L^2(\mathbb{R})$ and that the multiresolution
analysis conditions (MRA1) and (MRA4)–(MRA6) hold. (In the proof
that the V_j are closed subspaces and the proof of (MRA6), Exercise 4.28
may be useful.)

7.13 Confirm the remark made following Theorem 7.4 that if the value of the
integral in the theorem is allowed to be ± 1, then the conditions of the
theorem are necessary as well as sufficient

7.14 Compute and plot by hand the graphs of the first three approximations
$\phi^{(0)}$, $\phi^{(1)}$ and $\phi^{(2)}$ to $\phi = {}_2\phi$ given by Theorem 7.8.

7.15 (a) Verify the calculation summarised in Subsection 7.5.3 of the values
of $\phi = {}_2\phi$ at the points 1 and 2.

(b) Convert the two linear equations from part (a) into a matrix equation,
and confirm that the length-2 vector whose entries are $\phi(1)$ and $\phi(2)$
is an eigenvector of the coefficient matrix corresponding to the
eigenvalue 1 and that the normalisation of the eigenvector provided
by the relation $\phi(1) + \phi(2) = 1$ derived from Theorem 6.18 yields the
values of $\phi(1)$ and $\phi(2)$ found in the main text.

7.16 (a) Using the calculations of the previous exercise, confirm the symbolic
expressions found for the values of $\phi = {}_2\phi$ at the half-integer points
in $[0, 3]$.

(b) Similarly calculate in symbolic form the values of ϕ at the quarter-
integer points in $[0, 3]$.

(c) Use the numerical values of the scaling coefficients to plot the
graphs of the first three piecewise-linear approximations to ϕ, as in
Theorem 7.9.

7.17 (a) Treat the case of $\phi = {}_3\phi$ as in the previous two exercises, computing
the values of ϕ at the integer, half-integer and quarter-integer points

in isupp $\phi = [0, 5]$. (It is probably simpler now to do the calculations numerically rather than symbolically, using the values for the scaling coefficients given in Table 7.1.)

(b) Hence plot the graphs of the first three piecewise-linear approximations to ϕ.

7.18 Verify the claim in Theorem 7.9 that the sequence of approximations $\phi^{(n)}$ converges uniformly to ϕ as $n \to \infty$. (This result is really a general statement in analysis. A proof requires only the assumptions that the limit function is uniformly continuous, which is the case in particular if the function is continuous and has bounded support, and that successive piecewise-linear approximations are made over grids of points whose mesh-size approaches 0, where by the mesh-size we mean the maximum distance between neighbouring points).

7.19 This exercise and the next develop a few basic ideas about the dyadic and related numbers, and introduces a useful scheme for carrying out inductive arguments over the dyadics.

Denote the set of all dyadic numbers by $\mathbb{D}$. Also, for $j \in \mathbb{Z}$, denote by $\mathbb{D}_j$ the set of dyadic numbers that can be expressed in the form $k/2^j$ for $k \in \mathbb{Z}$. We will call $\mathbb{D}_j$ the set of dyadic numbers of **level** j.

(a) Show that $\mathbb{D}$ is closed under addition, subtraction and multiplication, but not division. What are the closure properties of $\mathbb{D}_j$?

(b) Show that $\mathbb{D}_j \subseteq \mathbb{D}_{j+1}$ for all j and that $\mathbb{D} = \bigcup_{j \in \mathbb{Z}} \mathbb{D}_j$.

(c) Show that $x \in \mathbb{D}$ if and only if the binary decimal expansion of x is terminating. Characterise $\mathbb{D}_j$ in similar terms.

(d) Show that $1/3 \notin \mathbb{D}$. What is the binary decimal expansion of $1/3$?

(e) Show that $\mathbb{D}$ is dense in $\mathbb{R}$.

(f) Show that $\mathbb{D}$ is of measure 0.

7.20 Denote by $\mathbb{D}(\sqrt{3})$ the set of numbers of the form $x + y\sqrt{3}$, where $x, y \in \mathbb{D}$. Also, for $x, y \in \mathbb{D}$, write

$$\overline{x + y\sqrt{3}} = x - y\sqrt{3},$$

and call this number the **conjugate of** $x + y\sqrt{3}$ (note that this conjugate has nothing directly to do with the complex conjugate of a complex number).

(a) Show that $\mathbb{D}(\sqrt{3})$ is closed under addition, subtraction and multiplication, but not division.

(b) Show that $\mathbb{D}(\sqrt{3})$ is closed under conjugation.

(c) Show that if $x \in \mathbb{D}$ and $\phi = {}_2\phi$, then $\phi(x) \in \mathbb{D}(\sqrt{3})$. (Use induction on the dyadic level of the argument x. Confirm that the formulae (7.11) give $a_0, a_1, a_2, a_3 \in \mathbb{D}(\sqrt{3})$, and then use the scaling equation to show that if x is dyadic of level $j \geq 0$, then $\phi(x)$ is dyadic of level $j + 1$.)

7.21 This exercise investigates one way of explaining the formal symmetry in the values of $\phi = 2\phi$ at dyadic points that was noted in the discussion just preceding Theorem 7.9.

Set $a = a_0$ and $b = a_3 = \overline{a_0}$ (see Exercise 7.20 for the notation), and note that $a_1 = 1 - b$ and $a_2 = 1 - a$. (Observe that the interchange of a_0 with a_3 and a_1 with a_2 corresponds just to the interchange of a and b.) Write the scaling equation for ϕ in the form

$$\phi(x) = \sum_{i=1}^{4} c_i(a, b)\phi(f_i(x)),$$

where $c_1(a, b) = a$ and $f_1(x) = 2x$, $c_2(a, b) = 1 - b$ and $f_2(x) = 2x - 1$, and so on.

(a) Show that

$$\phi(3 - x) = \sum_{i=1}^{4} c_i(b, a)\phi(3 - f_i(x)).$$

(b) Use this to argue inductively that at all dyadic points x, the value of $\phi(3 - x)$ is an expression in a and b that can be obtained from the expression for $\phi(x)$ by the interchange of a and b.

(c) Verify the result of part (b) specifically for the values of ϕ at the half-integers.

7.22 The previous exercise examined the symmetry in the values of $\phi = 2\phi$ at dyadic points in symbolic terms – the numerical values of the scaling coefficients or of ϕ at dyadic points played no role in the discussion. An alternative way of understanding the symmetry is in terms of the numerical values. Specifically, prove by induction over the dyadic levels that $\phi(3 - x) = \overline{\phi(x)}$ for all $x \in \mathbb{D}$. Verify this result directly for the values of ϕ at the half-integers.

7.23 Prove Lemma 7.10, the result on the partition of the identity, using induction over the dyadic levels.

7.24 Use the partition of unity and partition of the identity properties (Theorem 6.18 and Lemma 7.10) to prove Lemma 7.11.

7.25 Use Lemma 7.11 and the scaling equation to prove Lemma 7.12.

7.26 Use Lemma 7.12 to find expressions for $\phi = 2\phi$ on each of the quarter-integer intervals within $[0, 3]$ in terms of its values in $[0, 1]$.

7.27 Use induction over the dyadic levels to prove Theorem 7.13, with the omission of the condition that $\alpha \neq 0$. (Proving that $\alpha \neq 0$ as part of the induction seems difficult, and a proof of this condition by different means is left for the next exercise.) Check that the induction argument can be refined to yield the extra information that $\alpha, \beta, \gamma \in \mathbb{D}(\sqrt{3})$.

7.28 (a) Show that $\phi = {}_2\phi$ cannot be a linear function on any dyadic interval within isupp $\phi = [0, 3]$. (Assume that ϕ is linear on some such dyadic interval I, taken for concreteness to lie within $[0, 1]$ (the other possibilities can be dealt with similarly). Use the properties of partition of unity and partition of the identity to find a linear combination of $\phi(x)$, $\phi(x + 1)$ and $\phi(x + 2)$ that is identically 0 on I, and show that this contradicts the findings of Exercise 6.28.)

   (b) Deduce that the constants α in Theorem 7.13 and Exercise 7.27 must be non-zero for all j and k.

7.29 Show that $\phi = {}_2\phi$ is differentiable from the left at 3, with derivative 0. (It is required to prove that

$$\lim_{x \to 3^-} \frac{\phi(3) - \phi(x)}{3 - x} = 0.$$

Consider x such that $3 - 1/2^j \leq x < 3 - 1/2^{j+1}$, show that

$$\left| \frac{\phi(3) - \phi(x)}{3 - x} \right| = \left| \frac{\phi(x)}{3 - x} \right| \leq M \, |a_3|^j \, 2^{j+1}$$

for some constant M that is independent of j, and deduce the result.)

7.30 The graph of $\phi = {}_2\phi$ makes it impossible to doubt that $\phi(x)$ is strictly positive for all x such that $\frac{1}{2} \leq x \leq 1$, but since this information is needed in the next exercise, here is the outline of a formal argument to prove it.

   (a) Consider the sequence of approximations $\phi^{(0)}, \phi^{(1)}, \phi^{(2)}, \ldots$ to ϕ produced by Theorem 7.9. Observe that

$$\max_{0 \leq x \leq 3} \left| \phi^{(0)} \right| = \phi^{(0)}(1) = 2a_0 \leq 2,$$

   use Lemma 7.12 and induction to show that $\max_{0 \leq x \leq 3} \left| \phi^{(m)}(x) \right| \leq 2$ for all $m \geq 0$, and deduce that $\max_{0 \leq x \leq 3} \left| \phi(x) \right| \leq 2$. (This is a rather weak upper bound on ϕ, but it is enough for our purposes here.)

   (b) Use a similar argument to show that $\max_{0 \leq x \leq 3} \left| \phi^{(m)}(x) - \phi(x) \right| \leq 4a_0^m$ for all m.

   (c) Confirm that $4a_0^m \leq 1/4$ for $m \geq 6$, show by a direct argument that $\phi^{(6)}(x) \geq 2a_0^2$ whenever $\frac{1}{2} \leq x \leq 1$, and deduce that $\phi(x) \geq a_0^2$ for $\frac{1}{2} \leq x \leq 1$.

7.31 Show that the right-hand derivative of $\phi = {}_2\phi$ at 0 is infinite. More precisely, show that $\lim_{x \to 0^+} \phi(x)/x$, whose value, if defined, would be the right-hand derivative of ϕ at 0, diverges to ∞. (Use an argument much like that for Exercise 7.29, making use of the fact from Exercise 7.30 that there is a positive constant c such that $\phi(x) \geq c$ for all $x \in [1/2, 1]$.)

7.32 Use the scaling equation for $\phi = {}_2\phi$ to find an explicit expression for $\phi(3 - 1/2^j)$ for $j \geq 0$. Deduce that ϕ changes sign infinitely often in the interval $(3 - \epsilon, 3]$ for every $\epsilon > 0$.

7.33 The comment was made in Subsection 7.7.1 that the marked asymmetry of the Daubechies scaling functions diminishes a little in the corresponding wavelets. Investigate qualitatively why this is the case, by considering the structure of the wavelet equation given by Corollary 6.14.

8

Wavelets in the Fourier Domain

8.1 Introduction

We have noted in previous chapters that the usual approach to the theory of wavelets makes use of the Fourier transform. Application of the transform allows all the problems we want to solve to be 'translated' into a different form, where a different range of techniques can usually be applied to help solve them; and since there is also a 'translation' back in the other direction, solutions obtained in this way give us solutions to the original problems.

Anything approaching a full discussion of this approach is impossible in this book. If the Fourier transform is to be discussed in any depth, a substantial body of other theory (Lebesgue integration theory, in particular) needs to be developed first. A proper discussion of the transform itself is then a non-trivial application of this background work, and the application to wavelets itself is again non-trivial. While all this work certainly repays the effort, this book is not the place to attempt it.

Nevertheless, the Fourier transform approach is the subject of this chapter. We aim to explore the structure of wavelet theory with the extra insight given by use of the Fourier transform, though necessarily without many proofs. As in the previous chapters, it is possible to proceed usefully in this way – as, for example, in our discussion of Lebesgue integration earlier, where we introduced just enough of the ideas and results to allow our discussion to proceed. Thus although the going will be somewhat easier for a reader with a deeper level of background knowledge than that relied on in earlier chapters, we do not assume such knowledge.

8.2 The Complex Case

Our attention so far in this text has been exclusively on real-valued structures: our vector and inner product space theory has been for spaces over the field of real numbers, and the specific space $L^2(\mathbb{R})$ which has been the location for our wavelet theory is a space of real-valued functions. However, the Fourier transform is intrinsically an operator on *complex-valued* functions, so it is necessary for us to spend some time systematically shifting our focus from the real to the complex case. Thus we will need to discuss vector spaces and (especially) inner product spaces over the field of complex numbers, and the space of complex-valued square-integrable functions on $\mathbb{R}$. Further, we would like to be able to use in this chapter results from the wavelet theory of earlier chapters, so we must discuss the changes that are needed to allow the transfer of results from the real theory into the complex theory. Fortunately, as will become clear quickly, doing all of this is largely a straightforward exercise.

8.2.1 Complex-Valued Functions

We consider complex-valued functions of a real variable, that is, functions $f \colon \mathbb{R} \to \mathbb{C}$. Any such function is made up of a **real part** and an **imaginary part**. In fact, since $f(x)$ is a complex number for each fixed $x \in \mathbb{R}$, we can write $f(x) = u(x) + iv(x)$ for some real numbers $u(x)$ and $v(x)$, and then the functions $u, v \colon \mathbb{R} \to \mathbb{R}$ are the real and imaginary parts of f, respectively. (Remember, despite the name, that the imaginary part of a complex number is a real number; the imaginary part of a complex-valued function is likewise a real-valued function.)

We did not attempt to give a precise or formal definition of the Lebesgue integrability of real-valued functions in earlier chapters, and neither will we do so here for complex-valued functions. However, nothing essentially new is needed in the complex case, because *a complex-valued function $f = u + iv$ is Lebesgue-integrable if and only if its real and imaginary parts u and v are Lebesgue-integrable, and*

$$\int_{-\infty}^{\infty} f = \int_{-\infty}^{\infty} u + i \int_{-\infty}^{\infty} v.$$

The algebraic rules that are familiar from the real case carry over straightforwardly to the complex case. For example, the last equation above generalises to the following statement: *if f and g are integrable complex-valued functions and α and β are complex constants, then $\alpha f + \beta g$ is integrable, and*

$$\int_{-\infty}^{\infty} \alpha f + \beta g = \alpha \int_{-\infty}^{\infty} f + \beta \int_{-\infty}^{\infty} g.$$

Consequently, in our later work, it will not be necessary to explain or justify manipulations of integrals in any more detail in the complex case than we have so far done in the real case.

Although we have only discussed functions defined and integrated over the whole real line above, everything we have said applies equally well to complex-valued functions on, for example, intervals I of bounded length – most notably, the interval $I = [-\pi, \pi]$.

Many of the standard procedures for evaluating specific integrals also carry over to the complex case. As an illustration, we have

$$\int_0^\pi e^{ix}\,dx = (1/i)\left[e^{ix}\right]_0^\pi = -i\left[e^{i\pi} - 1\right] = -i[-1-1] = 2i.$$

Note that we could have split the complex exponential here into real and imaginary parts, using Euler's formula, but this is unnecessary and actually makes the problem more complicated:

$$\int_0^\pi e^{ix}\,dx = \int_0^\pi \cos x + i\sin x\,dx = \left[\sin x - i\cos x\right]_0^\pi = [i - (-i)] = 2i.$$

8.2.2 Complex Vector Spaces

If we wish to work with vector spaces over the complex field, the axioms for a vector space in Section 2.1 require no change beyond the obvious replacement of references to real numbers by references to complex numbers. Further, the subspace criterion given in Theorem 2.1 holds without change in the complex case; see Exercise 8.1.

Our collection of standard vector spaces also remains much as it was in the real case, with a few more or less obvious changes. Briefly, the complex counterparts of our earlier real spaces are as follows.

- For each $n \in \mathbb{N}$, the collection $\mathbb{C}^n$ of all n-tuples of complex numbers is a vector space over the field $\mathbb{C}$ if vector addition and scalar multiplication are defined entry by entry (see Exercise 8.2).
- The set $\mathbb{C}^\infty$, defined in exact analogy to $\mathbb{R}^\infty$, is a vector space over $\mathbb{C}$. Further, the collection

$$\ell^2 = \left\{(a_1, a_2, a_3, \ldots) \in \mathbb{C}^\infty : \sum_{n=1}^\infty |a_n|^2 \text{ converges}\right\}$$

of complex square-summable sequences forms a vector subspace of $\mathbb{C}^\infty$. Note that the absolute value signs are essential in the complex case (indeed, the collection would not form a vector space without them; see Exercise 8.4), though we can of course think of them as being present implicitly in the real case too. The most convincing justification of their presence, however, comes below, when we discuss ℓ^2 as a complex inner product space.

- The set $F(I)$ of all complex-valued functions on an interval I is a vector space if addition and scalar multiplication are defined pointwise, and $C(I)$, the subset of continuous complex-valued functions, is a subspace.
- Most significant for the work of this chapter, however, is $L^2(I)$, the vector space of square-integrable complex-valued functions on an interval I, that is, the functions $f: I \to \mathbb{C}$ such that

$$\int_I |f|^2 < \infty.$$

Not unexpectedly, the most important case for us will be that of $L^2(\mathbb{R})$. Note that, much as for ℓ^2, the integrand here is no longer the square of f as in the real case, but the square of the absolute value of f. The reasons for this are moreover essentially the same as in the case of ℓ^2.

Note that in the cases of ℓ^2 and $L^2(I)$, we are reusing the notation for the real case in the complex case. In this chapter, the notation will always refer to the complex case unless noted otherwise. If we need to consider the space of real-valued square-integrable functions $L^2(I)$, we will refer to it as *the real space* $L^2(I)$, and for emphasis or clarity we will also sometimes refer explicitly to *the complex space* $L^2(I)$.

8.2.3 Complex Inner Product Spaces

The definition of an inner product space V over the complex numbers, in contrast to the definition of a complex vector space, does require a significant change from the real case. The first of the inner product space axioms in Definition 3.1, the axiom of symmetry, must be changed to the property of **conjugate symmetry**:

$$\langle x, y \rangle = \overline{\langle y, x \rangle} \quad \text{for all } x, y \in V.$$

This new form of the axiom gives $\langle x, x \rangle = \overline{\langle x, x \rangle}$ for all $x \in V$, so it follows immediately that $\langle x, x \rangle$ is a real number for all $x \in V$. This last assertion is often added explicitly to the fourth inner product space axiom, since the first part of that axiom says that $\langle x, x \rangle \geq 0$, and numerical comparison of two complex numbers is impossible unless both are real (see Exercise 8.10).

Given that $\langle x, x \rangle$ is real and non-negative, we can define the norm of x exactly as in the real case by the formula $\|x\| = \langle x, x \rangle^{1/2}$, and of course the value of the norm itself is also a non-negative real number.

Most of the elementary results from Chapter 3 carry through to the complex case with little or no change. In particular, the very important Cauchy–Schwarz inequality (Theorem 3.7) and the three standard properties of a norm (Theorem 3.8), including the triangle inequality, are unchanged in the complex case, though minor changes are needed in some of the proofs. (See Exercise 8.7. Also, see Exercise 8.8 for a result that does not carry over to the complex case without change.)

Further, all the standard complex vector spaces of Subsection 8.2.2 form complex inner product spaces, provided we make one simple and systematic change in the definitions of the corresponding inner products. We list all of these briefly, although the final one is of by far the most importance here.

- For $x = (x_1, x_2, \ldots, x_n)$ and $y = (y_1, y_2, \ldots, y_n)$ in $\mathbb{C}^n$, we define

$$\langle x, y \rangle = \sum_{i=1}^{n} x_i \overline{y_i}.$$

- For $x = (x_1, x_2, x_3, \ldots)$ and $y = (y_1, y_2, y_3, \ldots)$ in ℓ^2, we similarly define

$$\langle x, y \rangle = \sum_{n=1}^{\infty} x_n \overline{y_n}.$$

- If I is a closed, bounded interval, we define an inner product on $C(I)$ by specifying that for $f, g \in C(I)$,

$$\langle f, g \rangle = \int_I f \overline{g}.$$

- Similarly, but now for an arbitrary interval I, we define an inner product on the complex space $L^2(I)$ by setting

$$\langle f, g \rangle = \int_I f \overline{g}$$

for $f, g \in L^2(I)$. Note that the problem with the fourth inner product space axiom that arose in the real case (see Subsection 3.2.3) arises again in the complex case, and is overcome in an identical way: by passing to a suitable space of equivalence classes.

In Chapters 3 and 4, we developed a significant body of theory about inner product spaces, all on the basis that the spaces were real spaces. At various points in the present chapter, we need to call upon those earlier results, but this time in the complex case rather than the real case. One notable example concerns the two main theorems on projections and Fourier series in Hilbert spaces, Theorems 4.38 and 4.39. Although we noted earlier that there are a few minor differences between the real and complex cases, all of the significant results, such as the two just cited, are valid in both cases. In this chapter, we will therefore make use of the complex versions of real results as needed and without special comment. (In any doubtful case, the reader can of course work through the proof and rewrite it in complex form.)

8.2.4 Fourier Series in the Complex Domain

We show briefly in this subsection how reformulation of the theory of trigonometric Fourier series in the broader context of complex-valued functions makes the theory simpler and more uniform. (There is a sense in which the complex domain is really the 'right' domain for this theory, and this is even clearer in the case of the Fourier transform.) Many details of the discussion are left as exercises, which are straightforward in all cases (see Exercises 8.11 and 8.12).

We know that in the real space $L^2([-\pi, \pi])$ the collection consisting of the constant function $1/\sqrt{2\pi}$ and the trigonometric functions $(1/\sqrt{\pi}) \cos nx$ and $(1/\sqrt{\pi}) \sin nx$ for $n \in \mathbb{N}$ forms an orthonormal basis, which we will denote by B for the present discussion. As a consequence, every function f in the real space $L^2([-\pi, \pi])$ has a Fourier series expansion with respect to B, in which the Fourier coefficients are the real numbers obtained by projection of f onto the respective elements of B.

Turning our attention to the complex space $L^2([-\pi, \pi])$, we observe almost immediately that the same collection of functions B remains an orthonormal basis. (The basis functions are of course real-valued, but $\mathbb{R} \subseteq \mathbb{C}$, so it is also technically correct to refer to them as complex-valued.) Checking the orthonormality of B is completely straightforward: the inner products that need to be evaluated are identical to those of the real case because the complex conjugation that is involved is here applied to real arguments. Checking the completeness of B is also easy, given its completeness in the real case. In fact, let $f = u + iv$ be in $L^2([-\pi, \pi])$ and suppose that for each function $b \in B$ we have $\langle f, b \rangle = 0$. Then

$$\int_{-\pi}^{\pi} fb = \int_{-\pi}^{\pi} (u + iv)b = \int_{-\pi}^{\pi} ub + i \int_{-\pi}^{\pi} vb = 0,$$

which gives

$$\int_{-\pi}^{\pi} ub = 0 \quad \text{and} \quad \int_{-\pi}^{\pi} vb = 0$$

for all $b \in B$. But from the completeness of the collection B in the real space $L^2([-\pi, \pi])$ it follows that $u = v = 0$, and so $f = u + iv = 0$, proving that B is also complete in the complex space $L^2([-\pi, \pi])$.

Thus, every function f in the complex space $L^2([-\pi, \pi])$ has a Fourier series expansion with respect to the basis B; the coefficients are defined, as in the real case, by projection of f onto each element of B, but the projection is now carried out by use of the complex inner product. Of course, if f is non-trivially complex-valued (that is, not merely real-valued), then at least some of its Fourier coefficients must turn out to be non-trivially complex-valued as well; the fact that the same collection B now spans a bigger collection of functions is only possible because we now have a bigger collection of scalars available as coefficients – all of $\mathbb{C}$, rather than just $\mathbb{R}$.

In the complex case, however, we can find a complete orthonormal basis which, while very closely related to the trigonometric basis, represents functions in $L^2(\mathbb{R})$ more uniformly and 'efficiently'. This is through the use of the complex exponential function and Euler's formula

$$e^{ix} = \cos x + i \sin x$$

for real x (the formula is in fact valid for all complex arguments, but the real values are the ones of interest here, since we are dealing with trigonometric functions of a real argument). Our new orthonormal basis consists of the functions

$$\frac{1}{\sqrt{2\pi}} e^{inx} \quad \text{for } n \in \mathbb{Z}.$$

Note that we can reconstruct all the members of the trigonometric basis B easily from the exponential basis: taking $n = 0$ in the exponent gives us the constant function $1/\sqrt{2\pi}$, while suitable linear combinations of e^{inx} and e^{-inx} for $n \in \mathbb{N}$ yield the trigonometric functions in B.

It is a simple integration exercise to verify that the above collection of exponential functions is an orthonormal set, and it is also easy to deduce the completeness of the collection from that of the trigonometric functions: any function in $L^2([-\pi, \pi])$ that is orthogonal to all the exponentials must be orthogonal to all the trigonometric functions, and hence must be the zero function.

Every function f in the complex space $L^2([-\pi, \pi])$ therefore has a Fourier series with respect to the complex exponentials, in which the Fourier coefficient corresponding to $(1/\sqrt{2\pi})e^{inx}$ is the component of f on $(1/\sqrt{2\pi})e^{inx}$, that is,

$$\left\langle f, (1/\sqrt{2\pi})e^{inx} \right\rangle = \frac{1}{\sqrt{2\pi}} \int_{-\pi}^{\pi} f(x)\overline{e^{inx}}\, dx = \frac{1}{\sqrt{2\pi}} \int_{-\pi}^{\pi} f(x)e^{-inx}\, dx.$$

(Note that in a complex inner product space, the inner product is only conjugate symmetric, not symmetric, so, in contrast to the real case, the order in which we form the inner product of the two elements matters.) If the Fourier coefficient is denoted by c_n for each $n \in \mathbb{Z}$, the Fourier series for f is therefore

$$f(x) = \frac{1}{\sqrt{2\pi}} \sum_{n=-\infty}^{\infty} c_n e^{inx},$$

where

$$c_n = \frac{1}{\sqrt{2\pi}} \int_{-\pi}^{\pi} f(x) e^{-inx} \, dx$$

for each $n \in \mathbb{Z}$.

➤ Much as in the case of the trigonometric Fourier series, this procedure gives rise to two factors of $1/\sqrt{2\pi}$, one from the Fourier coefficient and one from the corresponding basis element, though once we substitute the expression for c_n into the summation, of course, a single multiplicative factor of $1/2\pi$ will result. In many circumstances, it is more natural and convenient to work with the orthogonal but unnormalised basis elements e^{inx}, and to redefine the Fourier coefficients so as to produce the correct multiplicative constant $1/2\pi$. In the specific context of this book, however, we have good reasons for wanting to deal with bases that are orthonormal, and so we are willing to tolerate the minor inconvenience of having two separate factors of $1/\sqrt{2\pi}$.

As for an ordinary infinite series, a doubly infinite sum such as $\sum_{n=-\infty}^{\infty} c_n e^{inx}$ must be defined as the limit of a sequence of partial sums, but we have not so far specified the summation range for the partial sums in the doubly infinite case. Our convention is that the partial sums are *symmetric*: the kth partial sum is defined to be

$$\sum_{n=-k}^{k} c_n e^{inx},$$

and the infinite sum is the limit of these finite sums as $k \to \infty$. (However, as noted in the real case in the comments just preceding Subsection 4.6.1, the order of summation of a complex Fourier series cannot affect the sum, so if a different convention for the partial sums were adopted, it would make no difference to the sum of the series.)

8.2.5 Trigonometric Polynomials

Sums of the form $\sum_{n=-k}^{k} c_n e^{inx}$ are known as **trigonometric polynomials**; here, k is any non-negative integer and $c_n \in \mathbb{C}$ for all n. Thus, from the discussion just above, we could have defined trigonometric polynomials as the partial sums of Fourier series. Alternatively, we could say that trigonometric polynomials are Fourier series that have only a finite number of non-zero Fourier coefficients; note that we can always express the summation in such a Fourier series over a symmetric summation range, by including some terms with zero Fourier coefficients if necessary.

We will see in Sections 8.4 and 8.5 below that trigonometric polynomials arise in a very natural way when we discuss wavelets of bounded support as viewed from the Fourier domain. We do not require much of their theory here, though at least one significant result about them would be required for a complete account of the work of later sections. Several fairly straightforward but interesting results about trigonometric polynomials are left for the exercises.

➤ Depending on the context, it is convenient to be able to switch between thinking of a trigonometric polynomial as a function with domain $[-\pi, \pi]$ and as a function with domain $\mathbb{R}$. Regarded as a function on $\mathbb{R}$, a trigonometric polynomial is of course periodic of period 2π and its values outside of $[-\pi, \pi]$ are obtained by periodic extension of its values inside $[-\pi, \pi]$.

Note that as a function on $[-\pi, \pi]$ a trigonometric polynomial belongs to $L^2([-\pi, \pi])$, while as a function on $\mathbb{R}$ a trigonometric polynomial cannot be in $L^2(\mathbb{R})$ unless it is identically zero (see Exercise 8.15).

8.2.6 Wavelet Theory in the Complex Case

In Chapters 6 and 7, we developed a significant body of the general theory of wavelets in the real space $L^2(\mathbb{R})$, with a focus on the case of wavelets of bounded support. All of this theory carries over to the complex case with few changes (much as with the theory of inner product spaces, as noted at the end of Subsection 8.2.3). The only changes required, in fact, are those needed to reflect the changed definition of the inner product in the complex space $L^2(\mathbb{R})$: complex conjugate signs are needed at certain points. Since we wish to be able to draw on already established results from the real theory and use them in the complex case, we will briefly survey the work of Chapters 6 and 7, noting a few specific changes that are needed.

First, the definition of a multiresolution analysis, Definition 6.1, requires no formal change. (Of course, the context for the definition is now the complex space $L^2(\mathbb{R})$, with its complex inner product, but this difference is not explicit in the definition – the inner product only enters into the definition implicitly in the form of orthonormality relations, for example.) Similar comments apply to Theorem 6.6, which gives a condition on a function and a multiresolution analysis that is sufficient to ensure that the function is a wavelet. The definition of a wavelet itself is also formally unchanged.

The first point at which a formal change is required is in the discussion of the scaling equation and the scaling coefficients in Subsection 6.3.3. The scaling equation reads

$$\phi(x) = \sum_{k \in \mathbb{Z}} a_k \phi(2x - k)$$

as before, but the scaling coefficients, since they are explicitly defined as inner products in the complex space $L^2(\mathbb{R})$, now acquire a complex conjugate sign:

$$a_k = 2 \int_{-\infty}^{\infty} \phi(x) \overline{\phi(2x - k)} \, dx \quad \text{for } k \in \mathbb{Z}.$$

An analogous change is required in Theorem 6.9, which constructs a wavelet from the scaling equation and scaling coefficients; the formula for the wavelet is now

$$\psi(x) = \sum_{k \in \mathbb{Z}} (-1)^k \overline{a_{1-k}} \, \phi(2x - k).$$

Finally, the fundamental relations of Section 6.4 require a similar change. Working with the relations in the form given in (6.12), the relations now take the form

$$\left. \begin{array}{c} \displaystyle\sum_{k \in \mathbb{Z}} a_k \, \overline{a_{k+2\ell}} = 2\delta_{0,\ell} \quad \text{for all } \ell \in \mathbb{Z}, \\[2ex] \displaystyle\sum_{k \in \mathbb{Z}} a_k = 2. \end{array} \right\}$$

(For the form of the fundamental relations given in (6.13), with explicit summation bounds, no separate discussion is needed here, since the bounds are identical in the complex case.)

➤ If a scaling function is real-valued, the relevant definitions show immediately that the corresponding scaling coefficients and wavelet are also real-valued, and all the complex conjugate signs introduced above could therefore be omitted. However, we are working with the complex theory in this chapter in order to study the application of the Fourier transform, for which the complex domain is the most natural one, and so we have no need to single out the real case in any special way in the discussion. (Note that wavelets can indeed be genuinely complex-valued – see Exercise 8.18 – even though those in practical use are almost invariably real-valued.)

In Section 7.4, we noted that the solutions to the equations that govern the scaling coefficients in the Daubechies case have complex solutions when $N \geq 3$ as well as the desired real solutions. However, these complex solutions are of no more relevance now, in the complex case, than they were in the real case, because the systems of equations to which they are solutions are those appropriate to the real case, not the complex case, in which complex conjugation operations are present; the complex solutions are spurious in both cases.

8.3 The Fourier Transform

The Fourier transform is a certain mapping $\mathcal{F} \colon L^2(\mathbb{R}) \to L^2(\mathbb{R})$ which is a **Hilbert space isomorphism**. This means that $\mathcal{F}$ has the following properties.

- $\mathcal{F}$ is one-to-one and onto.
- $\mathcal{F}$ is linear; that is, $\mathcal{F}(\alpha f + \beta g) = \alpha \mathcal{F} f + \beta \mathcal{F} g$ for all $f, g \in L^2(\mathbb{R})$ and all $\alpha, \beta \in \mathbb{C}$.
- $\mathcal{F}$ preserves the inner product of $L^2(\mathbb{R})$; that is, $\langle \mathcal{F} f, \mathcal{F} g \rangle = \langle f, g \rangle$ for all $f, g \in L^2(\mathbb{R})$.

Observe that these three conditions tell us that $\mathcal{F}$ respects the structure of $L^2(\mathbb{R})$ in all its aspects: its structure as a set in the first condition, as a vector space in the second and as an inner product space in the third. Since the norm is an integral feature of an inner product space, we would hope that the norm is also preserved (that is, that $\|\mathcal{F}f\| = \|f\|$ for all f), and this is indeed the case (see part (c) of Exercise 8.8).

This description of $\mathcal{F}$ is significant as far as it goes, but of course it is not a definition of $\mathcal{F}$: there are many Hilbert space isomorphisms from $L^2(\mathbb{R})$ to $L^2(\mathbb{R})$, including the identity mapping. Let us now turn to the question of the definition.

Although we ultimately wish to define the Fourier transform as a function with domain $L^2(\mathbb{R})$, the natural space on which to define it initially is the space $L^1(\mathbb{R})$. Recall that if I is a non-trivial interval on the real line, then $L^1(I)$ is the vector space of complex-valued functions f on I with the property that $\int_I |f| < \infty$. This space was introduced briefly (in the real case) in Exercise 4.42, and we note two facts from that exercise (valid also in the complex case) that are relevant here. The first is that neither of $L^1(\mathbb{R})$ and $L^2(\mathbb{R})$ is contained in the other. The second is that if I is a bounded interval, then $L^1(I)$ is not contained in $L^2(I)$, but $L^2(I)$ is contained in $L^1(I)$.

Definition 8.1   Let $f \in L^1(\mathbb{R})$. Then the **Fourier transform of** f is the complex-valued function $\hat{f}$ given by

$$\hat{f}(\xi) = \frac{1}{\sqrt{2\pi}} \int_{-\infty}^{\infty} f(x) e^{-i\xi x}\, dx$$

for $\xi \in \mathbb{R}$.

The reason for the imposition here of the condition that $f \in L^1(\mathbb{R})$ is straightforward in essence: the modulus of the integrand above,

$$\left| f(x) e^{-i\xi x} \right| = |f(x)|,$$

must have finite integral over $\mathbb{R}$ to give a well-defined function $\hat{f}$, and this is the condition that $f \in L^1(\mathbb{R})$.

➤ For a specific function f, the discussion above allows us to denote the Fourier transform of f either by $\mathcal{F}f$ or by $\hat{f}$, but we will mostly use the notation $\hat{f}$. (When the expression for the function involves more than a single symbol, we may for purely typographical reasons use this notation in a variant form, as in $(f+g)\hat{\,}$, for example.) We generally reserve the use of the symbol $\mathcal{F}$ for situations where we are thinking of the properties of the Fourier transform as a mapping in its own right from $L^2(\mathbb{R})$ to $L^2(\mathbb{R})$ (and its use in this chapter is confined to the present section).

We analogously define the inverse Fourier transform as follows.

Definition 8.2 Let $g \in L^1(\mathbb{R})$. Then the **inverse Fourier transform of** g is the complex-valued function given by

$$\frac{1}{\sqrt{2\pi}} \int_{-\infty}^{\infty} g(\xi) e^{ix\xi} \, d\xi$$

for $x \in \mathbb{R}$.

When necessary, we will denote the function just defined by $\mathcal{F}^{-1}g$, but we will not use any notation similar to the notation $\hat{f}$ used for $\mathcal{F}f$.

➤ Observe the strong similarity between the formulae defining the two transforms: they differ only in the sign of the exponent. In fact, a simple change of variable shows that the inverse Fourier transform of g is the Fourier transform of the function h defined by setting $h(\xi) = g(-\xi)$.
   A helpful implication of this is that any result proved about the Fourier transform has an obvious counterpart, identical or almost identical in form, for the inverse transform.

Both the name and the notation used above suggest that we should have $\mathcal{F}\mathcal{F}^{-1}f = f$ and $\mathcal{F}^{-1}\mathcal{F}f = f$, but this is not the case without severe restrictions, at least until we extend the transforms so that they apply to all functions in $L^2(\mathbb{R})$. This is because although both transforms, as so far defined, are applied to L^1 functions as input, they do not in general produce L^1 functions as output (see Exercise 8.19). If we make the additional assumption that both f and $\hat{f}$ are in $L^1(\mathbb{R})$, then the two transforms do indeed behave as inverses, in the following sense.

Theorem 8.3 *Suppose that $f \in L^1(\mathbb{R})$ and $\hat{f} \in L^1(\mathbb{R})$. If $g = \mathcal{F}^{-1}\mathcal{F}f$, then then $f = g$ almost everywhere.*

Our aim now is to find definitions of the Fourier transform and the inverse Fourier transform on $L^2(\mathbb{R})$, and we present two results that lead straightforwardly to this outcome. The key is to focus initially on those functions that are in *both $L^1(\mathbb{R})$ and $L^2(\mathbb{R})$*, that is, in $L^1(\mathbb{R}) \cap L^2(\mathbb{R})$ (we noted just prior to Definition 8.1 that neither of $L^1(\mathbb{R})$ and $L^2(\mathbb{R})$ contains the other). The first result shows that the Fourier transform as so far defined on $L^1(\mathbb{R})$ behaves especially well when restricted to $L^1(\mathbb{R}) \cap L^2(\mathbb{R})$; the second result shows that the transform can be extended from $L^1(\mathbb{R}) \cap L^2(\mathbb{R})$ to the whole of $L^2(\mathbb{R})$ in such a way as to preserve the good behaviour.

Theorem 8.4 *Let $f \in L^1(\mathbb{R}) \cap L^2(\mathbb{R})$. Then $\hat{f} \in L^2(\mathbb{R})$ and $\|f\| = \|\hat{f}\|$.*

Let us see what information we now have about the transform $\mathcal{F}$, considered as a mapping from $L^1(\mathbb{R}) \cap L^2(\mathbb{R})$ to $L^2(\mathbb{R})$. First, the map is linear: it is in fact linear on the whole of $L^1(\mathbb{R})$, simply by the linearity of integration. Second, the map is one-to-one. To see this, suppose that we have $f, g \in L^1(\mathbb{R}) \cap L^2(\mathbb{R})$ such that $\mathcal{F}f = \mathcal{F}g$. Then $\mathcal{F}(f - g) = (f - g)\hat{} = 0$ by linearity, so $\left\|(f - g)\hat{}\right\| = 0$. Hence, by the theorem, $\|f - g\| = 0$, and so $f = g$ (with equality as usual in the L^2 sense, which we know is really equality of functions almost everywhere.) Third, $\mathcal{F}$ preserves not only the norm but, by part (b) of Exercise 8.8, also the inner product of $L^2(\mathbb{R})$.

Theorem 8.5 *The collection $L^1(\mathbb{R}) \cap L^2(\mathbb{R})$ is dense in $L^2(\mathbb{R})$. In more detail, for every $f \in L^2(\mathbb{R})$ there is a sequence of functions $f_n \in L^1(\mathbb{R}) \cap L^2(\mathbb{R})$ such that $f_n \to f$ in $L^2(\mathbb{R})$ as $n \to \infty$, that is, $\|f - f_n\| \to 0$ as $n \to \infty$.*

The proof of this result is not hard, but requires a little Lebesgue integration theory, so we will not discuss it here. However, we do wish to note one easily specified sequence of functions $\{f_n\}$ in $L^1(\mathbb{R}) \cap L^2(\mathbb{R})$ that converges to a given f in $L^2(\mathbb{R})$. We define f_n by 'truncating' f to the interval $[-n, n]$ for each n; more precisely, we define f_n by setting

$$f_n(x) = \begin{cases} f(x), & -n \le x \le n, \\ 0, & \text{otherwise.} \end{cases}$$

(Note that using the notation for characteristic functions, we could also write $f_n = f\chi_{[-n,n]}$.) These truncated functions are in $L^1(\mathbb{R}) \cap L^2(\mathbb{R})$ and converge to f in $L^2(\mathbb{R})$, as required. (For the first of these assertions, see Exercise 8.20.)

It is now straightforward to define the Fourier transform for all functions in $L^2(\mathbb{R})$. Given $f \in L^2(\mathbb{R})$, choose a sequence $\{f_n\}$ such that $f_n \in L^1(\mathbb{R}) \cap L^2(\mathbb{R})$ for all n and $f_n \to f$ in $L^2(\mathbb{R})$ as $n \to \infty$. Now since the sequence $\{f_n\}$ converges, it is a Cauchy sequence. By Theorem 8.4, the Fourier transform on $L^1(\mathbb{R}) \cap L^2(\mathbb{R})$ preserves the norm, so it follows that the sequence $\{\hat{f}_n\}$ is also a Cauchy sequence:

$$\left\|\hat{f}_m - \hat{f}_n\right\| = \left\|(f_m - f_n)\hat{}\right\| = \|f_m - f_n\| \to 0$$

as $m, n \to \infty$. But $L^2(\mathbb{R})$ is a Hilbert space, that is, is complete, and so the sequence $\{\hat{f}_n\}$ has a limit in $L^2(\mathbb{R})$, and we define this limit to be $\hat{f}$. For this to be a satisfactory definition, it must be the case that $\hat{f}$ does not depend on the particular sequence $\{f_n\}$ converging to f that we chose at the start, and this turns out to be easy to prove.

Perhaps the simplest and most obvious choice for the sequence $\{f_n\}$ is the one discussed just above, and we use that for our formal definition.

Definition 8.6 Let $f \in L^2(\mathbb{R})$. For each $n \in \mathbb{N}$, let f_n be the 'truncation' of f to the interval $[-n, n]$. Then $\hat{f}$, the **Fourier transform of** f, is the limit of the sequence $\{\hat{f}_n\}$ in $L^2(\mathbb{R})$.

Some technical work that we will not go into (but see Exercise 8.21) shows that the extended Fourier transform $\mathcal{F}$ given by this definition is a Hilbert space isomorphism from $L^2(\mathbb{R})$ to $L^2(\mathbb{R})$, as claimed at the start of the section. Finally, the inverse Fourier transform $\mathcal{F}^{-1}$ is defined on $L^2(\mathbb{R})$ in complete analogy to $\mathcal{F}$, and it can be shown that the two transforms are indeed mutual inverses: $\mathcal{F}^{-1}\mathcal{F}f = f$ and $\mathcal{F}\mathcal{F}^{-1}f = f$ for all $f \in L^2(\mathbb{R})$.

➤ Three common conventions are in use for the definition of the Fourier transform and the inverse Fourier transform, and one needs to be careful when looking in the literature to find out which convention is in use. Many books, for example, give tables of Fourier transforms, and the formulae for the transforms will differ according to the convention.

The reason for the difference is the factor of $1/\sqrt{2\pi}$ in the expressions for both the Fourier transform and the inverse Fourier transform of an L^1 function in our formulation. This factor must be present in some form and must be carried through every calculation involving the transforms, but it need not be incorporated in the formulae in exactly the way we have done. (Recall that similar issues arose earlier in the discussion of Fourier series in the complex space $L^2([-\pi, \pi])$.)

Our convention, as above, is to include a factor of $1/\sqrt{2\pi}$ in both transforms. The second convention simplifies the Fourier transform by, in effect, putting both of the $1/\sqrt{2\pi}$ factors into the inverse transform. Thus, on this convention, we would have expressions

$$\int_{-\infty}^{\infty} \cdots \quad \text{and} \quad \frac{1}{2\pi} \int_{-\infty}^{\infty} \cdots$$

for the transform and the inverse transform, respectively, where the integrals themselves are exactly as in our formulation. The third convention eliminates the multiplicative factors altogether by complicating the integrands slightly, using the formulae

$$\int_{-\infty}^{\infty} f(x)e^{-2\pi i \xi x}\, dx \quad \text{and} \quad \int_{-\infty}^{\infty} g(\xi)e^{2\pi i x \xi}\, d\xi,$$

respectively, for the transform and the inverse transform.

8.3.1 Scaling, Dilation and Translation

Scaling, dilation and translation are at the heart of the theory of wavelets, and we will need to know shortly how the Fourier transform interacts with these operations.

To express the interactions, we introduce some notation (which we in fact introduced briefly much earlier, in Exercise 1.11). Let f be a complex-valued function on $\mathbb{R}$ and let $b, c \in \mathbb{R}$ with $b \neq 0$. Then we write $D_b f$ to denote the dilation of f by b, that is, $(D_b f)(x) = f(bx)$, and we write $T_c f$ to denote the translate of f by c, that is, $(T_c f)(x) = f(x - c)$. (We will not need a special notation for scaling.)

The following result summarises the information we need, and we refer to Exercise 8.26 for proofs.

Theorem 8.7   *Let $f \in L^2(\mathbb{R})$ and let $a, b, c \in \mathbb{R}$ with $b \neq 0$. Then*

(i) $(af)\widehat{}(\xi) = a\hat{f}(\xi)$,

(ii) $(D_b f)\widehat{}(\xi) = (1/b)\hat{f}(\xi/b)$ *and*

(iii) $(T_c f)\widehat{}(\xi) = e^{-i\xi c}\,\hat{f}(\xi)$.

➤ Strictly, we should add the qualification 'for almost all ξ' to the three equations of the theorem. Although these are expressed as equations between function values, this is for notational convenience only (especially in the case of part (iii)), and the relations in reality are relations of equality in the L^2 sense between functions.

8.4 Applications to Wavelets

The Fourier transform, as a mapping from $L^2(\mathbb{R})$ to $L^2(\mathbb{R})$, is linear, one-to-one and onto and preserves the inner product and the norm, and the inverse Fourier transform has exactly the same properties. It follows, as noted earlier, that any statement about $L^2(\mathbb{R})$ can in principle be translated into a different but equivalent statement about $L^2(\mathbb{R})$ by application of the transform; the statements will be equivalent since the translation process can be reversed by application of the inverse transform. We speak loosely of the translation being into *the Fourier domain*, even though this domain is just $L^2(\mathbb{R})$, the same as the domain that we are mapping from.

➤ In the field of mathematics known as harmonic analysis, the Fourier transform is defined on a wide class of L^2 spaces other than $L^2(\mathbb{R})$, and in general the two domains involved are then different. For example, in this general theory, complex Fourier *series* are just a particular type of Fourier *transform*, defined on $L^2([-\pi, \pi])$ rather than $L^2(\mathbb{R})$. Since the formation of its Fourier series transforms a function in $L^2([-\pi, \pi])$ to a square-summable sequence of complex numbers indexed by $\mathbb{Z}$, the space $\ell^2(\mathbb{Z})$ is the Fourier domain in this case. (In this general setting, $\ell^2(\mathbb{Z})$ could correctly be denoted by $L^2(\mathbb{Z})$. See Subsection 2.3.2 for a brief discussion of $\ell^2(\mathbb{Z})$, though we are dealing now with the complex space rather than the real space of Chapter 2.)

Our aim here is to examine the form taken by some of the most fundamental elements of the theory of wavelets under the application of the Fourier transform. We will see that their translated versions are strikingly different in character.

Given a multiresolution analysis consisting of spaces $\{V_j : j \in \mathbb{Z}\}$ and a scaling function ϕ, the crucial property of ϕ (condition (MRA6) from Chapter 6) is that the translates $\{\phi(x - k) : k \in \mathbb{Z}\}$ form an orthonormal basis for V_0. We begin by noting what form the orthonormality condition takes when translated to the Fourier domain.

Theorem 8.8   *If $\phi \in L^2(\mathbb{R})$, then the collection of translates $\{\phi(x - k) : k \in \mathbb{Z}\}$ is an orthonormal set if and only if*

$$\sum_{k\in\mathbb{Z}}\left|\hat{\phi}(\xi+2k\pi)\right|^2 = \frac{1}{2\pi} \quad \textit{for almost all } \xi.$$

Orthonormality of translates is the first important property of a scaling function ϕ. The second is that ϕ satisfies a scaling equation. We next examine the form that the scaling equation takes under translation to the Fourier domain.

The original form of the scaling equation is

$$\phi(x) = \sum_{k\in\mathbb{Z}} a_k\phi(2x-k),$$

where (refer to Subsection 8.2.6)

$$a_k = 2\int_{-\infty}^{\infty} \phi(x)\,\overline{\phi(2x-k)}\,dx.$$

In Exercise 8.22, we observe that the Fourier transform 'passes through' infinite sums in $L^2(\mathbb{R})$, and in Theorem 8.7 we analysed how the Fourier transform interacts with scaling, dilation and translation. It follows, using this information and applying the transform to both sides of the scaling equation, that

$$\hat{\phi}(\xi) = \frac{1}{2}\sum_{k\in\mathbb{Z}} a_k e^{-i(\xi/2)k}\hat{\phi}(\xi/2) = m_0(\xi/2)\hat{\phi}(\xi/2),$$

where

$$m_0(\xi/2) = \frac{1}{2}\sum_{k\in\mathbb{Z}} a_k e^{-i(\xi/2)k}.$$

Replacing $\xi/2$ by ξ throughout (purely for notational convenience), we therefore have the following.

Theorem 8.9 *The scaling equation in the Fourier domain has the form*

$$\hat{\phi}(2\xi) = m_0(\xi)\hat{\phi}(\xi),$$

where

$$m_0(\xi) = \frac{1}{2}\sum_{k\in\mathbb{Z}} a_k e^{-i\xi k}.$$

The function m_0 is often called the **scaling filter** (where the term *filter* is borrowed from the theory of signal processing).

Observe that m_0 is periodic, with period 2π, because each term $e^{-i\xi k}$ has this property. Further, the restriction of m_0 to the interval $[-\pi, \pi]$ is in $L^2([-\pi, \pi])$, and in fact the series expression for m_0 given in the theorem is (after minor rearrangement) the Fourier series for this restriction with respect to the basis of exponential functions of Subsection 8.2.4.

This follows from the fact that the coefficients in the series form a square-summable sequence, which in turn is the case because the a_k were originally defined as Fourier coefficients – the Fourier coefficients of $\phi(x/2)$ with respect to the basis $\{\phi(x - k) : k \in \mathbb{Z}\}$ of V_0. Note that the coefficients actually form a doubly-infinite square-summable sequence, since their indexing is over $\mathbb{Z}$, and so the sequence belongs to the complex space $\ell^2(\mathbb{Z})$. (For $\ell^2(\mathbb{Z})$, see the brief remark near the start of this section, and see Subsection 2.3.2 for a brief discussion of the corresponding real space $\ell^2(\mathbb{Z})$.)

▶ The original scaling equation $\phi(x) = \sum_{k \in \mathbb{Z}} a_k \phi(2x - k)$, though expressed in terms of function values for convenience, is really an equality in the L^2 sense, since it is the expression of the equality between a function and its Fourier series in a subspace of $L^2(\mathbb{R})$. The Fourier transform preserves this L^2 equality, and so the transformed scaling equation $\hat{\phi}(\xi) = \frac{1}{2} \sum_{k \in \mathbb{Z}} a_k e^{-i(\xi/2)k} \hat{\phi}(\xi/2)$ again represents an L^2 equality.

However, after we manipulate the last equation into the form $\hat{\phi}(2\xi) = m_0(\xi)\hat{\phi}(\xi)$, we can no longer claim L^2 equality. The reason for this is simple: the function m_0 is not in $L^2(\mathbb{R})$. It belongs to $L^2([-\pi, \pi])$, as noted above, but since it is not the zero function its norm in $L^2([-\pi, \pi])$ is non-zero, and since it is periodic the integral that would define its norm in $L^2(\mathbb{R})$ must therefore be infinite. (See also Exercise 8.15.)

It is nevertheless correct to say that the equation $\hat{\phi}(2\xi) = m_0(\xi)\hat{\phi}(\xi)$ holds for almost all ξ: since the equation $\hat{\phi}(\xi) = \frac{1}{2} \sum_{k \in \mathbb{Z}} a_k e^{-i(\xi/2)k} \hat{\phi}(\xi/2)$ holds in the L^2 sense, it holds almost everywhere, and the expressions $\frac{1}{2} \sum_{k \in \mathbb{Z}} a_k e^{-i(\xi/2)k} \hat{\phi}(\xi/2)$ and $m_0(\xi/2)\hat{\phi}(\xi/2)$ are equal pointwise, simply by definition of $m_0(\xi)$.

From Chapter 6 and (especially) Chapter 7, we know that to find a scaling function and wavelet all we need to do is to find a suitable sequence of scaling coefficients. It is therefore important to note that this problem is equivalent to the problem of finding a suitable scaling filter m_0. (This is just another instance of the principle noted earlier that the Fourier transform 'translates' information into equivalent information in the Fourier domain.) Theorem 8.9 above tells us explicitly that the sequence of scaling coefficients $\{a_k : k \in \mathbb{Z}\}$ of a scaling equation gives rise to m_0, but the reverse is also true, because the series expression for m_0 is a Fourier series, and the coefficients of a Fourier series with respect to a given orthonormal basis can be recovered from the series by calculation of its inner products with the elements of the basis. Thus the main issue we face when using Fourier transform methods to find wavelets is to construct suitable scaling filters m_0. (Of course, just as in the analysis of Chapter 7, there are non-trivial questions about what 'suitable' means in the Fourier domain, and results below give information on this.)

The wavelet equation is derived in an exactly parallel way to the scaling equation: the wavelet ψ lies in V_1, and this gives rise to a Fourier series expansion

$$\psi(x) = \sum_{k \in \mathbb{Z}} b_k \phi(2x - k),$$

where

$$b_k = 2 \int_{-\infty}^{\infty} \psi(x) \overline{\phi(2x - k)} \, dx.$$

We can then translate these equations into the Fourier domain much as in the case of the scaling equation, as follows.

Theorem 8.10 *The wavelet equation in the Fourier domain has the form*

$$\hat{\psi}(2\xi) = m_1(\xi)\hat{\phi}(\xi),$$

where

$$m_1(\xi) = \frac{1}{2} \sum_{k \in \mathbb{Z}} b_k e^{-i\xi k}.$$

The function m_1 is often called the **wavelet filter**. Again in parallel with the scaling function case, m_1 is periodic with period 2π and the series for m_1 is the Fourier series for the restriction of m_1 to $[-\pi, \pi]$.

From Chapter 6, we know that the coefficients a_k and b_k are far from independent; in fact each sequence determines the other through the equation $b_k = (-1)^k \overline{a_{1-k}}$, which is a reformulation of the conclusion of Theorem 6.9 in the complex case. Because of this close relation between the a_k and the b_k, we should expect to find that there is also a close relation between m_0 and m_1, and this relation is expressed in the following result.

Theorem 8.11 *Suppose that ϕ is the scaling function of a multiresolution analysis and that ψ is the corresponding wavelet. Let m_0 and m_1 denote the scaling filter and wavelet filter, respectively. Then the following equations are satisfied for almost all ξ:*

$$\left| m_0(\xi) \right|^2 + \left| m_0(\xi + \pi) \right|^2 = 1,$$

$$\left| m_1(\xi) \right|^2 + \left| m_1(\xi + \pi) \right|^2 = 1,$$

$$m_0(\xi) \overline{m_1(\xi)} + m_0(\xi + \pi) \overline{m_1(\xi + \pi)} = 0.$$

➤ We note in passing that the functions m_0 and m_1 are often referred to together (and sometimes separately) as **quadrature mirror filters**. The use of the term *filter* has already been commented on. The introduction of the word *quadrature* is related to the squaring operation involved in the first two equations in the theorem, while the word *mirror* is a reflection of the fact that $e^{i\xi}$ and $e^{i(\xi+\pi)}$ are at mirror-image positions on the unit circle in the complex plane.

In the case of a scaling function and wavelet of bounded support, only a finite number of the scaling and wavelet coefficients are non-zero, and it follows immediately from their definitions above that *the scaling and wavelet filters m_0 and m_1 are both trigonometric polynomials in the case of bounded support.*

In Subsections 7.5.1 and 7.5.3, we discussed results on the existence of scaling functions in the bounded support case and results that imply algorithms for their construction. We now introduce a result in the Fourier domain that gives roughly parallel information.

Theorem 8.12 *Let $m(\xi)$ be a trigonometric polynomial such that*

$$|m(\xi)|^2 + |m(\xi + \pi)|^2 = 1 \quad \text{for all } \xi \in \mathbb{R}, \tag{8.1}$$

$$m(0) = 1 \tag{8.2}$$

and

$$m(\xi) \neq 0 \quad \text{for } -\pi/2 \leq \xi \leq \pi/2. \tag{8.3}$$

Then the formula

$$\hat{\phi}(\xi) = \frac{1}{\sqrt{2\pi}} \prod_{j=1}^{\infty} m(2^{-j}\xi) \tag{8.4}$$

defines a function $\hat{\phi} \in L^2(\mathbb{R})$, and $\hat{\phi}$ is the Fourier transform of a function $\phi \in L^2(\mathbb{R})$ which is of bounded support and is the scaling function of a multiresolution analysis of which $m(\xi)$ is the scaling filter.

Theorem 7.8 showed how the scaling equation can be used to generate a scaling function of bounded support by a method of successive approximation. Theorem 8.9 shows that the scaling equation in the Fourier domain takes the form

$$\hat{\phi}(2\xi) = m_0(\xi)\hat{\phi}(\xi), \tag{8.5}$$

and it is interesting to observe that the infinite product formula (8.4) for the Fourier transform of a scaling function given by Theorem 8.12 arises naturally as a translation of the successive approximation process into the Fourier domain.

Specifically, if we apply the formula (8.5) repeatedly, we obtain

$$\hat{\phi}(\xi) = m_0(\xi/2)\,\hat{\phi}(\xi/2)$$
$$= m_0(\xi/2)\,m_0(\xi/4)\,\hat{\phi}(\xi/4)$$
$$= m_0(\xi/2)\,m_0(\xi/4)\,m_0(\xi/8)\,\hat{\phi}(\xi/8)$$
$$\vdots$$

and, in general,

$$\hat{\phi}(\xi) = \left(\prod_{j=1}^{n} m_0(\xi/2^j)\right)\hat{\phi}(\xi/2^n)$$

for all $n \in \mathbb{N}$. It can be shown that under appropriate assumptions, such as those in the theorem, we can pass from this sequence of finite products to the infinite product expression for $\hat{\phi}$ in Equation (8.4) of the theorem, though the argument is non-trivial.

➤ Although it does not help much with the main difficulty in the proof of the theorem, it is worth explaining the disappearance of the factor involving $\hat{\phi}$ from the right-hand side of the finite product above and the appearance in its place of the factor $1/\sqrt{2\pi}$ in the infinite product of the theorem.

Consider $\phi \in L^2(\mathbb{R})$. If ϕ is also in $L^1(\mathbb{R})$, or in particular if ϕ has bounded support, then it is a standard result from Fourier transform theory that $\hat{\phi}$ is continuous (it is in fact uniformly continuous). It follows that the value of $\hat{\phi}(\xi/2^n)$ in the finite product converges to $\hat{\phi}(0)$ as $n \to \infty$. Part (d) of Exercise 8.18 shows that if we wish ϕ to be a scaling function, we can assume that $\int_{-\infty}^{\infty} \phi = 1$, and then from part (a) of Exercise 8.28 it follows that $\hat{\phi}(0) = 1/\sqrt{2\pi}$, the factor appearing in the infinite product.

Note that we have not formally defined infinite products or developed any of their theory, and we will not do so, since we do not need them other than for the present brief discussion and one or two of the exercises. In close analogy to the case of infinite sums, an infinite product, if it exists, is defined to be the limit of the corresponding sequence of finite *partial products*. With some care, much of the discussion of infinite products can be reduced to a discussion of the infinite sums of the (complex) logarithms of the terms involved, since the logarithm function converts (finite) products into (finite) sums.

8.5 Computing the Daubechies Scaling Coefficients

Finding trigonometric polynomials satisfying the three conditions of Theorem 8.12, and perhaps especially condition (8.1), might appear to be difficult, and it turns out to be useful to break the problem into two steps. The first step involves looking for trigonometric polynomials that satisfy a significantly simpler set of conditions, and the second involves a process that converts these functions into trigonometric polynomials that satisfy the conditions that we really want.

Roughly speaking, the first step amounts to finding $|m(\xi)|^2$ and the second step involves finding $m(\xi)$ itself. More precisely, in the first step we seek a trigonometric polynomial $M(\xi)$ that is real-valued and non-negative and satisfies the conditions

$$M(\xi) + M(\xi + \pi) = 1 \quad \text{for all } \xi \in \mathbb{R}, \tag{8.6}$$
$$M(0) = 1 \tag{8.7}$$

and

$$M(\xi) \neq 0 \quad \text{for } -\pi/2 \leq \xi \leq \pi/2. \tag{8.8}$$

In the second step, given $M(\xi)$, we seek a trigonometric polynomial $m(\xi)$ such that $|m(\xi)|^2 = M(\xi)$; the new polynomial $m(\xi)$ is often referred to, somewhat inaccurately, as a 'square root' of the original polynomial $M(\xi)$. It is clear (see Exercise 8.35) that such a square root $m(\xi)$ must satisfy the three original conditions (8.1), (8.2) and (8.3).

There are several systematic ways of handling the first step. We will examine just one, chosen because it leads specifically to the Daubechies scaling functions and wavelets; other procedures lead to other wavelet families. Given the relative simplicity of condition (8.6) in comparison with condition (8.1), it is not too hard to find suitable functions $M(\xi)$.

We start with the identity

$$1 = \cos^2(\xi/2) + \sin^2(\xi/2).$$

Fix an integer $N \geq 1$. Then using the binomial theorem, we have

$$1 = \left(\cos^2(\xi/2) + \sin^2(\xi/2)\right)^{2N-1}$$

$$= \sum_{k=0}^{2N-1} \binom{2N-1}{k} \cos^{4N-2-2k}(\xi/2) \sin^{2k}(\xi/2). \tag{8.9}$$

Since $2N - 1$ is odd, the last sum contains an even number of terms, $2N$ terms, specifically. We define $M(\xi)$ to be the sum of *the first half* of those terms:

$$M(\xi) = \sum_{k=0}^{N-1} \binom{2N-1}{k} \cos^{4N-2-2k}(\xi/2) \sin^{2k}(\xi/2). \tag{8.10}$$

We claim that the three conditions (8.6), (8.7) and (8.8) hold for $M(\xi)$. The argument for the first condition depends on the simple observation that

$$\sin(\alpha + \pi/2) = \cos\alpha \qquad \text{and} \qquad \cos(\alpha + \pi/2) = -\sin\alpha.$$

Using these identities, it is straightforward to check that substitution of $\xi + \pi$ for ξ in the expression for $M(\xi)$ converts the expression into the sum of *the second half* of the terms in the expression (8.9). Since the terms in (8.9) add to 1, condition (8.6) follows. The checking of conditions (8.7) and (8.8) (and the detailed checking of condition (8.6)) is left for Exercise 8.36. We note also that $M(\xi)$ can be expressed as a trigonometric polynomial, and that $M(\xi)$ is real-valued and non-negative (see Exercises 8.16 and 8.37).

We now need to find a 'square root' of $M(\xi)$, to give the function $m(\xi)$ required for Theorem 8.12. There is a general method for finding square roots of non-negative real-valued trigonometric polynomials, but rather than presenting this here, we will apply ad hoc methods to deal with the cases $N = 1, 2, 3$. We will see that these generate the first three of the Daubechies scaling functions and wavelets, beginning with the Haar case. Most of the calculations will be presented in summary form only, and the details are left as exercises (see Exercise 8.42).

8.5.1 The Case $N = 1$

When $N = 1$, the general expression (8.10) for $M(\xi)$ reduces to

$$M(\xi) = \cos^2(\xi/2),$$

and by the discussion above, the three conditions (8.6), (8.7) and (8.8) required for $M(\xi)$ hold.

The obvious square root in this case is the function

$$\cos(\xi/2) = \tfrac{1}{2}(e^{i\xi/2} + e^{-i\xi/2}),$$

but this function, while satisfying the three conditions (8.1), (8.2) and (8.3), is not a trigonometric polynomial. (The last claim is immediately plausible, given the fractional exponents on the right-hand side, but can be confirmed formally; see Exercise 8.41.) However, it is possible (at least in principle) to modify a candidate square root to produce a function that satisfies all of the required conditions. Specifically, if we are given a function that satisfies conditions (8.1), (8.2) and (8.3), the validity of these conditions will be preserved if we multiply it by any function $\mu(\xi)$ that is periodic with period 2π and satisfies $|\mu(\xi)| = 1$ and $\mu(0) = 1$; if moreover we are able to choose $\mu(\xi)$ so that the process results in a trigonometric polynomial, then we will have a square root with all of the required properties.

In the present case, we choose $\mu(\xi) = e^{-i\xi/2}$. It is easy to check that this function has the properties just stated; in particular,

$$m(\xi) = \cos(\xi/2)\mu(\xi)$$

is a trigonometric polynomial, because multiplication by $\mu(\xi)$ clears the problematic fractional exponents from the expression above for $\cos(\xi/2)$:

$$m(\xi) = \cos(\xi/2)\mu(\xi) = \tfrac{1}{2}(e^{i\xi/2} + e^{-i\xi/2})e^{-i\xi/2} = \tfrac{1}{2}(1 + e^{-i\xi}).$$

Thus all the required conditions for $m(\xi)$ are now met, so we have a function with the properties required to apply Theorem 8.12.

Now the formula for the scaling filter $m_0(\xi)$ in Theorem 8.9 is

$$m_0(\xi) = \frac{1}{2} \sum_{k \in \mathbb{Z}} a_k e^{-i\xi k}.$$

Matching this against the expression for $m(\xi)$ and using the fact that the coefficients in a Fourier series are uniquely determined (see Exercise 4.30), we find that

$$a_0 = a_1 = 1$$

and that $a_k = 0$ for all $k \neq 0, 1$. These are of course the familiar scaling coefficients for the Haar wavelet.

8.5.2 The Case $N = 2$

Using $N = 2$ in the formula (8.10), we obtain the expression

$$M(\xi) = \cos^6(\xi/2) + 3\cos^4(\xi/2)\sin^2(\xi/2)$$
$$= \cos^4(\xi/2)\left[\cos^2(\xi/2) + 3\sin^2(\xi/2)\right].$$

Now the term in brackets here is the sum of the squares of two real numbers, and we can therefore express it as the square of the modulus of a suitable complex number, giving

$$M(\xi) = \cos^4(\xi/2)\left|\cos(\xi/2) + \sqrt{3}i\sin(\xi/2)\right|^2.$$

An obvious candidate square root is therefore

$$\cos^2(\xi/2)\left[\cos(\xi/2) + \sqrt{3}i\sin(\xi/2)\right].$$

However, when we convert the trigonometric terms here into exponential form in an attempt to express the function as a trigonometric polynomial, we encounter the same problem as in the case $N = 1$: we obtain odd integer powers of the expression $e^{i\xi/2}$. We therefore apply the idea we used for $N = 1$, of multiplying through by a function $\mu(\xi)$ that is periodic of period 2π, satisfies $|\mu(\xi)| = 1$ and $\mu(0) = 1$ and is chosen so as to eliminate the fractional powers of $e^{-i\xi}$. A suitable function turns out to be given by $\mu(\xi) = e^{-3i\xi/2}$, and after some simplification, we obtain

$$m_0(\xi) = \frac{1}{8}\left[\left(1 + \sqrt{3}\right) + \left(3 + \sqrt{3}\right)e^{-i\xi} + \left(3 - \sqrt{3}\right)e^{-2i\xi} + \left(1 - \sqrt{3}\right)e^{-3i\xi}\right].$$

Comparison of this with the general expression for $m_0(\xi)$ yields

$$a_0 = \frac{1 + \sqrt{3}}{4}, \qquad a_1 = \frac{3 + \sqrt{3}}{4}, \qquad a_2 = \frac{3 - \sqrt{3}}{4}, \qquad a_3 = \frac{1 - \sqrt{3}}{4},$$

with all the other a_k equal to 0, and these are the scaling coefficients for the second Daubechies wavelet (see the equations (7.11) of Section 7.4). (Note that in choosing an 'obvious' square root above, we deliberately ignored the existence of another almost equally obvious square root; see Exercise 8.39 for an investigation of this point.)

8.5.3 The Case $N = 3$

We can deal with the case $N = 3$ in a similar way, though the detailed calculations are significantly more complicated. We find that

$$M(\xi) = \cos^{10}(\xi/2) + 5\cos^{8}(\xi/2)\sin^{2}(\xi/2) + 10\cos^{6}(\xi/2)\sin^{4}(\xi/2)$$

$$= \cos^{6}(\xi/2)\left[\cos^{4}(\xi/2) + 5\cos^{2}(\xi/2)\sin^{2}(\xi/2)) + 10\sin^{4}(\xi/2)\right].$$

For $N = 2$, the expression for $M(\xi)$ contained a sum of two squares, which we wrote as the square of the modulus of a complex number. In the present case, the bracketed term is a sum of three squares, but after some manipulation we can convert it to a sum of two squares. Carrying out a process akin to completion of the square, we find that

$$M(\xi) = \cos^{6}(\xi/2)\left[\left(\cos^{2}(\xi/2) - \sqrt{10}\sin^{2}(\xi/2)\right)^{2}\right.$$

$$\left. + \left(5 + 2\sqrt{10}\right)\cos^{2}(\xi/2)\sin^{2}(\xi/2)\right]$$

$$= \cos^{6}(\xi/2)\left|\left(\cos^{2}(\xi/2) - \sqrt{10}\sin^{2}(\xi/2)\right)\right.$$

$$\left. + i\sqrt{5 + 2\sqrt{10}}\cos(\xi/2)\sin(\xi/2)\right|^{2},$$

and our first try at a square root is therefore the function

$$\cos^{3}(\xi/2)\left[\left(\cos^{2}(\xi/2) - \sqrt{10}\sin^{2}(\xi/2)\right) + i\sqrt{5 + 2\sqrt{10}}\cos(\xi/2)\sin(\xi/2)\right].$$

As in both of the previous cases, multiplication by a suitably chosen function $\mu(\xi)$ is needed to yield the form of a trigonometric polynomial. We find that the function $\mu(\xi) = e^{-5i\xi/2}$ is as required, and calculation eventually gives the values

$$a_0 = \frac{1}{16}\left(1 + \sqrt{10} + \sqrt{5 + 2\sqrt{10}}\right) = 0.4704672078\ldots,$$

$$a_1 = \frac{1}{16}\left(5 + \sqrt{10} + 3\sqrt{5 + 2\sqrt{10}}\right) = 1.1411169158\ldots,$$

$$a_2 = \frac{1}{8}\left(5 - \sqrt{10} + \sqrt{5 + 2\sqrt{10}}\right) = 0.6503650005\ldots,$$

$$a_3 = \frac{1}{8}\left(5 - \sqrt{10} - \sqrt{5 + 2\sqrt{10}}\right) = -0.1909344156\ldots,$$

$$a_4 = \frac{1}{16}\left(5 + \sqrt{10} - 3\sqrt{5 + 2\sqrt{10}}\,\right) = -0.1208322083\ldots,$$

$$a_5 = \frac{1}{16}\left(1 + \sqrt{10} - \sqrt{5 + 2\sqrt{10}}\,\right) = 0.0498174997\ldots,$$

with all other a_k equal to 0, and these are the scaling coefficients for the third Daubechies wavelet. (We did not derive analytic expressions for these coefficients in the body of Chapter 7, but see Exercise 7.10 and compare the numerical values in Table 7.1.)

As for the case $N = 2$, there are other possibilities for the square root above; see Exercise 8.40.

8.5.4 Higher Values of N

The first of the two steps in the construction of $m_0(\xi)$, that is, the construction of the function $M(\xi)$, proceeds in the same way for all N, as already outlined; the expressions obtained become increasingly complicated, but there is no difficulty in principle in continuing through higher and higher values of N.

However, as N increases, the second of the steps, the formation of a suitable square root of $M(\xi)$, requires different and more systematic methods, since the ad hoc arguments used for $N = 1, 2, 3$ become impractical. It also becomes necessary to use numerical approximations rather than exact expressions for the functions $m_0(\xi)$ and their coefficients, the Daubechies scaling coefficients. This of course parallels our experience in Chapter 7 in computing the Daubechies scaling coefficients directly from the fundamental relations, where numerical methods were essential for all but the smallest values of N.

As we noted earlier in this section, there is a general theorem showing that the formation of a suitable square root is always possible in the circumstances required, and the proof of the result implicitly describes an algorithm for their computation – numerically rather than analytically in general. (The technique is known in the signal processing literature as **spectral factorisation**.) However, we cannot pursue these ideas further here.

Exercises

8.1 Confirm that Theorem 2.1 and the two lemmas used in its proof remain valid in the complex case.

8.2 (a) Verify that the (complex) vector space axioms hold in $\mathbb{C}^n$.
     (b) Verify that the real vector space axioms also hold in $\mathbb{C}^n$.
     (c) Show that $\mathbb{C}^n$ has dimension n as a vector space over $\mathbb{C}$, and write down a basis.

(d) Show that $\mathbb{C}^n$ has dimension $2n$ as a vector space over $\mathbb{R}$, and write down a basis.

8.3 Prove that $\mathbb{C}^\infty$ is a complex vector space and that the collection ℓ^2 of square-summable complex sequences is a vector subspace of $\mathbb{C}^\infty$. (Most of the ingredients of the proof of the real case in Theorem 2.4 are relevant here, but modifications are needed.)

8.4 Show that the set of complex sequences $\{a_n\}$ such that $\sum_{n=1}^\infty a_n^2$ converges is not closed under entry-by-entry addition, and therefore is not a vector space over $\mathbb{C}$.

8.5 For any interval I, show that $F(I)$ is a vector space and that $C(I)$ and $L^2(I)$ are subspaces of $F(I)$.

8.6 Show that the (complex) inner product space axioms hold in $\mathbb{C}^n$.

8.7 Rework the proofs of the Cauchy–Schwarz inequality and the three standard properties of a norm, including the triangle inequality, confirming that the results hold unchanged in the complex case even though minor changes are needed in the arguments.

8.8 (a) Prove that the following so-called **polarisation identity** holds for any vectors x and y in a complex inner product space:

$$\langle x, y \rangle = \tfrac{1}{4}(\|x + y\|^2 - \|x - y\|^2 + i\|x + iy\|^2 - i\|x - iy\|^2).$$

(b) Show that the similar but simpler formula found in the real case in Exercise 3.13 does not hold in the complex case.

(c) Suppose that V and W are complex inner product spaces and that $T: V \to W$ is a linear function (that is, $T(\alpha x + \beta y) = \alpha T(x) + \beta T(y)$ for all $x, y \in V$ and all $\alpha, \beta \in \mathbb{C}$). Show that T preserves the norm ($\|T(x)\| = \|x\|$ for all x) if and only if T preserves the inner product ($\langle T(x), T(y) \rangle = \langle x, y \rangle$ for all x and y).

8.9 The definition of the convergence of a sequence to a limit is as follows: $\lim_{n\to\infty} a_n = \ell$ (or $a_n \to \ell$ as $n \to \infty$) if for all $\epsilon > 0$, there exists $N \in \mathbb{N}$ such that $n > N$ implies $|a_n - \ell| < \epsilon$. Note that this definition holds equally well for convergence of sequences of complex numbers as for sequences of real numbers, since $|x - y|$ is geometrically the distance between x and y whether $x, y \in \mathbb{R}$ or $x, y \in \mathbb{C}$.

(a) Show that if $a_n = b_n + ic_n$ and $r = s + it$ (where b_n, c_n, s and t are real), then $a_n \to r$ in $\mathbb{C}$ if and only if $b_n \to s$ and $c_n \to t$ in $\mathbb{R}$.

(b) Show that $e^{i/n} \to 1$ in $\mathbb{C}$.

8.10 Show that it is not possible to define a consistent numerical ordering on the complex numbers that obeys reasonable rules. Specifically, suppose we had an ordering $<$ satisfying the following for all $z, w \in \mathbb{C}$:

(i) if z and w are distinct, then either $z < w$ or $w < z$;

(ii) if $z < w$ then $-w < -z$; and

(iii) if $z > 0$ and $w > 0$ then $zw > 0$.

Show that it follows both that $1 > 0$ and that $1 < 0$.

8.11 Confirm that in the complex space $L^2([-\pi, \pi])$, as claimed in Subsection 8.2.4, the elements of the standard basis of trigonometric functions can be expressed as linear combinations of the elements of the standard basis of exponential functions, and vice versa.

8.12 Consider the collection of functions $(1/\sqrt{2\pi})e^{inx}$, for $n \in \mathbb{Z}$, in the complex inner product space $C([-\pi, \pi])$ or $L^2([-\pi, \pi])$.

(a) Use integration of complex-valued functions to show that the collection forms an orthonormal set.

(b) Prove the orthonormality of the collection instead by reducing the problem to the real case (see Exercise 8.11).

(c) Show that the collection is an orthonormal basis.

8.13 Show that the set of trigonometric polynomials is dense in $L^2([-\pi, \pi])$. (Here, and in the other exercises on trigonometric polynomials, it is useful to keep in mind the observations in the first paragraph of Subsection 8.2.5.)

8.14 Show that two trigonometric polynomials are equal as functions if and only if their coefficients are equal. Deduce that a trigonometric polynomial is identically 0 if and only if its coefficients are all 0.

8.15 Show that a trigonometric polynomial, regarded as a function on $[-\pi, \pi]$, belongs to $L^2([-\pi, \pi])$, but, regarded as a function on $\mathbb{R}$, does not belong to $L^2(\mathbb{R})$ unless it is identically 0. (For the second assertion, see the remark between Theorems 8.9 and 8.10.)

8.16 Consider a trigonometric polynomial $p(x) = \sum_{n=-k}^{k} c_n e^{inx}$, where $c_n \in \mathbb{C}$ for all n.

(a) Show that $p(x)$ is real-valued if and only if $c_{-n} = \overline{c_n}$ for all n.

(b) Deduce that $p(x)$ is real-valued if and only if it can be expressed in the form $\sum_{n=0}^{k}(a_n \cos nx + b_n \sin nx)$, where $a_n, b_n \in \mathbb{R}$ for all n.

8.17 By inspecting the relevant arguments from Chapter 6, confirm that the elements of wavelet theory in the complex case as presented in Subsection 8.2.6 are correct.

8.18 (a) Show that if $\phi \in L^2(\mathbb{R})$ is the scaling function for a multiresolution analysis, then the function $e^{i\theta}\phi$ is also a scaling function for the same multiresolution analysis, for any fixed real number θ.

(b) Show that if $\psi \in L^2(\mathbb{R})$ is a wavelet (whether associated with a multiresolution analysis or not), then so is the function $e^{i\theta}\psi$ for any fixed real number θ. Deduce that genuinely complex-valued wavelets exist.

(c) Suppose that $\phi \in L^2(\mathbb{R})$ is the scaling function for a multiresolution analysis, and write $I = \int_{-\infty}^{\infty} \phi$. The first part of Theorem 6.16 states that $I = \pm 1$. However, this result is for the real case, and the conclusion differs slightly in the complex case. Specifically, show that the correct conclusion in the complex case is that $|I| = 1$. (Exercise 6.18 outlines a proof that $I = \pm 1$ in the real case. Confirm that the entire argument remains valid in the complex case except for the final step, given in part (f), where from the fact that $|I| = 1$ it is deduced that $I = \pm 1$. In the complex case, we cannot take this last step, and so our conclusion must be that $|I| = 1$.)

(d) Deduce from parts (a) and (c) that when we are seeking a scaling function in the complex case we lose no generality in restricting attention to functions ϕ such that $\int_{-\infty}^{\infty} \phi = 1$. (See the remarks following Theorem 7.4 for a comparable statement in the real case.)

8.19 In relation to Theorem 8.4, prove the following statements, by finding suitable examples or otherwise.

(a) If $f \in L^2(\mathbb{R})$, then $\hat{f}$ need not belong to $L^1(\mathbb{R})$.

(b) If $f \in L^1(\mathbb{R}) \cap L^2(\mathbb{R})$, then $\hat{f}$ need not belong to $L^1(\mathbb{R})$. (Any example for part (b) is obviously also an example for part (a), but it may be interesting to think about the problems separately.)

(c) If $f \in L^2(\mathbb{R})$, then the integral $\int_{-\infty}^{\infty} f(x)e^{-i\xi x}\,dx$ may be infinite for all ξ.

(d) If $f \in L^1(\mathbb{R})$, then $\hat{f}$ need not belong to $L^2(\mathbb{R})$.

8.20 Following Theorem 8.5, it was asserted that if $f \in L^2(\mathbb{R})$, then the 'truncation' f_n of f to the interval $[-n, n]$ lies in $L^1(\mathbb{R}) \cap L^2(\mathbb{R})$ for each n. Use part (d) of Exercise 4.42 to give a proof.

8.21 Show directly from Definition 8.6 that the Fourier transform on $L^2(\mathbb{R})$ is a linear mapping.

8.22 (a) Use the fact that the Fourier transform on $L^2(\mathbb{R})$ preserves the norm to show that it 'preserves the convergence of sequences'; that is, that if $f_n \to f$ as $n \to \infty$ for a sequence $\{f_n\}$ in $L^2(\mathbb{R})$ and $f \in L^2(\mathbb{R})$, then also $\hat{f}_n \to \hat{f}$ as $n \to \infty$. (This result and the result of part (b) have a lot of similarity to the results of Subsection 4.3.2. Moreover, a comment in that subsection shows that the fact that the Fourier transform preserves convergence is equivalent to the statement that it is a continuous mapping from $L^2(\mathbb{R})$ to itself.)

(b) Deduce that the transform also 'passes through infinite summations' in $L^2(\mathbb{R})$, after first formulating the statement in precise terms.

8.23 It can be shown using a little of the Lebesgue theory that if $f \in L^1(\mathbb{R})$ then $\hat{f}$ is uniformly continuous. What information from Section 8.3 allows us to conclude trivially that this result does not carry over to all $f \in L^2(\mathbb{R})$?

8.24 (a) For $a < b$, show that $\chi_{[a,b]} \in L^1(\mathbb{R}) \cap L^2(\mathbb{R})$ and compute $\hat{\chi}_{[a,b]}(\xi)$.

(b) Deduce that for $a > 0$,

$$\hat{\chi}_{[-a,a]}(\xi) = \sqrt{\frac{2}{\pi}} \frac{\sin a\xi}{\xi}.$$

(c) Show that $\hat{\chi}_{[-1,1]}$ is not in $L^1(\mathbb{R})$. (Several approaches are possible. One that is reasonably straightforward in principle, though it requires some technical care, is to compare the relevant integral with the harmonic series.)

(d) Show that $\hat{\chi}_{[-1,1]} \in L^p(\mathbb{R})$ for all $p > 1$. (Refer to Exercise 4.42 for the definition and a brief discussion of $L^p(\mathbb{R})$ in the real case.)

(e) For $a < b$, show that $\hat{\chi}_{[a,b]} \notin L^1(\mathbb{R})$ and that $\hat{\chi}_{[a,b]} \in L^p(\mathbb{R})$ for all $p > 1$. (It is possible to do this either by repeating earlier arguments in a generalised form or by using earlier results and suitably applying Theorem 8.7.)

8.25 (a) From the previous exercise, the Fourier transform of the function $\chi_{[-1,1]}$ is the function whose value at ξ is $\sqrt{2/\pi} (\sin \xi)/\xi$. Compute the inverse Fourier transform of this last function, and confirm that, as Theorem 8.3 guarantees, it is almost everywhere equal to $\chi_{[-1,1]}$. (Since the function $\sqrt{2/\pi} (\sin \xi)/\xi$ is in $L^2(\mathbb{R})$ but is not in $L^1(\mathbb{R})$, the definition requires its inverse transform to be computed as the limit in $L^2(\mathbb{R})$ of the sequence of inverse transforms of its 'truncations', that is, as the limit of the sequence

$$\frac{1}{\sqrt{2\pi}} \int_{-n}^{n} \sqrt{\frac{2}{\pi}} \frac{\sin \xi}{\xi} e^{i\xi x} \, d\xi$$

as $n \to \infty$. Show that the last expression equals

$$\frac{1}{\pi} \int_{-n}^{n} \frac{\sin \xi \cos \xi x}{\xi} \, d\xi, \tag{8.11}$$

use trigonometric formulae and integral substitutions to reduce the problem to the calculation of the improper integral

$$\int_{-\infty}^{\infty} \frac{\sin t}{t}\, dt$$

and apply the fact that this integral has the value π.)

(b) Use a computer algebra system to make plots of the functions in the sequence (8.11) over a suitable domain and for a suitable range of values of n.

8.26 Consider the three parts of Theorem 8.7.

(a) Prove that the statements hold when $f \in L^1(\mathbb{R})$. (Confirm that the scaled, dilated and translated copies of f are again in $L^1(\mathbb{R})$, and work directly with the integrals that define their Fourier transforms.)

(b) Prove that the statements hold for all $f \in L^2(\mathbb{R})$.
The following three established facts are helpful here:

(i) the fact that a sequence in an inner product space can have at most one limit (part (i) of Theorem 4.3);

(ii) the fact, from Exercise 8.22, that the Fourier transform preserves the convergence of sequences; and

(iii) the fact, noted just before Definition 8.6, that for $f \in L^2(\mathbb{R})$, the definition of $\hat{f}$ can be taken to be $\lim_{n\to\infty} \hat{f}_n$ for *any* sequence of functions $f_n \in L^1(\mathbb{R}) \cap L^2(\mathbb{R})$ for which $\lim_{n\to\infty} f_n = f$.

To prove the second statement of the theorem (for example), let $\{f_n\}$ be a sequence in $L^1(\mathbb{R}) \cap L^2(\mathbb{R})$ that converges to f. Show that $D_b f_n \to D_b f$, observe that $D_b f_n \in L^1(\mathbb{R}) \cap L^2(\mathbb{R})$, and deduce that $(D_b f_n)\hat{\ } \to (D_b f)\hat{\ }$. Also show independently that $(D_b f_n)\hat{\ }$ converges to the function whose value at ξ is $(1/b)\hat{f}(\xi/b)$, and draw the required conclusion. (The arguments for the other two statements are similar.)

8.27 Use Theorem 8.7 (and perhaps the result of part (b) of Exercise 1.11) to find an expression for the Fourier transform of the scaled, dilated, translated function $\phi_{j,k}$, where $\phi \in L^2(\mathbb{R})$ and $j, k \in \mathbb{Z}$.

8.28 (a) Show that if $f \in L^1(\mathbb{R})$, then $\hat{f}(0) = (1/\sqrt{2\pi})\int_{-\infty}^{\infty} f$. Observe that this holds in particular if $f \in L^2(\mathbb{R})$ and has bounded support.

(b) Deduce using part (c) of Exercise 8.18 that if $\phi \in L^2(\mathbb{R})$ has bounded support and is the scaling function of a multiresolution analysis, then $|\hat{\phi}(0)| = 1/\sqrt{2\pi}$. From Theorem 8.8, deduce further that $\hat{\phi}(2\pi k) = 0$ for every non-zero integer k.

8.29 For use in the next two exercises, show that

$$e^{iz} + 1 = 2\cos(z/2)e^{iz/2} \quad \text{and} \quad e^{iz} - 1 = 2i\sin(z/2)e^{iz/2}$$

for all $z \in \mathbb{C}$.

8.30 Let ϕ be the scaling function for the Haar wavelet.

(a) Show that

$$\hat{\phi}(\xi) = \frac{i}{\sqrt{2\pi}} \frac{e^{-i\xi} - 1}{\xi} = \sqrt{\frac{2}{\pi}} \frac{\sin(\xi/2)e^{-i\xi/2}}{\xi}$$

when $\xi \neq 0$, and also compute $\hat{\phi}(0)$.

(b) Confirm the results of Exercise 8.28 for ϕ.

8.31 Let m_0 be the scaling filter for a multiresolution analysis.

(a) What is the value of $m_0(\pi n)$ when n is an even integer?

(b) If the scaling function has bounded support, what is the value of $m_0(\pi n)$ when n is an odd integer?

8.32 (a) Show that the scaling filter for the Haar wavelet is given by

$$m_0(\xi) = \tfrac{1}{2}(e^{-i\xi} + 1) = \cos(\xi/2)e^{-i\xi/2}.$$

(Compare this with the findings of Subsection 8.5.1.)

(b) Confirm that the scaling equation in the Fourier domain given in Theorem 8.9 holds for m_0.

(c) Confirm that the first of the quadrature mirror filter equations of Theorem 8.11 holds for m_0.

(d) Confirm that conditions (8.1), (8.2) and (8.3) of Theorem 8.12 hold for m_0.

8.33 Let ϕ and m_0 be the scaling function and the scaling filter, respectively, for the Haar wavelet. Show that the infinite product formula of Theorem 8.12, applied when $m = m_0$, yields $\hat{\phi}$, as computed in Exercise 8.30.

The identity

$$\prod_{j=1}^{\infty} \cos(2^{-j}x) = \frac{\sin x}{x}$$

will probably be needed. (This result was proved by Euler, though a special case was known to the French mathematician François Viète in the late sixteenth century. See the next exercise for a suggested proof.) One or two general facts about the manipulation of infinite products will also be needed. (These are straightforward analogues of results for infinite series, and formulating them and working through their proofs would make an interesting exercise in its own right.)

8.34 Prove the infinite product formula quoted in the previous exercise. (Use the standard identity $\sin 2\alpha = 2\cos\alpha\,\sin\alpha$ to find a simple expression for the nth partial product, and let n tend to ∞.)

8.35 Confirm that if a function $M(\xi)$ satisfies conditions (8.6), (8.7) and (8.8), and if $|m(\xi)|^2 = M(\xi)$, then $m(\xi)$ satisfies conditions (8.1), (8.2) and (8.3). Also confirm that if a function satisfying conditions (8.1), (8.2) and (8.3) is multiplied by a function $\mu(\xi)$ that is periodic of period 2π and satisfies $|\mu(\xi)| = 1$ and $\mu(0) = 1$, then the product function again satisfies conditions (8.1), (8.2) and (8.3).

8.36 Confirm that conditions (8.6), (8.7) and (8.8) of Section 8.5 hold for the function $M(\xi)$ of equation (8.10). In the case of (8.8), prove the stronger statement that $M(\xi) = 0$ if and only if ξ is an odd multiple of π.

8.37 Show that the function $M(\xi)$ of Equation (8.10) can be expressed as a trigonometric polynomial and that it is real-valued and non-negative.

8.38 In the case $N = 1$ of Subsection 8.5.1, verify directly that conditions (8.6), (8.7) and (8.7) hold and that $M(\xi)$ is a trigonometric polynomial.

8.39 In the case $N = 2$ of Subsection 8.5.2, a choice of $\pm$ sign was possible in the formation of the 'obvious candidate square root' of $M(\xi)$, although we ignored it at the time. There are therefore two candidate square roots. Investigate what happens to the rest of the calculation if we adopt the other choice of sign, and hence the other square root. Relate your findings to the work of Chapter 7 and to Exercise 7.9 in particular.

8.40 As in Exercise 8.39, choices are possible in the formation of the square root of $M(\xi)$ in the case $N = 3$ of Subsection 8.5.3. This time, however, there are two *independent* choices of $\pm$ sign, giving four square roots. Find out exactly where the choices of sign arise, and investigate what happens to the rest of the calculation if we use the other choices. Again relate your findings to the work of Chapter 7.

8.41 It was claimed in Subsection 8.5.1 that $\cos(\xi/2)$ is not a trigonometric polynomial. This is intuitively plausible, but we can prove it formally, as follows. Each function in $L^2([-\pi, \pi])$ has a uniquely determined Fourier series with respect to a given orthonormal basis (see Exercise 4.30), and hence with respect to the basis of exponentials $(1/\sqrt{2\pi})e^{in\xi}$ in particular. Now an expression for a function in $L^2([-\pi, \pi])$ as a trigonometric polynomial is a Fourier series for that function, one with only finitely many non-trivial terms. Therefore, if a function in $L^2([-\pi, \pi])$ has a Fourier series with infinitely many non-trivial terms, that function is not expressible as a trigonometric polynomial. Apply this observation to prove the claim.

8.42 Carry out in detail the calculations given in summary form in Subsections 8.5.1, 8.5.2 and 8.5.3.

8.43 Repeat the work of Section 8.5 using expansions of $\cos^2(\xi/2) + \sin^2(\xi/2)$ to even powers instead of odd powers. (The main reason for using odd powers was simply convenience: an expansion to an odd power has an even number of terms, and using precisely the first half of them gives a suitable polynomial $M(\xi)$. If we expand to an even power $2N$, then there will be an odd number, $2N + 1$, of terms in the expansion, and we need to decide which terms to use for $M(\xi)$. The trick is to form $M(\xi)$ from the first N terms and *one half* of the $(N + 1)$th term.)

Appendix
Notes on Sources

Given the largely introductory nature of this book, the decision was made early on not to include explicit references to the literature in the main text. However, for the reader who wants pointers to sources, this appendix provides them.

It should be said that the coverage of the literature here is quite selective, and the appendix is very far from forming a literature survey. Roughly, three types of references are included. First, there are references of a general nature on certain topics, which are introduced not so much as sources for particular results in the main text but as accounts of bodies of theory as a whole. Second, there are references that are used as sources for specific results, though our formulations may be less general than those referred to. Third, there are references to advanced material that is mentioned or hinted at in the text but is beyond the scope of the text.

Some references are undergraduate-level mathematical texts, some are texts at a more advanced level, and some are expository or research articles from mathematical journals. We generally do not list sources in detail for material that is essentially standard and that is therefore covered by many texts, since the reader will have no difficulty locating suitable references. Since the focus thus tends to be on the less standard material, a consequence is that the density of our coverage of sources tends to increase from chapter to chapter.

When there are several relevant texts on a topic, the selection made here is often a personal one, and failure to mention a particular text should not be taken as implicit disparagement. Any reader who does not have access to the specific texts mentioned or prefers to look for other styles of presentation will find that numerous standard texts are available which cover virtually all the topics of Chapters 2, 3 and 4.

Chapters 1 to 4: general comments. Since Chapter 1 is the book in microcosm, all its topics are treated in more detail and with more rigour later in the book, and so we give no references specifically for this chapter. Neither do we list references for linear algebra, the subject of Chapter 2. The elementary theory of linear algebra is assumed known here, and in any case the emphasis of Chapter 2 is on examples of vector spaces, especially those of infinite dimension, rather than on development of the theory, and the examples of Chapter 2 are dealt with in one source or another among those listed for Chapters 3 and 4.

There are also relatively few detailed references for Chapters 3 and 4, since all the material in these chapters, other than the small quantity of material on wavelets, is standard and can be found in many texts on linear algebra, measure and integration, real analysis and functional analysis. Most texts on measure and integration and on functional analysis, and some on linear algebra, deal with spaces over the complex field from the start, whereas in this book we deal only with the real case, except in Chapter 8. However, as noted in that chapter, the differences are not of great significance, at least at the level of our discussion in this book, and the texts mentioned are ones that should be useful for any reader of this book.

The books of Debnath and Mikusiński (1999), Sen (2014) and Young (1988) all cover (among many other things) most of the following topics of Chapters 3 and 4: function spaces such as $C(I)$ and $L^2(I)$ for intervals I in $\mathbb{R}$; pointwise convergence and convergence with respect to the L^2 norm; inner product spaces; completeness and completion; orthonormality, orthogonality and the Gram–Schmidt process; best approximation; orthogonal projection and direct sums; orthonormal bases; Hilbert spaces; and Fourier series, both in the case of a general Hilbert space and in the classical case of trigonometric Fourier series in $L^2([-\pi, \pi])$. The book of Saxe (2002) provides accessible introductions to most of the above topics and to others as well, such as topology, measure theory and integration, and it is also of note for its interesting historical asides on various mathematical theories and on some of the mathematicians whose names are associated with them. As general references on real analysis, see, for example, Bartle (1976), Berberian (1994), Royden and Fitzpatrick (2010) and Rudin (1987); for measure and integration, see Rudin (1987) and Stroock (1999); and for metric spaces and topology, see Simmons (1963).

Chapter 3. We introduced orthogonal polynomials in Subsection 3.5.4 (and discussed them at various later points in Chapters 3 and 4) by way of the Legendre polynomials, and we noted in passing that there are other such orthogonal families, such as the Hermite, Laguerre and Chebyshev polynomials. These and other such families are treated much more fully in the book of Davis (1975).

Chapter 3: exercises. For the Cantor set and its properties, mentioned in Exercise 3.8, see Royden and Fitzpatrick (2010) and Hong et al. (2004). For the Heine–Borel Theorem, used in Exercise 3.9, see any book on real analysis. For the very important topic of convex sets, introduced merely in passing in Exercise 3.16, see any text on functional analysis or specifically on convexity. For Gram matrices and determinants, introduced in Exercises 3.31 and 3.32, see Davis (1975).

Chapter 4. Theorem 4.32, on the completeness of $L^2(I)$ for an interval I, is a special case of Theorem 3.11 of Rudin (1987). Zorn's lemma and the Axiom of Choice are mentioned in the comment following Theorem 4.34. For these statements and a discussion of their logical status, see, for example, Simmons (1963). For the fact (Theorem 4.35) that the standard collection of trigonometric functions forms an orthonormal basis in $L^2([-\pi, \pi])$, see, for example, Section 4.24 of Rudin (1987). For the comment near the beginning of Section 4.6 about the existence of possibly uncountable orthonormal bases, see, for example, the somewhat more general statement of Theorem 4.22 of Rudin (1987). For the comment on rearrangement of the order of summation in Fourier series that immediately precedes Subsection 4.6.1, see, for example, results in Heil (2011) on unconditional convergence of series in Banach spaces. The completely satisfactory analysis of pointwise convergence in $L^2([-\pi, \pi])$ referred to in the comments just preceding Theorem 4.40 was given in a famous paper of Carleson (1966), which shows among other things that the Fourier series of a function $f \in L^2([-\pi, \pi])$ converges to f almost everywhere. Carleson's paper is notoriously difficult, and although there are now other proofs of his result that are somewhat less difficult, no known proof is easy. Theorem 4.40, giving a sufficient condition for the pointwise convergence of a Fourier series, can found in many sources (and in several variant forms); see, for example, Section 4.6 of Bachman et al. (2000) and Chapter 1, Section 10 of Tolstov (1962). Kolmogorov's example, referred to in the comments just following Theorem 4.40, is Theorem II.3.6 of Katznelson (1968).

Chapter 4: exercises. For the Weierstrass Approximation Theorem, used in Exercise 4.19, see, for example, Simmons (1963). Greatly generalised versions of the Weierstrass theorem have been discovered, and can be found in many of the more advanced analysis and functional analysis texts; see, for example, the Stone–Weierstrass theorem in Simmons (1963), and other results in still more advanced texts. The author is indebted to Rodney Nillsen for the idea of Exercise 4.22. The density theorems of Exercise 4.19 and part (a) of Exercise 4.23 are special cases of the fundamental result in the theory of Lebesgue measure and integration that for any interval I the continuous functions with bounded support are dense in $L^2(I)$; see Theorem 3.14 of Rudin (1987) for an even more general result. The density theorem of part (b) of Exercise 4.23 can be deduced from that of part (a); in fact, so can the much stronger statement that the infinitely differentiable functions of bounded support are dense in $L^2(\mathbb{R})$. (In rough outline, let f be a continuous function with bounded support. The Weierstrass Approximation Theorem allows f to be approximated arbitrarily closely on its support by a polynomial g. Then g can be 'truncated' to have bounded support while retaining infinite differentiability and without disturbing the closeness of the approximation too much, by suitably employing what are sometimes called 'bump functions', for which see Chapter IV, §1 of Lang (1997). Alternatively, the result is a very special case of Theorem 2.16 of Lieb and Loss (2001).) For the normed spaces and Banach spaces of Exercise 4.42, which generalise inner product spaces and Hilbert spaces, respectively, see any text on functional analysis, such as Debnath and Mikusiński (1999). I do not know the origin of the elegant example in Exercise 4.36 involving rearrangement of the alternating harmonic series. However, Exercise K.75 of Körner (2004) gives a similar example based on the alternating harmonic series which Körner attributes to Dirichlet (see also Example 3.53 of Rudin (1976)).

Chapters 5 to 8: general comments. Numerous texts that are about wavelets, or that include substantial sections on wavelets, have been useful in the writing of the present book. We note just a handful of these here, focusing on those that readers of this book might find most useful to reinforce or complement our account: Bachman et al. (2000), Boggess and Narcowich (2009), Chui (1992), Daubechies (1992), Hernández and Weiss (1996), Kaiser (1994), Nievergelt (2013), Ruch and Van Fleet (2009), Walnut (2002) and Wojtaszczyk (1997). The book of Hubbard (1998) also deserves mention. It is probably the only one of its kind, and is highly successful in its own terms. It is written by a popular science writer, not a mathematician, but gives an account of what wavelets are, where they come from and how they are used that is informed and informative but at the same time essentially non-technical.

The orthonormal basis for $L^2(\mathbb{R})$ generated from what is now called the Haar wavelet was introduced in Haar (1910) (Haar's construction was on the interval $[0, 1]$ rather than on the whole real line $\mathbb{R}$, but it contains the essential idea of the modern construction). Although modern wavelet theory began to take shape only in the 1980s, the Haar system was not forgotten in the intervening years, and it received mention from time to time in the literature; see, for example, Goffman and Pedrick (1965). A considerable amount of other work was also done in those years that can be seen in retrospect as foreshadowing aspects of the theory that emerged in the 1980s. Interesting accounts of this 'prehistory' of wavelet theory can be found in Meyer (1993) and Kahane and Lemarié-Rieusset (1995). The modern theory, and in particular the central concept of a multiresolution analysis, can be said to have made its first appearance in the published literature in Mallat (1989).

We now turn to comments and sources for Chapters 6, 7 and 8; no discussion is needed specifically on Chapter 5 beyond the comments just made on the Haar wavelet, since all the main ideas of the chapter are repeated later in a more general form.

Chapter 6. Most of the material of Chapter 6 is standard in the wavelets literature and can be found in various, often more general, forms in many wavelets texts. The wavelet construction of Daubechies mentioned in Section 6.1 is in the long article Daubechies (1988), much of the content of which appeared later in the book Daubechies (1992). Section 6.2 and the first two subsections of Section 6.3 are an elaboration in detail of parts of Section 5.1 of Daubechies (1992). I am indebted to Rodney Nillsen for comments which stimulated substantial rewriting and improvement of the exposition of this material. Theorem 6.8 is standard; see, for example, Proposition 6.4.23 of Pinsky (2002) and Equation (5.1.39) of Daubechies (1992). The proof of Theorem 6.12 spells out in detail an argument found in numerous sources, such as Theorem 4.5 of Wojtaszczyk (1997). The results of Theorem 6.16 are again standard, and hold under more general conditions. The various parts of the result can be found, for example, in Theorem 7.51, Corollary 7.52 and Theorem 8.13 of Walnut (2002), Proposition 2.16 of Wojtaszczyk (1997) and Proposition 6.4.23 of Pinsky (2002). (See also the comment on Exercise 6.18.) It is noted just before Theorem 6.18 that any given orthonormal set in a Hilbert space can be enlarged to an orthonormal basis; for this, see Theorem 4.22 of Rudin (1987).

Chapter 6: exercises. For part (a) of Exercise 6.3, see Exercise 2.2 of Wojtaszczyk (1997); for part (b), see Example 6.4.4 and Subsection 6.4.1 of Pinsky (2002). The outline argument given in Exercise 6.18 for part (i) of Theorem 6.16 adapts an idea from the proof of Proposition 6.4.33 of Pinsky (2002), though Pinsky's argument is devoted to the proof of a different result.

For the case of scaling functions of unbounded support mentioned at the end of Exercise 6.22, see Lemma 4.4 and Remark 4.5 of Wojtaszczyk (1997). For Exercises 6.23 to 6.29, see Lemarié-Rieusset and Malgouyres (1991). For Exercises 6.30 to 6.32, see, for example, Chapter 5, and especially Example 5.1, of Wojtaszczyk (1997) and Section 6.7, and especially Subsection 6.7.1, of Pinsky (2002).

Chapter 7. For the geometric properties and parametrisations of A_N mentioned in Section 7.2, see the important unpublished paper Pollen (1989), and also Wells (1993), Lina and Mayrand (1993), Schneid and Pittner (1993) and Regensburger (2007). In the discussion preceding Theorem 7.2, it was noted that the wavelet coefficients of a smooth function tail off rapidly with respect to a wavelet that has many vanishing moments; for this, see Section 7.4 of Daubechies (1992) and Subsection 9.1.2 of Walnut (2002). For results from which Theorem 7.2 follows, see Proposition 3.1 of Wojtaszczyk (1997) and Theorem 5.5.1 and Corollary 5.5.2 of Daubechies (1992). Statements equivalent to Theorem 7.3 can be found in many sources; see, for example, Theorem 9.11 of Walnut (2002). For the conversion of the augmented system (7.7) to the augmented system (7.8) discussed in Subsection 7.3.1, see Regensburger (2007). The account given of the exact solutions to the augmented system (7.8) in Section 7.4 and in Section 8.5 for the cases $N = 1, 2, 3$ is extended to the cases $N = 4, 5$ in Shann and Yen (1999) and Shann and Yen (1997). For the expression for the number of (real) solutions to the system (7.8) given in Section 7.4, see Subsection 8.1.1 of Daubechies (1992). For discussion of the extremal phase choice defined in Section 7.4, see Section 5.6 of Oppenheim and Schafer (1989), Section 4.8 of Percival and Walden (2000) and Section 6.4 of Daubechies (1992), and for the reference to front-loading, see Section 11.10 of Percival and Walden (2000). For a result implying Theorem 7.4, see Theorem 6.4.27 of Pinsky (2002). The results of Lawton (1990) yield all the statements of Theorem 7.5 except those pertaining specifically to the Daubechies scaling coefficients; for the Daubechies case, see Section 6.4 of Daubechies (1992). For Theorem 7.7, refer to the sources listed below for the more detailed statements of Section 7.7. For Theorem 7.8, see Theorem 5.23 of Boggess and Narcowich (2009). For Theorem 7.9, see the comments on Exercise 7.18. For the algorithm discussed briefly near the end of Subsection 7.5.3, see Section 6.5 of Daubechies (1992). The main ideas of Section 7.6 come from the paper of Pollen (1992), in which it is also shown how the continuity of the second Daubechies scaling function and wavelet can be proved by elementary arguments. (The important information in Theorem 7.13 that $\alpha \neq 0$ is not noted in Pollen's paper, and it is unclear whether it can be obtained within that framework. This is part of the reason for the inclusion of Exercise 7.28 and the exercises in Chapter 6 on which it depends.) For the detailed smoothness information given on $_N\phi$ and $_N\psi$ in Section 7.7 for small N and, asymptotically, for large N, see page 226 and pages 238–239 of Daubechies (1992), Section 5.A of Daubechies and Lagarias (1992) and Theorem 5.1 of Daubechies and Lagarias (1991). Families of wavelets of bounded support other than the Daubechies family are discussed briefly in Subsection 7.7.1; for the least asymmetric wavelets, see Sections 6.4 and 8.1 of Daubechies (1992); for attempts to maximise the smoothness of wavelets see Section 7.3 of Daubechies (1992) and Lang and Heller (1996); and for the coiflets mentioned at the end of the subsection, see Section 8.2 of Daubechies (1992).

Chapter 7: exercises. For part (c) of Exercise 7.3, see Theorem 3 of Lawton (1990) (this paper was already referred to in the comments above on Theorem 7.5), Theorem 3 of Lawton (1991), Section 2 of Wells (1993) and Section 4 of Regensburger (2007). For the ideas used in Exercise 7.7, see Regensburger (2007). Note that the analytical result underlying Exercise 7.18 is roughly a converse of the much more important result that the uniform limit of a sequence of continuous functions on a closed and bounded interval is continuous (for which see any real analysis text). For the dyadic numbers introduced systematically in Exercises 7.19 and 7.20, see also Section 5.1 and Subsection 9.1.2 of Nievergelt (2013).

Chapter 8. The first four subsections of Section 8.2 require no detailed listing of sources; as noted earlier, most texts that cover these topics do so in the complex case. Similar comments apply to the sixth subsection, on wavelet theory. There are innumerable references for the Fourier transform theory of Section 8.3, and we mention only the old but excellent book of Rudin (1987). The main results of Section 8.4, namely, Theorems 8.8 to 8.12, are in essence standard results in the Fourier-analytic treatment of wavelet theory, and can be found explicitly or implicitly in many sources. For Theorem 8.8 see Corollary 2.9 of Wojtaszczyk (1997) and Corollary 6.4.9 of Pinsky (2002); for Theorem 8.9, see equations (2.16) and (2.17) of Wojtaszczyk (1997) and equations (6.4.15) and (6.4.16) of Pinsky (2002); for Theorem 8.10, see equations (6.4.28) and (6.4.30) of Pinsky (2002); for Theorem 8.11 see Proposition 6.4.36 of Pinsky (2002); and for Theorem 8.12, see Theorem 4.1 of Wojtaszczyk (1997) and Theorem 6.5.2 of Pinsky (2002). The field of harmonic analysis is mentioned briefly in the comment near the start of Section 8.4. The book of Katznelson (1968) is an approachable introduction to this topic. The theory of infinite products is mentioned in the comment at the end of Section 8.4 and in Exercises 8.33 and 8.34. For a brief discussion of the theory, see Section 8.4 of Walnut (2002). For the two-stage process for the construction of the trigonometric polynomial $m(\xi)$ in Section 8.5 (finding $M(\xi) = |m(\xi)|^2$ first, and then the square root $m(\xi)$ itself), see Subsection 6.5.2 of Pinsky (2002) and the article of Strichartz (1993). For the specific procedure used to find $M(\xi)$, involving the expansion of powers of $\cos^2(\xi/2) + \sin^2(\xi/2)$, and then the procedure used to extract the square root in the cases $N = 1, 2, 3$, see Strichartz (1993) again. For the spectral factorisation mentioned at the end of Subsection 8.5.4, see Section 6.1 of Daubechies (1992) and Lemma 4.6 of Wojtaszczyk (1997).

Chapter 8: exercises. For the value of the improper integral $\int_{-\infty}^{\infty}(\sin t)/t\, dt$ in Exercise 8.25, see, for example, Problem 19-42 of Spivak (1994). On the infinite product formula of Exercise 8.34, compare Note 5 on page 211 of Daubechies (1992).

References

Bachman, George, Narici, Lawrence, and Beckenstein, Edward. 2000. *Fourier and wavelet analysis*. Universitext. Springer-Verlag, New York.

Bartle, Robert G. 1976. *The elements of real analysis*. Second edn. John Wiley & Sons, New York, London and Sydney.

Berberian, Sterling K. 1994. *A first course in real analysis*. Undergraduate Texts in Mathematics. Springer-Verlag, New York.

Boggess, Albert, and Narcowich, Francis J. 2009. *A first course in wavelets with Fourier analysis*. Second edn. John Wiley & Sons Inc., Hoboken, NJ.

Carleson, Lennart. 1966. On convergence and growth of partial sums of Fourier series. *Acta Math.*, **116**, 135–157.

Chui, Charles K. 1992. *An introduction to wavelets*. Wavelet Analysis and Its Applications, vol. 1. Academic Press, Inc., Boston, MA.

Daubechies, Ingrid. 1988. Orthonormal bases of compactly supported wavelets. *Comm. Pure Appl. Math.*, **41**(7), 909–996.

Daubechies, Ingrid. 1992. *Ten lectures on wavelets*. CBMS-NSF Regional Conference Series in Applied Mathematics, vol. 61. Society for Industrial and Applied Mathematics (SIAM), Philadelphia, PA.

Daubechies, Ingrid, and Lagarias, Jeffrey C. 1991. Two-scale difference equations. I. Existence and global regularity of solutions. *SIAM J. Math. Anal.*, **22**(5), 1388–1410.

Daubechies, Ingrid, and Lagarias, Jeffrey C. 1992. Two-scale difference equations. II. Local regularity, infinite products of matrices and fractals. *SIAM J. Math. Anal.*, **23**(4), 1031–1079.

Davis, Philip J. 1975. *Interpolation and approximation*. Dover Publications, Inc., New York. Re-publication, with minor corrections, of the 1963 original, with a new preface and bibliography.

Debnath, Lokenath, and Mikusiński, Piotr. 1999. *Introduction to Hilbert spaces with applications*. Second edn. Academic Press, Inc., San Diego, CA.

Goffman, Casper, and Pedrick, George. 1965. *First course in functional analysis*. Prentice-Hall, Inc., Englewood Cliffs, NJ.

Haar, Alfred. 1910. Zur Theorie der orthogonalen Funktionensysteme. *Math. Ann.*, **69**(3), 331–371.

Heil, Christopher. 2011. *A basis theory primer*. Expanded edn. Applied and Numerical Harmonic Analysis. Birkhäuser/Springer, New York.

Hernández, Eugenio, and Weiss, Guido. 1996. *A first course on wavelets*. Studies in Advanced Mathematics. CRC Press, Boca Raton, FL.

259

Hong, Don, Gardner, Robert, and Wang, Jianzhong. 2004. *Real analysis with an introduction to wavelets and applications*. Academic Press, San Diego, CA.

Hubbard, Barbara Burke. 1998. *The world according to wavelets*. Second edn. A K Peters Ltd, Wellesley, MA.

Kahane, J.-P., and Lemarié-Rieusset, P.-G. 1995. *Fourier series and wavelets*. Gordon and Breach Science.

Kaiser, Gerald. 1994. *A friendly guide to wavelets*. Birkhäuser Boston, Inc., Boston, MA.

Katznelson, Yitzhak. 1968. *An introduction to harmonic analysis*. John Wiley & Sons, Inc., New York, London and Sydney.

Körner, T. W. 2004. *A companion to analysis*. Graduate Studies in Mathematics, vol. 62. American Mathematical Society, Providence, RI.

Lang, Markus, and Heller, Peter N. 1996. The design of maximally smooth wavelets. Pages 1463–1466 of: *Proceedings of the IEEE International Conference on Acoustics, Speech, and Signal Processing*, vol. 3. IEEE.

Lang, Serge. 1997. *Undergraduate analysis*. Second edn. Undergraduate Texts in Mathematics. Springer-Verlag, New York.

Lawton, Wayne M. 1990. Tight frames of compactly supported affine wavelets. *J. Math. Phys.*, **31**(8), 1898–1901.

Lawton, Wayne M. 1991. Necessary and sufficient conditions for constructing orthonormal wavelet bases. *J. Math. Phys.*, **32**(1), 57–61.

Lemarié-Rieusset, Pierre-Gilles, and Malgouyres, Gérard. 1991. Support des fonctions de base dans une analyse multi-résolution. *C. R. Acad. Sci. Paris Sér. I Math.*, **313**(6), 377–380.

Lieb, Elliott H., and Loss, Michael. 2001. *Analysis*. Second edn. Graduate Studies in Mathematics, vol. 14. American Mathematical Society, Providence, RI.

Lina, Jean-Marc, and Mayrand, Michel. 1993. Parametrizations for Daubechies wavelets. *Phys. Rev. E (3)*, **48**(6), R4160–R4163.

Mallat, Stéphane G. 1989. Multiresolution approximations and wavelet orthonormal bases of $L^2(\mathbf{R})$. *Trans. Amer. Math. Soc.*, **315**(1), 69–87.

Meyer, Yves. 1993. *Wavelets*. Society for Industrial and Applied Mathematics (SIAM), Philadelphia, PA. Algorithms and applications. Translated from the French and with a foreword by Robert D. Ryan.

Nievergelt, Yves. 2013. *Wavelets made easy*. Modern Birkhäuser Classics. Birkhäuser/ Springer, New York. Second corrected printing of the 1999 original.

Oppenheim, Alan V., and Schafer, Ronald W. 1989. *Discrete-time signal processing*. Prentice-Hall, Inc., Englewood Cliffs, NJ.

Percival, Donald B., and Walden, Andrew T. 2000. *Wavelet methods for time series analysis*. Cambridge Series in Statistical and Probabilistic Mathematics, vol. 4. Cambridge University Press, Cambridge.

Pinsky, Mark A. 2002. *Introduction to Fourier analysis and wavelets*. Brooks/Cole Series in Advanced Mathematics. Brooks/Cole, Pacific Grove, CA.

Pollen, David. 1989 (May). *Parametrization of compactly supported wavelets*. Company Report, Aware, Inc., AD890503.1.4.

Pollen, David. 1992. Daubechies' scaling function on [0, 3]. Pages 3–13 of: *Wavelets*. Wavelet Anal. Appl., vol. 2. Academic Press, Inc., Boston, MA.

Regensburger, Georg. 2007. Parametrizing compactly supported orthonormal wavelets by discrete moments. *Appl. Algebra Engrg. Comm. Comput.*, **18**(6), 583–601.

Royden, H. L., and Fitzpatrick, Patrick. 2010. *Real analysis*. Fourth, revised edn. Prentice Hall.

Ruch, David K., and Van Fleet, Patrick J. 2009. *Wavelet theory.* John Wiley & Sons, Inc., Hoboken, NJ.

Rudin, Walter. 1976. *Principles of mathematical analysis.* Third edn. International Series in Pure and Applied Mathematics. McGraw-Hill Book Co., New York, Auckland and Düsseldorf.

Rudin, Walter. 1987. *Real and complex analysis.* Third edn. McGraw-Hill Book Co., New York.

Saxe, Karen. 2002. *Beginning functional analysis.* Undergraduate Texts in Mathematics. Springer-Verlag, New York.

Schneid, J., and Pittner, S. 1993. On the parametrization of the coefficients of dilation equations for compactly supported wavelets. *Computing,* **51**(2), 165–173.

Sen, Rabindranath. 2014. *A first course in functional analysis: theory and applications.* Anthem Press, London.

Shann, Wei-Chang, and Yen, Chien-Chang. 1997. *Exact solutions for Daubechies orthonormal scaling coefficients.* Technical Report 9704, Department of Mathematics, National Central University.

Shann, Wei-Chang, and Yen, Chien-Chang. 1999. On the exact values of orthonormal scaling coefficients of lengths 8 and 10. *Appl. Comput. Harmon. Anal.,* **6**(1), 109–112.

Simmons, George F. 1963. *Introduction to topology and modern analysis.* McGraw-Hill Book Co., Inc., New York, San Francisco, CA, Toronto and London.

Spivak, Michael. 1994. *Calculus.* Third edn. Publish or Perish, Inc., Houston, Texas.

Strichartz, Robert S. 1993. How to make wavelets. *Amer. Math. Monthly,* **100**(6), 539–556.

Stroock, Daniel W. 1999. *A concise introduction to the theory of integration.* Third edn. Birkhäuser Boston, Inc., Boston, MA.

Tolstov, Georgi P. 1962. *Fourier series.* Translated from the Russian by Richard A. Silverman. Prentice-Hall, Inc., Englewood Cliffs, NJ.

Walnut, David F. 2002. *An introduction to wavelet analysis.* Applied and Numerical Harmonic Analysis. Birkhäuser Boston, Inc., Boston, MA.

Wells, Jr., Raymond O. 1993. Parametrizing smooth compactly supported wavelets. *Trans. Amer. Math. Soc.,* **338**(2), 919–931.

Wojtaszczyk, P. 1997. *A mathematical introduction to wavelets.* London Mathematical Society Student Texts, vol. 37. Cambridge University Press, Cambridge.

Young, Nicholas. 1988. *An introduction to Hilbert space.* Cambridge Mathematical Textbooks. Cambridge University Press, Cambridge.

Index